Benjamin Wolf, PhD

The Fertile Triangle
The Interrelationship of Air, Water, and Nutrients in Maximizing Soil Productivity

"**T**he combination of practical observations and technical guidance in *The Fertile Triangle* provides readers with the capability of replacing guesswork with clear, factual, proven guidelines. It is important to recognize that crops require a necessary balance of variable inputs. This balance is critical for establishing a base for producing a quality and nutritious crop of fruits or vegetables for competitive world markets.

For the past fifteen years, Dr. Wolf's guidance with Central American Produce has been a major factor in the successful production and marketing development programs in Central America, the Caribbean Islands, and South America, which has supplied fruits and vegetables (i.e., cantaloupes, honeydews, specialty melons, asparagus, and specialty vegetables) for world markets.

Dr. Wolf has been the catalyst in providing technical guidance to technicians in a developing country with a limited background or capability in agriculture. He supplied the basic fundamentals to these technicians that corrected serious negative technical inputs. At the present time, Dr. Wolf continually evaluates the production inputs with field observations combined with water, soil, and plant analyses through A & L Southern Agricultural Laboratories.

The practical and technical information in this publication will provide you with the basis of eliminating guesswork with clear, definite guidelines for successful production and marketing development programs."

David N. Warren
President,
Central American Produce, Inc.
Pompano Beach, FL

Food Products Press
An Imprint of The Haworth Press, Inc.

The Fertile Triangle

The Interrelationship of Air, Water, and Nutrients in Maximizing Soil Productivity

FOOD PRODUCTS PRESS
New, Recent, and Forthcoming Titles
of Related Interest

The Fertile Triangle
The Interrelationship of Air, Water, and Nutrients in Maximizing Soil Productivity

Benjamin Wolf, PhD

Food Products Press
An Imprint of The Haworth Press, Inc.
New York • London • Oxford

Published by

Food Products Press®, an imprint of The Haworth Press, Inc., 10 Alice Street, Binghamton, NY 13904-1580

Cover design by Marylouise E. Doyle.

Library of Congress Cataloging-in-Publication Data

Wolf, Benjamin, 1913-
    The fertile triangle : the interrelationship of air, water, and nutrients in maximizing soil productivity / Benjamin Wolf.
        p.   cm.
    Includes bibliographical references and index.
    ISBN 1-56022-878-4 (alk. paper)
    1. Soil productivity. 2. Soils—Composition. I. Title.
S596.7.W65 1999
631.4'22—dc21   98-38131
                                                                                     CIP

This book is dedicated to the fond memory of Professor Frank G. Helyar, former Director of Resident Instruction at the College of Agriculture, Rutgers University, New Brunswick, New Jersey. It was through his help of providing assisted living quarters that I and many other students were able to complete our undergraduate studies. This was especially beneficial during the Depression years, when I was an undergraduate, but still is of great benefit to students today in the form of Helyar House, which provides cooperative living quarters for many worthwhile students.

In addition to helping me complete my undergraduate studies, it was his recommendation that provided a graduate assistantship in the Soils Department, allowing me to fulfill my graduate work for the MS and PhD degrees. The recommendation was especially important since my undergraduate work had trained me as a dairy bacteriologist, with little training in soils. His recommendation was based on my knowledge of chemistry in addition to bacteriology but was motivated by a great deal of faith. I am forever thankful for this faith, which made it possible for me to obtain the advanced degrees and enjoy a very interesting and rewarding career in soils.

ABOUT THE AUTHOR

Benjamin Wolf, PhD, has over sixty years of research and consulting experience in soil, plant, and water analysis and crop production. Throughout his career, Dr. Wolf has made innovative analyses and discoveries. He developed practical methods of soil and plant analysis, showed that rapid soil tests can be used to evaluate fertilizer and lime requirements, used plant analyses to diagnose nutritional problems as early as 1945, used the nitrogen test as a basis for applying nitrogen fertilizers as early as the mid-1940s, introduced micronutrients into fertilizer programs in several states and foreign countries, introduced foliar feeding for cooperative growers as early as 1949, introduced the practice of fertigation to cooperative growers in 1952, and was largely responsible for the development of the cut flower business in Columbia. Most of Dr. Wolf's present consulting work is done in Guatemala, Honduras, Costa Rica, and the Dominican Republic. His research and activities have greatly increased crop yields and quality, while lifting general farming in many areas to much higher productive levels. He has published three books and over forty articles in trade magazines and scientific journals.

CONTENTS

Acknowledgments

The author wishes to acknowledge the help of the following in the preparation of this book:

Dr. George Snyder, Soil Scientist, IFAS, EREC, Belle Glade, Florida, who provided many valuable changes in the manuscript.

In addition to Dr. Snyder, many other soil scientists, as well as agronomists, horticulturists, and plant physiologists who over the years have provided the basis of understanding the functions of air, water, and nutrients in soils and other media for crop production. Some of their work is highlighted in the many references or additional readings.

Mr. David Warren, President of Central American Produce Company, growers and importers of offshore fruits and vegetables, who over a number of years has been willing to apply on a large scale practices that maximize air, water, and nutrients in soil. Good yields with outstanding quality, which is especially needed for offshore shipping, have resulted when the three components were in balance. I have used the results of these practices to confirm my understanding of their importance, and these results have prompted me to write this book.

Introduction

The health and yields of crops are largely dependent on three constituents of soil or other media—air, water, and nutrients. No crop can produce satisfactory yields—of both sufficient quantity and quality—unless satisfactory amounts of all three items are present. While there are many ways in which soils or other media affect crop production, it is essentially their ability to provide air, water, and nutrients that defines their productivity. The amounts necessary for large yields of good quality vary with different crops and even with different stages of the same crop.

A correct balance between air, water, and nutrients is essential to obtain desired yields. Larger quantities of water need to be present to obtain the full value of added nutrients. The addition of water requires more nutrients—especially if water washes out some essential nutrients from the root zone. The effectiveness of both nutrients and water are greatly reduced as soil air falls below certain levels—the desirable level varying primarily with the crop, its stage of growth, and soil temperature. Since air in soil or other media usually decreases as water is increased, the addition of large quantities of water for maximum growth needs to be carefully controlled lest it lower air below optimum levels.

The interdependency of the components is most strikingly exemplified in the relationship between air and water. Air and water occupy about equal volumes in a fertile mineral soil (see Figure A). Rain or irrigation temporarily increases the water content at the expense of air. If sufficient water is added, air is depleted to the point that many microbiological soil functions (nitrification, nitrogen fixation, and organic matter decomposition) and the uptake of water and nutrients come to a halt. In most fertile media, the excess water quickly drains away, restoring adequate levels of air but taking some valuable nutrients with it. As water is used by plants or is lost to evaporation, its reduction allows an increase in the amount of air. The increase in air volume is about equal to water volume lost in most artificial media and in sands and light loams. Because clays shrink upon drying, their increase in air volume is less than the decrease of water volume except for certain soils that develop cracks upon drying. If cracks develop, the final volume of air can far exceed the original volume of water.

FIGURE A. The Composition of a Fertile Mineral Soil Illustrating the Relationship of Air and Water

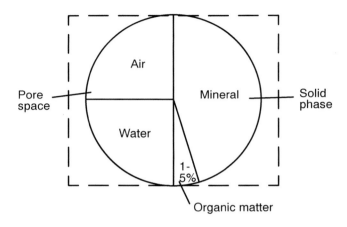

Note: In most mineral soils, air increases as water decreases and vice versa. An ideal relationship exists when they are about equal.

The interdependency of the three constituents can be diagrammed as an equilateral triangle, with the relationship of one to the other two constituents represented by sides of equal length. The area of the triangle then would correspond to the crop yield, which increases as the length of the sides is increased. Maximum area or yield is obtained as the integrity of the equilateral triangle is maintained by increasing the three sides—not necessarily equally but in equal proportions. The length of the sides is limited by the soil or medium, which in turn can be depicted as a circle surrounding the triangle (see Figure B).

The maximum area or yield is obtained as the points of the triangle representing sides of equal length touch the circle, and lessened as any leg is shortened to less than its ideal potential. Small differences in length are usually not significant, but large differences, particularly if one side dominates the triangle, can seriously affect crop production (Figure C represents an excess of water).

Satisfactory yields are still possible even with shortened lengths if the sides are always maintained above certain minima by rapid replacement as needed, and providing a sufficient balance is maintained between them.

FIGURE B. The Relationship of Balanced Air, Nutrition, and Water in Fertile Soil

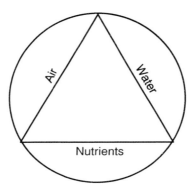

Note: Air, nutrient, and water sides are extended to their full potential and are balanced in a fertile soil.

FIGURE C. The Unbalanced Relationship of Air, Nutrients, and Water in Soil

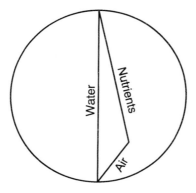

Note: An example of unbalanced air, nutrients, and water. In this case one side (water) is dominating the triangle, greatly shortening the air side and limiting the nutrient side, resulting in poor crop production.

A fertile triangle would have the three sides at about equal length, but in most cases fertility increases as the sides increase to certain limits. The upper limits vary with different soils, being greatest most of the time in loam soils because of their ability to hold large amounts of air, water, and nutrients without unduly affecting other parameters of crop production.

It is desirable to improve the length of the sides in all media and soils including loams to the maximum permitted by the limits of the medium, the economics of the crop, and acceptable environmental stewardship. In many cases, changes in tillage methods, working of the soil, utilization of organic matter or water, selection of fertilizers, or methods of applying and timing fertilizers involve little or no extra cost but can improve air, water, and nutrient sides.

Improving basic characteristics of soil and media may require substantial increases in costs, but even these are often worthwhile for certain high-priced crops or special situations.

The maximum response to the three components is seldom attained in soils. This is one of the incentives motivating researchers and growers to seek substitutes for soil in the form of artificial media, in which higher levels of air, water, and nutrients can be maintained. The higher levels used in artificial systems (hydroponics and aeroponics) have achieved greater yields of a number of crops, but costs have kept their usage to relatively small production. Although restricted land availability due to the population explosion may usher in a period when such practices will play a more important role in crop production, our best hope for major increases in crop productivity at present lies in increasing the three sides to a balanced maximum in soils.

Theoretically, crop yields will continue to rise as sides of the triangle are increased as long as a desirable balance is maintained between them, but limits imposed by other factors, such as insufficient light or carbon dioxide, eventually restrict growth. These limiting factors can also be increased, making it possible to attain super-high yields, but not without additional costs, which at present limit these special treatments to highly specialized crops of high value.

The importance of each one of these basic requirements (sides of the triangle) and how each is maintained at ideal concentrations are considered in several sections. Section I deals with the three sides of the triangle—air, water, and nutrients—and their importance and usual concentrations in soil and other media; Section II with characteristics of soil and other media that affect the length of the triangle sides. Section III considers farm practices that indirectly modify the length of the sides. Several of these practices affect more than one side. Section IV examines the direct application of air, nutrients, or water, and Section V evaluates several approaches that have maximized one side of the triangle while attempting to adjust the others for optimum growth.

In all sections, the aim has been to provide enough material for a basic understanding of the subject as well as potentially for practical applications. My long association with varied commercial enterprises has given me an appreciation of the need for understanding how air, water, and nutrients can be used for maximum crop production, but unless the grower realizes the relationships between these important soil characteristics, changes needed to maximize them are often delayed to the point of severely impinging on profitable operation.

Since the book brings together the basics of maximizing a soil's potential, it can be a great aid for students of agriculture and related sciences in understanding the many facets of soil management needed for optimum crop production. As such, it can be helpful not only for those majoring in soil fertility or crop production, but others in horticulture, agronomy, animal husbandry, and related agricultural sciences who are interested in producing high yields of specialty crops. For maximum utilization by soil fertility students, the book needs to be introduced shortly after or in conjunction with an elementary course in soils. For others interested primarily in crop production, *The Fertile Triangle* could well be the primary soil text.

The book should also be highly useful to those who grow crops or those who assist growers. Farmers, farm managers, consultants, extension personnel, and those who sell various agricultural chemicals can profit from understanding the importance of air, water, and nutrients in the production of crops and how these items are interrelated. It is only by understanding their importance, their close relationships, and how they are affected by various farm practices that growers can utilize soil inputs to full advantage with minimal harmful effects on crops, soils, or the environment.

SECTION I:
SIDES OF THE TRIANGLE

This section deals with the importance of air, water, and nutrients present in a soil or other medium upon the production of crops, some of the interrelationships between the three components, amounts of each present in the soil at different times, and amounts needed for optimal health and growth.

Chapter 1

Air

Air in soil or other media is vitally important because plants and many microorganisms that affect crop production require a constant supply of oxygen for respiration, which provides needed energy. Air provided at the roots allows for water and nutrient uptake. All respiration involving oxygen (O_2) releases quantities of carbon dioxide (CO_2), the accumulation of which in the soil or other media above certain levels must be avoided in order that these vital processes continue.

Soil air is present both in gaseous (air-filled porosity) and liquid states (soil solution). The gaseous state, present in large soil pores, contributes the major portion of air requirements for plants and microorganisms by acting as a reservoir of needed O_2 and as a depository for CO_2 and other gases resulting from incomplete respiration or decomposition of organic matter (OM). Some of these gases formed from incomplete respiration or decomposition of OM may be toxic to plants. With sufficient large pores in well-drained soils, the spent air in soil pores can readily be exchanged for fresh air from the atmosphere.

The exchange of air between atmosphere and soil takes place in both the air-filled pores and pores partially filled with water. That taking place in pores partially filled with water decreases steadily as the water content increases, coming to a virtual halt as saturation approaches 90 percent.

As gas transport is halted due to flooding or slowed appreciably by wet or compact conditions, marked changes in plant growth take place. Root development can be seriously impaired as elongation and cell division quickly slow and may stop completely depending on the severity of the slowdown. The harmful effect of poor air exchange on plants is aggravated by changes in the soil microbial and animal populations if the slowdown is severe enough to cause anaerobic conditions. As indicated below, these changes severely affect several processes that are vital for a number of plants.

Although relatively small in volume, the exchange of air taking place in the soil solution is also of the utmost importance. The small amount of O_2

present in the soil solution is highly useful because it is readily absorbed by roots or utilized by soil microorganisms since the solution envelops the roots or soil particles as a film.

Oxygen in the film can be quickly exhausted and the solution overly saturated with CO_2 unless it can be readily replaced by O_2 from the pores. It has been estimated that the O_2 in a flooded clay soil can be exhausted in a single day whereas it may last about three days in a sandy soil with its large proportion of large pores. The actual exhaustion rate, largely reflecting the respiration rates of plants and microorganisms, depends on soil temperature, rising as temperatures increase.

The rapid replacement of air containing large amounts of CO_2 and possibly toxic gases with fresh air relatively rich in O_2 is no problem in well-drained soils containing adequate pore space. But the result can be disastrous if large pore space is reduced due to clay dispersal or compaction and if sufficient air is prevented from entering the pores because of overly wet conditions, compaction, or constricting layers.

The composition of air even in well-aerated soil differs from that directly above the soil. The differences in nitrogen (N_2) and O_2 contents are rather small and are not significant. For example, common concentrations are: about 79.1 percent N_2 in the air above the soil and 79.0 percent in the soil; 21.0 percent O_2 in air above the soil and 20.5 percent in the soil. The differences between CO_2 in the air above the soil and CO_2 in the soil are much greater, being 0.035 percent in air above but ranging from about 0.2 to 1.0 percent in the soil.

The amount of O_2 is lowered while that of CO_2 is simultaneously raised in deeper layers of the soil. This change is accentuated if the soil is subject to flooding or compaction. Oxygen contents even in dry soils can drop from about 20 percent at a 4-inch depth to about 15 percent at 3 feet. In wet soils, the range is closer to 13 percent at 4 inches and only about 7 percent at 3 feet. In flooded soils, the O_2 content can drop to almost zero. Corresponding changes in CO_2 would be 0.5 percent in a dry soil near the surface increasing to about 4 percent at the 3-foot depth. In a wet soil, the CO_2 content can be about 6 percent at the 4-inch level, increasing to about 10 percent at 3 feet. The increase in CO_2 and the decrease in O_2 in wet soils is greatly accentuated as temperatures increase.

Soil air may contain small amounts of other gases. Some of these, such as argon, are rather inert and not subject to appreciable change. Others, such as ammonia, ethylene, hydrogen sulfide, hydrogen, methane, and nitric and nitrous oxides, which are primarily due to anaerobic conditions, can fluctuate rather markedly and sometimes may have harmful effects on plants.

OXYGEN

Oxygen is the critical gas in the soil atmosphere since it is essential for respiration of all crop plants and most microorganisms. Most plants need adequate concentrations in contact with the roots. Some marsh plants, rice for example, can function with very little O_2 in the soil. Evidently, sufficient O_2 is transported from the atmosphere through leaves and stems to the roots for essential respiration. Some O_2 may be excreted from the roots and a desirable concentration maintained externally so that the production of anaerobic toxins capable of adversely affecting the roots is limited. For most plants, a renewable source of O_2 needs to be present at the root surfaces for sugars within the roots to be utilized as a source of energy. It is the energy produced by respiration that allows uptake of nutrients and water and permits various plant growth processes to continue.

As O_2 is reduced in the soil below optimum values for the particular species and its stage of development, growth of plant roots is slowed, with a corresponding loss in crop production. A further decrease in O_2 can begin to kill roots. The decomposition of dead roots can lead to formation of several substances toxic to plants. These toxins and those formed from reduction of chemicals (increased amounts of nitrites, iron, manganese, and hydrogen sulfide) in the soil can lead to the complete death of plants within relatively short periods.

Oxygen is essential for most soil microorganisms. It has been estimated that in a fertile soil more O_2 is consumed by microorganisms than by crop plants. Insufficient O_2 can slow down or halt the following important soil processes conducted by microorganisms:

1. The breakdown of organic matter with its subsequent release of nutrient elements. Halting or even slowing down the process can result in serious shortages of nitrogen, phosphorus, sulfur, and several micronutrients in many soils.
2. The formation of substances capable of binding individual soil particles into aggregates that increase soil porosity and increase resistance to wind and water erosion.
3. Symbiotic N_2 fixation by the legumes—alfalfa, beans, clovers, lupines, soybeans.
4. Nonsymbiotic fixation of N_2 by *Azotobacter* bacteria.
5. Nitrification of urea, ammonium, and nitrite forms of nitrogen. Although the formed nitrates are more subject to leaching than ammonium, nitrates are a readily available form of nitrogen and the presence of sufficient nitrates in relationship to ammoniacal forms helps ensure good quality of many crops.

In addition to slowing of beneficial bacterial processes, shortages of O_2 can lead to denitrification, whereby important amounts of nitrogen are lost for crop production as volatile nitrogen compounds (nitrous oxide and nitric oxide) are produced and leave the soil. The shortages of O_2 also can lead to outbreaks of serious plant diseases caused by soil-inhabiting pathogens. While growth of some soil pathogens is restricted as O_2 is reduced, the damage to plants is greatly increased by several organisms (*Fusarium, Pythium, Phytophthora, Sclerotinia sclerotium*) under these conditions.

Oxygen is also needed for large groups of soil fauna. These include a number of insects, nematodes, mites, spiders, slugs and snails, and earthworms. There may be little or no benefit for crop production from some of these, but there can be no doubt about the benefits of earthworms. By ingesting soil particles along with organic matter and coating them with calcium carbonate produced by their digestive glands, they produce casts that benefit soils. The material can greatly benefit the moisture-holding capacity of the lighter soils and improve the porosity of heavier soils. Earthworms survive well in moist soils but will be overcome in soils very low in O_2.

The ideal amount of O_2 varies with different plants and with different stages of the same plant. Uptake of nutrients by plants requiring large amounts of oxygen, such as tomato, tobacco, and papaya, is impaired when O_2 concentration in the medium falls below about 10 percent. Cereal crops or pastures need less but can be more demanding at critical stages such as germination and early development. In some early studies with orchards, it was found that 15 percent O_2 in the soil gave the best results; anything less than 15 percent resulted in a decreased uptake of mineral nutrients. While at least 12 percent was necessary for root initiation, and a concentration of 5 to 10 percent was needed for growth of existing root tips, only 0.1 to 3.0 percent was required for subsistence. However, at a concentration of only 1 percent, existing roots lost weight (Stolzy, 1974).

The amount of O_2 needed by plants is also affected by temperature, increasing as temperature rises. A satisfactory soil concentration of 10 percent O_2 for several plants grown at 65° F (18°C) might be insufficient at 85°F (29°C).

The increased demand for O_2 as temperature rises appears to be due to elevated rates of respiration at higher temperatures, although it has been wrongly attributed to lower solubility of O_2 in the soil solution as temperatures increase. The solubility of O_2 does decrease with rising temperatures, but only about 1.6 percent per °C, whereas the diffusion of O_2 increases at a rate of 3 to 4 percent per °C, indicating a net gain of O_2 in the solution with increasing temperatures. But the increased respiration at higher temperatures utilizes more of the O_2 than is gained by the elevated diffusion rate.

Some species survive low O_2 concentrations, particularly if the onset is rather slow, by development of adventitious roots near the well-aerated surface and/or by the development of aerenchyma tissue in their roots and shoots. Tomatoes, which readily produce adventitious roots, will survive slowly induced wet conditions, while tobacco, which cannot produce adventitious roots, is usually killed under such conditions.

Aerenchyma tissue, some of which is increased by slowly induced flooding or wet conditions, is formed by splitting, delamination, and separation of adjacent cell walls or the breakdown of cell walls. Voids, referred to as intercellular spaces, chambers, or lacuna, associated with aerenchyma tissue may aid gas exchange in roots and stems. While most water-tolerant plants have aerenchyma tissue in their roots and shoots, the amounts are quite variable (5 to 60 percent) (Allen, 1997).

SOIL SOLUTION OXYGEN

Relatively less seems to be known about O_2 concentrations in the soil solution compatible with root initiation and maintenance. It is the soil solution upon which the root depends for O_2. The O_2 can evidently be quickly depleted, but this is not serious if there is sufficient air exchange in the soil so that the needed levels are quickly replaced. From the standpoints of both number and length of roots, it appears as if a minimum of about 2.5 parts per million (ppm) of O_2 are needed in the soil solution for growth of some plants, with 5 to 10 ppm producing far better results. The amounts needed vary not only with different crop plants but even with cultivars of the same species.

Some idea of the need for O_2 in the soil solution is indicated by amounts beneficial in solution or gravel cultures, but even these results point to differing responses due to type of plant and cultural conditions. In a review of the literature by Hewitt (1966) it was found that O_2 tensions of less than about 0.7 ppm in solution cultures resulted in injury of avocado and citrus roots, with permanent cessation of growth by avocados but not citrus at 0.6 ppm. Root growth of soybean, tobacco, and tomato ceased if O_2 fell below 0.5 percent, but yields of tomato increased as O_2 content rose from 0.6 to 21 percent. The amount of O_2 required for good growth (at least for certain plants) appears to be greater if ammonium rather than nitrate-nitrogen is present or if manganese (Mn) or copper (Cu) is deficient. Tomato and apple, which have high requirements for O_2, did best with 16 ppm, while avocado, barley, citrus, and soybean, with moderate demand, did very well with only 6 to 8 ppm.

In gravel culture, a concentration of about 500 ppm of O_2 in the nutrient solution appears to be ample for carnations if continuous circulation of the solution is used. But ideal amounts are related to number of plants and time of day, being 0.4 to 0.55 milligrams per minute (mg/min) per plant in the daytime and 0.2 to 0.4 mg/min per plant during the night.

The solution culture work also shows that plants can be injured with an overabundance of O_2, but here too the results differ with different plants and conditions. Legumes seem to tolerate higher levels of O_2 than cereals. Barley and soybean were injured with excess O_2. While soybean showed iron (Fe) deficiency in the presence of high O_2 when nitrates were present in the solution, oats did not.

CARBON DIOXIDE

The concentration of CO_2 produced by respiration of plants, animals, and microorganisms and augmented by the decomposition of organic matter and diffusion from the atmosphere can well be ten times higher in the soil than in the atmosphere directly above it. The concentration in soil air increases with depth and moisture. It also increases with farm practices such as the use of plastic mulches, killing of sods, incorporation of manures, plowing down large quantities of plant residues, overwatering, or mechanical compaction of soils.

The presence of CO_2 can alter nutrition by changing soil pH, amounts of available nutrients (probably by pH changes), and carbon available to microorganisms and plants.

There have been reports of improved yields from additions of CO_2 injected as a gas or by the carbonation of irrigation water (Glenn and Welker, 1997). Unfortunately, injection of CO_2 has not provided consistent increases of crop yields, probably because of variable soil contents and/or a relatively narrow range between desirable and harmful levels.

In well-aerated soils, the production of large amounts of CO_2 is of no concern as soil air is readily exchanged with that of the atmosphere. It is only as the exchange is limited by factors such as soil compaction or overly wet conditions that its concentration may become a problem. Carbon dioxide is more soluble in water than O_2 and can displace diminished O_2 to the point of damaging plants and/or accelerating certain soilborne diseases.

Carbon dioxide readily combines with water to form carbonic acid (H_2CO_3). The acid can lower pH of the media. Its reduction in soils of high buffering capacity (clays, loams, peats) is of minor importance but can be a problem in lightly buffered soils (sands), and especially damaging in media used for hydroponics.

Acid formation from CO_2 can be helpful in calcareous soils where lowering of pH can increase availability of phosphorus (P), manganese (Mn), iron (Fe), and zinc (Zn). The reduction of pH takes place as partial pressures of CO_2 in the soil equal that of the atmosphere, but increases in pH can occur if partial pressures in the soil are greatly increased by respiration and organic matter degradation to levels much higher than the atmosphere. Whereas the pH of a calcareous soil may be about 7.2 with low partial pressures of CO_2, the pH can be 8.5 if high partial pressures of CO_2 exist. The increase in pH is accentuated with the use of high-bicarbonate irrigation water that greatly increases the partial pressure of CO_2.

The CO_2 content is substantially increased by additions of manure, green manure crops, or other large additions of organic matter. Such increases persist for only short periods in most soils as the extra CO_2 is removed by exchange with air above the soil. Some organics, such as manure, greatly increase the air space in the soil, thereby hastening the diffusion of the CO_2 to the air above the soil. Nevertheless, it is suspected that excess CO_2 generated by the incorporation of large amounts of organic matter may be partially responsible for the failure of certain seeds to germinate if planted soon after the incorporation.

Much of our knowledge of the harmful effects of excess CO_2 also has been gained from solution and sand cultures. These results indicate that negative effects of excess CO_2 appear to vary not only with the plant but also with the amounts of O_2 and bicarbonates (HCO_3^-) present. A concentration of 10 percent CO_2 was not harmful to corn if the solution contained 20 percent O_2, but as little as 5 percent CO_2 depressed potassium (K) uptake if O_2 was markedly reduced. Cotton also showed no growth inhibition of roots with 10 percent CO_2 if the solution contained 7.5 to 21 percent O_2, but root growth ceased when O_2 was less than 0.5 percent. With 21 percent O_2, 15 percent CO_2 did not adversely affect root elongation, but 30 to 45 percent CO_2 reduced growth and 60 percent totally repressed it.

The results obtained with gravel cultures are quite different. The reduction of pea, broad bean, or sunflower growth resulting from flushing pots after irrigation with as little as 2 percent CO_2 was about 20 percent. With 3 percent CO_2 the reduction was 50 percent and with 6 percent CO_2 it was 70 percent. Oats and barley were not as sensitive, showing only a 20 percent reduction at the 6 percent concentration of CO_2 (cited by Hewitt, 1966).

The pH and the presence of bicarbonates evidently influence the effects of CO_2 upon plant growth. In solution cultures, concentrations of CO_2 up to 100 ppm (0.1 percent) stimulated the growth of pea at pH 5.3, but additions of CO_2 at higher pH values may cause depressions in yields. In soils, depressions of yield associated with increases of CO_2 at high pH values are often due to iron deficiencies evidently induced by bicarbonates. The increases in bicarbonates result as calcium carbonate reacts with CO_2 and water as in the following equation:

$$CaCO_3 + CO_2 + H_2O \longrightarrow Ca^{2+} + HCO_3^-$$

The formation of bicarbonates is increased as CO_2, generated by respiration of roots and microorganisms and augmented by decomposition of organic matter, accumulates in soils that are compacted or overwatered by irrigation or rainfall. The increase is intensified with use of water high in bicarbonates or if soil pH is high, but lessened if soil is open, allowing good drainage and air exchange.

AIR EXCHANGE

To maintain good growing conditions for most crops, it will be necessary to exchange soil air, rich in CO_2 and possibly other potentially harmful gases but low in O_2, for air that contains satisfactory levels of O_2.

The actual exchange of soil air rich in CO_2 for air rich in O_2 takes place in the soil's pores (see the section, "Porosity"). Most of this exchange takes place in the air-filled pore space by the process of diffusion, although mass flow resulting from temperature or pressure gradient or from increases in volume of air-filled pore space as water is withdrawn may be responsible for some of the exchange.

Although much smaller in amount, there is an important exchange of O_2 and CO_2 in water-filled pores. The small amount of O_2 in water-filled pores is due to the poor solubility of O_2 in water (0.6 percent) as compared to the normal presence of 21 percent O_2 in air. Also, the diffusion of O_2 in water-filled pores is reduced by a factor of 10^4, making the movement of O_2 in water-filled pores rather limited. Despite the small amount of O_2 present in water-filled pores, its presence ensures sufficient O_2 for microorganisms and root hairs since these need to be bathed in a water film for maximum development. Failure to maintain sufficient O_2 tension in these films leads to decreased growth and possible cell death.

Just how much air needs to be exchanged to maintain good O_2 balance in the soil or medium varies with different conditions, plants, and microbial activity. But it has been estimated that the O_2 consumption per square meter of soil is about 10 liters per day on average. The 10 liters of O_2 are equivalent to about 50 liters of air or slightly more than 53,000 gallons of air per acre—a tremendous amount of air that needs to be exchanged to maintain healthy conditions for optimum crop production. If the air porosity of the soil is 20 percent, there will be 200 liters of air in a cubic meter of soil. If the 50-liter consumption rate were uniform to a depth of one meter, all the O_2 would be consumed in four days, requiring complete replacement of the 200,000+ liters (53,000 gal) in that time.

In most fertile soils with adequate pore space, the exchange of O_2 for built-up levels of CO_2 is readily accomplished. But the consequences are far different and result in greatly reduced crop production if air-filled porosity is compromised because of inadequate pore space due to compaction, deflocculation, or water that does not readily drain.

SOIL CAPACITY FOR HOLDING OR EXCHANGING AIR

Porosity

The amount of air present in a soil depends on its porosity, which also regulates the rapidity with which spent air can be replaced with oxygen-rich air. Porosity refers to that portion of a soil not occupied by solid particles, which in most fertile mineral soils is about 50 percent of the volume. Total porosity in mineral soils ranges from 0.28 to 0.75 m^3m^{-3}; that of organic soils from 0.55 to 0.94 m^3m^{-3}. The optimum volume is probably about 0.50 m^3m^{-3}. Ideally, about half of this volume ought to be air porosity. As a bare minimum, about $1/4$ of this porosity (0.12 m^3m^{-3}) ought to be present two to three days after irrigation or heavy rainfall.

A soil's pore space can be occupied by both gas and water. Different-sized pores make up a soil's porosity, with most of the air being confined to the larger pore spaces, and is designated as air-filled porosity. The space occupied by water is primarily confined to the smaller pores and is referred to as water-filled or capillary porosity.

The physical characteristics of a soil that permits rapid exchange of air and water and still retains enough water for good crop production depends on the soil having sufficient porosity or numbers of macropores (large pores) and micropores (small pores capable of capillary action).

Large amounts of air being held and readily exchanged depend on the soil having sufficient numbers of macropores (>100μm), which extend around the main structural soil units. These pores, largely dependent on the nature of the soil particles or aggregates, are also aided by channels or tunnels left by roots, rodents, insects, and earthworms and by the alternate shrinking and swelling of the clays. Macropores must be numerous and deep enough so that there is a rapid exchange of air in most of the root zone.

Sufficient numbers of macropores extending throughout the soil profile are also needed for water drainage. Rain or irrigation water needs to infiltrate rapidly into the soil but much of it also needs to drain rapidly so that the pores will regain sufficient air. Sufficient numbers and distribution of macropores ensures the removal of excess water and replacement with air. These pores are also instrumental in aiding the removal of air containing large amounts of CO_2 and little O_2 and replacing it with normal air containing much higher levels of O_2. Lack of O_2 in soils with insufficient large pore space can slow water and nutrient uptake and probably is becoming the limiting factor in many modern farm operations.

Retention and movement of water after initial drainage are made possible by micropores (small or fine pores <30μm), which extend through the interior of the soil structural units. There must be sufficient numbers of them to hold enough water against drainage forces and so located that they are readily reached by plant roots. The small pores or capillaries not only hold water but can move it upward from lower layers, and a sufficient number of them properly handled allows for dryland farming. Movement of water in the micropores is practically limited to capillary action. A soil with insufficient micropores tends to be droughty and unproductive unless there are ample rains, although the recent use of irrigation (especially drip irrigation) has enabled efficient use of such soils.

Mesopores (30 to 100μm) affect both air and water retention and movement, but to a lesser extent.

Optimum Porosity

Just how much air porosity is required for optimum growing conditions varies with different plants, stages of their growth, and whether they are grown in open soil or confined containers. The optimum volume in open soils is probably about 0.25 m^3m^{-3}. As a bare minimum, about half of this porosity (0.12 m^3m^{-3}) ought to be present two to three days after irrigation or heavy rainfall. It has been noted that the health of conifers is greatly reduced as soil air content falls below 4 percent, that of deciduous trees and fruit trees if it falls below 10 percent, and that of nurseries and some vegetables if the air content falls below 15 percent of the total porosity for relatively short periods.

The actual desirable O_2 content also varies with different plants. The critical content begins at about 5 to 10 percent by volume, with root growth of many plants being limited when O_2 is less than 0.1 percent of the pore space (Glinski and Lipiec, 1990).

The air-filled porosity is of more critical concern in pot cultures, because of the small volume of soil and a perched water table. Although some plants and short-term growing (seed germination) can be productive in media with rather low air porosity (2 to 5 percent of volume), some pot-grown plants (azaleas and orchids) require media with very high air porosities (20 percent of volume) and many of them require porosities of 10 to 20 percent (see Table 1.1).

TABLE 1.1. Desirable Air-Filled Porosities for Several Ornamentals

Very high >20%	High 20-10%	Intermediate 10-5%	Low 5-2%
Azalea	African violet	Camellia	Carnation
Bromeliad	Begonia	Chrysanthemum	Conifer
Fern	Daphne	Gladiolus	Geranium
Orchid, epiphytic	Foliage plants	Hydrangea	Ivy
	Gardenia	Lily	Palm
	Gloxinia	Poinsettia	Rose
	Heather		Stocks
	Orchid, terrestrial		Strelitzia
	Podocarpus		Turf
	Rhododendron		
	Snapdragon		

The upper limits in each category are more suitable for starting plants or repotting since there is a normal reduction in porosity with time as organic materials decompose and all materials tend to settle.

Source: White, J. W. 1973. Criteria for selection of growing media for greenhouse crops. *Penn. State Agric. Expt. Journal*, Series #4574:2.

Porosity of Different Soils

The amount of air-filled porosity is not closely related to total porosity because of the variable water content. Although clay loams and clays have a high total air space, air porosity suitable for rapid air exchange may be low. Much of the pore space is filled with water because pores or interconnecting passages are too small to allow for rapid and efficient drainage. On the other hand, despite lower porosity, air porosity and rapid air exchange are relatively high in most sands, where large particles allow for large air spaces that are readily drained and refilled with air.

Exceptions to high air porosity are noted in very fine sands or sands that may have restrictive subsoils limiting water drainage. Air spaces between very fine sand particles also may be small enough to limit water drainage.

The amount of air-filled porosity in the heavier soils can be quite variable. Unlike sand particles, which retain their identity, clay and silt particles tend to combine with other particles to form various aggregates. The spaces between aggregates comprise the pores or voids of the soil and in some cases can be quite large, substantially increasing the soil's degree of air-filled porosity, which largely depends on the nature of the aggregates and how well they can be maintained.

As will be covered in greater detail in other chapters, cultural practices greatly affect air-filled porosity. Generally, cultivation can benefit it temporarily but usually leads to smaller porosities. Tillage of soils when wet is highly counterproductive, but tillage of slightly frozen soils can aid in their drying, allowing earlier spring planting. The successful drying of frozen soil by tillage requires that the frozen layer be 1 to 4 inches deep without snow accumulation (Gooch, 1997).

As will be pointed out, some of the more important practices for improving porosity are (1) adding organic matter, either by growing it in place or bringing it in as manures, sludge, peat, or various wastes; and (2) the addition of calcium in the form of limestones or gypsum on the heavier soils, those containing substantial amounts of clay and silt. The lighter soils (sands) fail to show substantial porosity changes from limestone or gypsum additions.

Moisture and Air Exchange

Problems arise at times even though the soil or medium may have good porosity, primarily because a large proportion of the pores is filled with water from too much rain or irrigation. The problem may exist for a short time and usually will have no perceptible effect on crop yield if sufficient large pore spaces are present and drainage conditions are such that the excess water can readily move out of the root zone. If, on the other hand, large pore space is in short supply or if poor drainage conditions (compact subsoils, plow soles, hardened layers, or high water table) exist, water movement out of the root zone may be much too slow. The slow water movement out of the profile can seriously impinge on the replacement of

CO_2 and other gases harmful to respiration with air rich in O_2, needed for respiration.

Oxygen Diffusion Rate

The oxygen diffusion rate (ODR) is perhaps the best measurement of a soil's ability to replenish O_2 used by respiring plants or displaced by water. Generally, plant growth appears to be satisfactory if ODR remains above 40 g $\times$ 10^{-8}/cm^2/min to depths of about 15 inches, although the best growth of sugar beets, wheat, and citrus occurs if ODR is above 50 g $\times$ 10^{-8}/cm^2/min. Sugar beets appear to be very sensitive to poor ODR. Legumes, especially alfalfa, also are sensitive, while most grasses are more tolerant. Some critical ODR values for several plants are given in Table 1.2.

TABLE 1.2. Critical Oxygen Diffusion Rates (ODR) for Several Plants

Crop	ODR (μg m^{-2}s^{-1})
Barley	25
Corn	20-40
Cotton	33
Grasses	
Kentucky bluegrass	8-15
Ryegrass	17
Turfgrass	25
Snapdragon	33
Sunflower	33
Wheat	50

Source: Prepared from Glinski, J. and W. Stepniewski. 1985. *Soil Aeration and Its Role for Plants.* CRC Press Inc., Boca Raton, FL, p. 150.

Structure

Porosity and the ready exchange of O_2 for CO_2 is closely related to structure, which is more important in heavy soils rich in clay than in light or sandy soils. The latter have little or no colloidal content, which is so important in forming soil aggregates and maintaining their stability.

The type of aggregate and its size and shape are influenced by the kind of clay. If montmorillonite is the primary clay, prismatic angular structure results, but kaolinite and the hydrous oxides are apt to produce granular structures, which are the most productive.

Structure that allows good productivity of crops will combine sufficient numbers of large pores, necessary for rapid infiltration of water and exchange of gases, with sufficient small pores to retain enough water for plant growth. More about the factors affecting the formation of soil structure and its influence on crop growth is presented in Chapter 4.

Reducing Conditions in Soils

Reducing conditions can exist in soil or other media if O_2 falls to low levels for any reason (undue compaction, poor drainage, overwatering, excessive rainfall, etc.). Not only are certain vital processes (respiration, organic matter decomposition, and uptake of nutrients) inhibited or stopped, but certain important changes take place in the soil or media.

As soils are flooded, pronounced changes are brought about by the almost complete interruption of the normal gas exchange between soil and air. Flooding fills soil pores and almost stops oxygen infusion so that O_2 begins to decline rapidly. Oxygen depletion is hastened as water stimulates the activities of soil microorganisms, increasing the demand for O_2. If temperatures are high enough for rapid respiration, essential levels of O_2 can be depleted in hours.

While oxygen declines rapidly, other gases such as CO_2, N_2, hydrogen (H_2), and methane (CH_4) can accumulate, some of them to dangerous levels. The concentrations of these gases in submerged soils can range as follows: 10 to 95 percent N, 0 to 10 percent H_2, 1 to 20 percent CO_2, and 15 to 75 percent CH_4.

The amounts of reduced substances in relationship to oxidized substances are increased with certain fundamental changes in the medium. The redox potential, which measures the extent of reducing conditions, decreases as the amounts of O_2 become limiting. When measured by the voltage of a platinum electrode against a reference electrode, it shows a low voltage as O_2 is lowered and reducing conditions prevail.

A rather high voltage (+0.6 V) in the presence of ample O_2 allows for the formation of oxides of carbon (CO_2), hydrogen (H_2O), nitrogen (NO_3), sulfur (SO_4), iron (Fe_2O_3), and manganese (MnO_2). The absence of sufficient O_2, as indicated by a low potential, allows for reduction of these compounds, which may have negative results on crop production.

Some of the changes that can take place and affect crop production are the following: nitrate-nitrogen (NO_3-N) can be reduced to gaseous nitrogen (N_2) or nitrous oxide (N_2O), which may be lost to the atmosphere; oxidized iron (Fe^{3+}) and manganese (Mn^{4+} or Mn^{3+}) can be reduced to Fe^{2+} and Mn^{2+}. If only small amounts are reduced, there is no problem and plant growth may be stimulated, as the reduced forms are more available than the oxidized forms. On the other hand, in the presence of large amounts of these elements or prolonged flooding, toxic amounts of both elements may be formed, greatly inhibiting the growth process. Also, sulfates can be reduced to iron sulfide (FeS), pyrites (FeS_2), hydrogen sulfide (H_2S), and the anions bisulfide (HS^- and thiosulfate ($S_2O_3^-$). Relatively low levels of hydrogen sulfide, bisulfide, and thiosulfate can have marked negative effects on crop yields. Changes in several elements with time under flooding conditions are depicted in Figure 1.1.

The change in redox potential is pronounced in paddy soils used for rice production. These soils are flooded for periods of time. At first, nitrate, manganese, and iron reduction takes place. Usually these changes have little adverse effect on production unless iron and/or manganese are present in large amounts. But if flooding continues too long, negative redox potentials

FIGURE 1.1. Changes in Several Elements with Time Under Flooding Conditions

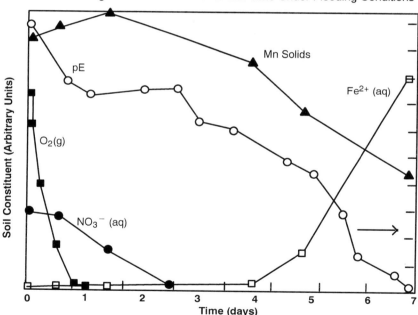

Source: Sposito, G. 1989. *The Chemistry of Soils.* Oxford University Press, New York, Oxford, based on data from F. T. Turner and W. H. Patrick. 1968. Chemical changes in waterlogged soils as a result of oxygen depletion. *Transactions IX Congress International Soil Science Society.* Adelaide, Australia, 4:53-65.

are present and sulfide, hydrogen, and methane can be formed, with serious negative effects on production. An outline of the progressive effects on redox potential and the accompanying changes taking place with increasing flooding are presented in Table 1.3, and in Figure 1.2.

TABLE 1.3. Changes in Redox Potential and Accompanying Chemical Changes with Increased Flooding

Step	Reaction	Redox potential (volts)	
	First stage:		
1	Disappearance of oxygen	+0.6	+0.5
2	Nitrate reduction	+0.6	+0.5
3	Formation of reduced manganese (Mn^{2+})	+0.6	+0.5
4	Formation of reduced iron (Fe^{2+})	+0.5	+0.3
	Second stage:		
5	Sulfate reduction and sulfide formation	0	−0.19
6	Hydrogen formation	−0.15	−0.22
7	Methane formation	−0.15	−0.19

Sources: Mengel, K. and E. A. Kirby. 1982. *Principles of Plant Nutrition.* International Potash Institute, Worblaufen-Bern, Switzerland; From a table by Takai, Y., T. Koyama, and T. Kamaru. 1957. Microbial metabolism of paddy soils. *J. Agr. Chem. Soc. Japan*, 31, 211-220.

It is more useful to express oxidizability of a soil with the term pE rather than voltage, where pE is the negative common logarithm of free electron activity, and is analogous to the expression of acidity as pH or the common logarithm of free proton activity. The pE value can be expressed quantitatively by the following equation:

$$pE = -\log (e-)$$

Large pE values indicate soil or other media with electron-poor (oxidized) species, while low pE values signify those with electron-rich (oxidized) species. The range of pE in soils varies from about +13.0 to −6.0. As pE drops below 11.0, O_2 as a gas can be reduced to liquid (H_2O). Nitrate reduction takes place at pE values below +8.0 as extra electrons

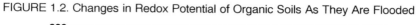

FIGURE 1.2. Changes in Redox Potential of Organic Soils As They Are Flooded

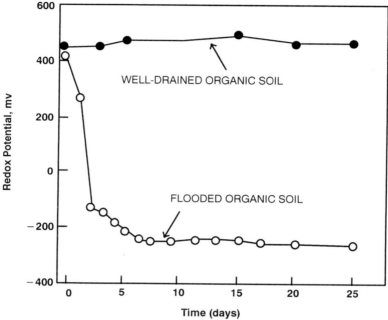

Source: Snyder, G. H. (Ed.). 1987. *Agricultural Flooding of Organic Soils.* Agricultural Experiment Station, University of Florida Institute of Food and Agricultural Science Bulletin 870 (Technical). Gainesville, Florida.

become available. At pE values of +7 to +5, reduction of Fe and Mn in solid phases can occur, but Fe reduction does not take place until O_2 and nitrate (NO_3^-) are depleted, while Mn reduction can take place in the presence of NO_3^-. Oxygen is consumed by aerobic microorganisms at pE values greater than +5, and it is not stable in neutral soils less than +5.0. A soil becomes anoxic (without oxygen) at a pE value of about +2, and at pE less than 0 enough electrons are present for sulfate to be reduced by anaerobic bacteria.

Not all of the changes taking place in a submerged or low-oxygen soil are harmful to plants. Besides the increase of available iron (Fe^{2+}) and manganese (Mn^{2+}), already noted, changes in pH, phosphate availability, and soluble cations can be positive. Beneficial changes in pH can occur especially as soils are flooded. Flooding tends to increase the pH of acid soils and decrease the pH of alkaline soils, with the final pH approaching the neutral point (pH 6.5 to 7.5). The increase in pH appears to be due to the capture of H^+ by free electrons present in the reduced environment. The rise can be beneficial if the soil is too acid. Fortunately the pH of calcareous and sodic soils decreases due to the solution of CO_2. Phosphate availability generally increases due to the decomposition of inositol hexaphosphate, an organic phosphorus compound, the decomposition of which is promoted by submergence. Phosphorus availability is also aided by the release of phosphates occluded by Fe^{3+} as it is reduced to Fe^{2+} on the oxide skin and also by hydrolysis of ferric hydroxide $Fe(OH)_3$. The increase of soluble Fe^{2+} and Mn^{2+} increases the cation exchange capacity (CEC) with a corresponding increase in soluble cations.

Despite these positive changes, the net result of reduced aeration is detrimental for most plants. Rice, because it can still obtain O_2 through the plant, actually does better at certain stages in a partially reduced soil atmosphere, provided that submersion is not overextended. Even the formation of hydrogen sulfide and its subsequent precipitation of Fe, Cu, Mn, and Zn can be beneficial for rice plants sometimes since the precipitation of some iron sulfide can reduce toxic amounts often present in submerged land to acceptable levels. But if overextended, the formation of methane and other toxic compounds can cause serious crop reduction. For the vast majority of plants that are unable to move O_2 into the root zone via the plant, even short periods of submersion carry multiple risks of inadequate O_2 for respiration, loss of volatile N compounds, plus deficits of micronutrients due to precipitation by sulfides as well as disruption of vital microbiological processes on which the plant depends.

SUMMARY

The maintenance of adequate amounts and quality of air in soils or other media is of vital importance in crop production, because of the effects of O_2 on plants and microorganisms. Both require O_2 for respiration in order to obtain energy for various life processes. Insufficient O_2 limits uptake of water and nutrients by plants and interferes with several functions of microorganisms (nitrification, nitrogen fixation, soil aggregate formation, breakdown of OM), which affect crop production.

The presence of sufficient amounts and quality of soil air depends on the ready removal of excess water from pore spaces, allowing air to replace much of the water. The quality of replaced air is gradually diminished as O_2 is gradually replaced with CO_2 resulting from respiration of both plants and microorganisms. The continuous renewal with sufficient

O_2 is only possible if there are sufficient large pore spaces, allowing for rapid drainage of excess water and exchange of spent air for that of the atmosphere rich in O_2. Factors that interfere with air movement (excess water, poor drainage, hardpans or compact soils, or lack of sufficient porosity) allow for the accumulation of CO_2 at the expense of O_2 in the soil pores. If there is very poor exchange of air, as happens with flooding, reducing conditions can prevail with the accumulation of harmful substances that can be toxic to plant roots and the loss of N in gaseous forms.

Chapter 2

Water

This chapter deals with water—its importance, plant requirements, and how it may be more efficiently used. No crop can be brought to full production without sufficient quantities of water, but the requirements for different plants vary tremendously—not only in the amounts needed but also the quality of water that can be tolerated. Also variable are the amounts required at different stages of growth. In this period of diminishing resources for growing crops, coupled with increasing demands for food and fiber, the intelligent use of water can easily make the difference between sufficiency or hunger for future world populations.

WATER AS PART OF THE FERTILE TRIANGLE

Unlike the pronounced inverse effect of water upon the aeration side of the triangle, that of water on nutrition is less obvious but usually more positively related. Generally, the addition of water benefits nutrition—from the standpoints of increased availability, uptake, and utilization. But excess water can reduce nutrition by reducing the amounts available through leaching and volatilization and by limiting the ability of the plant to take up that which remains.

The usual inverse relationship between the water and aeration sides of the triangle, which adversely affects nutrition as water is increased beyond a certain point, has been covered to some extent under "Oxygen" and "Reducing Conditions in Soils" in Chapter 1. The effect of water on increasing availability of nutrients, while broached in Chapter 1 as O_2 was considered, needs to be emphasized here. Not only is water necessary for the microbial release and fixation of nutrients, and the solution of nutrients from fertilizers and minerals, but also for the movement of nutrients into and upward throughout the plant and for the final incorporation of nutrients into vital components of the plant.

Vital microbial actions that release nutrient elements from organic materials and fix nitrogen from the air by both symbiotic and nonsymbiotic means are not possible without an adequate water supply. A more detailed description of microbial influence on availability of nitrogen is given in the section "Nitrogen" under "Essential Macronutrients" in Chapter 3. It is fortunate that soil–water relations which are ideal for most plant growth are also satisfactory for most of the microorganisms which provide various benefits for plants.

WATER FUNDAMENTALS

Water is essential for many vital functions of green plants, such as nutrient uptake and translocation, photosynthesis, and respiration. None of the myriad of chemical reactions taking place in the plant are possible without the presence of ample water. Except for the seed stage, water is the major portion of a plant's composition, being more than 90 percent of the mass of some water plants, but commonly 70 to 85 percent of fleshy plant material and even 50 percent of woody tissue.

Plant Water

Most cultivated plants require a continuous supply of water to produce good crops, and their growth rates are very dependent on water. Water is the major constituent of protoplasm. It is a major reactant in a number of metabolic processes and is the source of hydrogen atoms, which reduce carbon dioxide in the photosynthetic process. It is the universal solvent by which all ions and compounds are moved into or within the plant. Water helps to stretch the elastic cell walls to maintain full turgor, giving plants form and structure while promoting growth. It is also responsible for movement of many plant parts, which includes such important functions as swelling and shrinking of guard cells that regulate stomata opening and closing, the folding of leaflets, and the opening and closing of flowers.

Insufficient water can affect normal metabolism of plants, usually reducing vegetative and reproductive growth by affecting such important functions as photosynthesis, nutrient uptake and translocation, and transpiration. The harmful effects of such deficits largely depend on the severity of the deficit, its duration, and the stage of plant growth when it occurs. The changes in metabolism usually have multiple adverse effects on plants, which can range from restricting plant development and reproduction to making them more susceptible to disease and insect attack.

Some water is constantly being utilized to produce and maintain new cell development, but most of the absorbed water is lost from plants to evaporation from leaf surfaces by the process of transpiration. Plants can reduce but not stop transpirational losses by closing their stomata (small openings in leaf surfaces) through which CO_2 enters and water vapor escapes.

The amounts lost by transpiration vary with different plants and their stages of growth. Losses usually increase as plants grow in size and as solar radiation, air temperature, and wind velocity are increased, but are reduced as air humidity rises. If the plant loses water faster than it can be replaced, cells tend to shrink in volume, and if the deficit is continued, the protoplasm (the vital matter of the cell) tends to shrink away from the cell wall in a process known as plasmolysis. If severe enough, it can mean death of the cell and if enough cells are affected, the plant will go into a permanent wilt and die.

Long before this will occur or even before the plant begins to wilt, plant growth will be restricted because of such changes as reduced CO_2 intake due to closed stomata and by failure of new cells to develop or enlarge properly. In addition, the closing of the stomata, as a response to the water deficit, reduces transpiration, which in turn tends to increase canopy temperatures. The rise in temperature increases the respiration rate, hastening the depletion of carbohydrate reserves already lessened by reduced photosynthesis resulting from the closing of the stomata.

Some restriction in growth may not reduce yields, and in a few cases may increase them or produce better quality crops. Some plants require some restriction of growth to initiate or develop the fruit or seed production stage and to fully ripen fruits. For example, unrestricted water supplies during the fruit maturation period of cantaloupes, pears, and tomatoes produces fruit of poor quality. Such fruit will have lower solids, poorer color, and will ship and keep poorly.

Plant Requirements for Water

Amounts of water required for maximum economic yield (MEY) will vary with different plants grown under similar climatic conditions. Requirements for millets and sorghums are very low, corn and the small grains intermediate, and those of legumes and grasses high. Those of alfalfa are very high, in some years being almost twice those of small grains and four times those of sorghum or millet. The yields of broccoli, celery, endive, lettuce, and strawberry usually decline if the available water falls below 70 percent but yields of beet, rhubarb, sorghum, or watermelon are not lowered until the readily available water falls below 20 percent of the total.

Amounts of water required by crops at any moment vary with the age of the crop. While amounts needed vary with plants and climatic conditions, average values indicate high usage in the period when 40 to 80 percent of a plant's growth has been made. Maximum usage is during the period when about 55 to 70 percent of the crop's growth has been completed. If a plant takes about two months from germination to maturity, maximum use would be in the period 34 to 40 days from germination; for a plant requiring a four-month growing period, maximum use would be about 70 to 82 days following germination start.

The average amounts of water in acre-inches needed to produce several crops are given in Table 2.1. Although knowledge of average annual water requirements is useful, it is important to realize that the amount of water required to produce a crop varies not only with the crop but possibly in different years in the same location. The variation is due to genetic differences between crops but is greatly modified by climate. The climate affects not only intake and loss of water by the crop but also the length of time needed to produce the crop. The latter influences the effectiveness of the different forces. Table 2.2 shows some of the effects of different locations and time periods on water losses from several crops that received ample water in their development.

Reductions in Crop Yield Due to Insufficient Water

Critical Periods

There are critical periods in a plant's development when lack of water can be more serious than at other times. These may correspond to periods of maximum water usage but often coincide with an important stage of the plant's development.

TABLE 2.1. Average Annual Water Requirements to Produce Good Yields of Several Crops

Crop	Acre-inches of water
Truck crops	16
Cotton, small grains	18
Corn	20
Flowers, lawns, shrubs, sugar beets	24
Fruit trees	30
Alfalfa, irrigated pastures	36

Source: Donahue, R. L., R. H. Follett, and R. W. Tulloch. 1995. *Our Soils and Their Management,* Fourth Edition. The Interstate Printers and Publishers, Danville, IL, p. 314.

TABLE 2.2. Effect of Different Locations and Time Periods on the Water Losses
of Several Crops

Crop	Location	Crop duration	Evapotranspiration in	mm
Alfalfa	North Dakota	143 d (summer)	23.4	594
	Nevada	124 d (summer)	39.9	1,013
Grass	Canada	—	22.8	579
	Davis, CA	12 m	51.8	1,316
Barley	Wyoming	May-Aug.	15.2	386
	Mesa, AZ	Dec.-May	25.3	643
Beans	South Dakota	105 d	16.4	417
	Davis, CA	92 d	15.9	404
Corn	Ohio	124 d	18.5	470
	Bushland, TX	122 d	24.3	617
Potatoes	Alberta, Can.	—	19.9	505
	Phoenix, AZ	Feb.-June	24.3	617
Rice, flooded	Davis, CA	150 d	36.2	919
Sorghum	Kansas	—	21.7	551
	Mesa, AZ	July-Nov.	25.4	645
Wheat, hard	South Dakota	—	16.3	414
	Bushland, TX	Oct.-June	28.3	719
Sugar beets	Montana	Apr.-Sep.	22.5	571
	Kansas	Apr.-Nov.	36.5	927
Safflower	So. Idaho	Apr.-Nov.	25.0	635
Soybeans	South Dakota	—	15.7	399
Cotton	Arvin, CA	12 m	35.9	912
	Mesa, AZ	Apr.-Nov.	41.2	1,046
Cabbage, late	Mesa, AZ	Sep.-Mar.	24.9	632
Lettuce	Mesa, AZ	Sep.-Dec.	8.5	216
Peas, green	Alberta, Can.	—	13.4	340
Tomatoes	Alberta, Can.	—	14.4	366
	Davis, CA	May-Oct.	26.8	639
Apples	Wenatchee, WA	Apr.-Nov.	41.7	1,059
Oranges	Phoenix, AZ	12 m	39.1	993
Turf	Reno, NV	112 d	21.8	554

Note: d = days   m = months

Source: Miller, R. W. and R. L. Donahue, 1995. *Soils in Our Environment.* Prentice-Hall, Englewood Cliffs, NJ, p. 487. Reprinted by permission of Prentice-Hall, Inc., Upper Saddle River, NJ.

The point at which insufficient water affects yields (critical point) varies with different crops and stages of growth. For many plants, the period of seed germination and seedling development is critical. Relatively small reductions in the available water at transplanting, flowering, or the period of fruit development and enlargement can also seriously limit yields of many plants. Some critical periods for different crops are: silking and tasseling to soft dough for corn; boot to heading and heading to soft dough for sorghum; prior to antithesis and during enlargement of fruit for cucurbits; final swelling of peach fruits; head development in rice; seed filling of safflower; stolonization and tuber initiation of potato; and tillering and stem elongation of sugarcane. Critical periods of stress for several irrigated crops are summarized in Table 2.3.

TABLE 2.3. Critical Periods of Moisture Stress

Crop	Stress period*
Alfalfa	Soon after cutting or flower start for seed development
Apricots	Bud development and flowering
Barley	Early boot stage, tillering, and soft dough
Broccoli	Head formation and enlargement
Cabbage	Head formation and enlargement
Cantaloupe	Flowering to early fruit development
Citrus	Flowering to early fruit development
Corn	Silking and tasseling to soft dough
Olive	Just prior to flower start and fruit enlargement
Peach	Final swelling of fruit
Peanut	Flowering and seed development
Potato	Stolonization and tuber formation
Rice	Head development
Safflower	Seed filling
Sorghum	Boot to soft dough
Sugar beet	A few weeks after emergence
Sugar cane	Tillering and stem elongation
Tobacco	Knee high to flower start
Tomato	Flowering and period of rapid fruit enlargement
Watermelon	Flowering to harvest
Wheat	Booting, flowering, and early grain formation

* Seed germination is a stress period for most plants.

Source: Wolf, B. 1996. *Diagnostic Techniques for Improving Crop Production.* The Haworth Press, Inc., Binghamton, NY, p. 169.

Plant sensitivity to water shortages at different periods has been mathematically expressed as the crop susceptibility factor (CS), which can be written as:

$$CS = \frac{(x - x_1)}{x}$$

where x = marketable yield obtained from a control treatment that received sufficient water through its entire development, and x_1 = yield of an area that received ample water except for a particular growth stage. The crop susceptibilities for several important irrigated crops are given in Table 2.4.

Yield reductions that can be expected due to water shortages will vary with the importance of water at different stages of plant growth (see Table 2.5).

Plant Water Movement

Water moves into the plant through root hairs, which grow into pore spaces of the soil and are surrounded by water film. Water moves into the root due to two forces. (1) The difference between the high osmotic concentration of the cell solutes (4 to 10 atmospheres) and that of the soil solution (<1 atmosphere) separated by a semipermeable membrane in root hair cells favors water movement into the cells. The semipermeable membrane allows water to pass but restricts most ions. Water moves into the cells to balance out the higher osmotic concentration. (2) The pull of a film of water moving through the xylem upward through the plant as it is evaporated from leaf stomata creates a negative pressure or tension, which pulls water into the root cells. The negative pressure is also largely responsible for the movement of water up and throughout the plant.

The continued uptake of water into roots is favored by the movement of water films from adjacent soil particles to replace the water removed by the root hairs and the extension of the root forming new root hairs in untapped moist soil.

Problems may arise as the soil water is reduced, making it more difficult for the plant to obtain sufficient quantities. Much of the problem is due to increased water suction or tension by which water is held in the soil as water levels decrease. Also, increase in plant size may greatly increase the amounts needed to maintain vital functions. Deficits are more likely to result if root formation has been limited because of poor soil conditions or damage to roots by disease, nematodes, or insects. Climatic factors (high temperatures, low humidity, windy conditions) that favor water losses from the leaves by transpiration can cause the problem or greatly aggravate the seriousness of existing problems.

TABLE 2.4. Crop Susceptibilities (CS) to Water Shortages for Several Irrigated Crops

Crop	Growth stage	CS
Corn	Vegetative	0.25
	Silking and tasseling to soft dough	0.50
	After soft dough	0.21
Cotton	Prior to flowering	0.00
	Early flowering	0.21
	Peak flowering	0.32
	Late flowering	0.20
Grain sorghum	Vegetative six- to eight-leaf stage	0.25
	Boot to heading	0.36
	Heading to soft dough	0.45
	After soft dough	0.25
Peanuts	Vegetative to peak flowering and early pegging (30-50 days after emergence)	0.36
	Peak pegging and nut development (60-80 days)	0.24
	Late nut development and maturation (90-110 days)	0.12
Rice	Vegetative	0.17
	Reproductive and ripening	0.30
Southern peas	Vegetative—prior to flowering	0.13
	Flowering and early pod formation	0.46
	Pod development to maturation	0.43
Soybeans	Vegetative	0.12
	Early-to-peak flowering	0.24
	Late flowering and early pod development	0.35
	Late pod development to maturation	0.13

Source: Hiler, E. A. and T. A. Howell. 1983. Irrigating options to avoid critical stress: An overview. In H. M. Taylor, W. R. Jordan, and T. R. Sinclair (Eds.), *Limitations to Efficient Water Use in Crop Production*. American Society of Agronomy, Inc., Crop Science Society of America, Inc., and Soil Science Society of America, Inc., Madison, WI, p. 485.

FACTORS AFFECTING WATER STORAGE AND USE

Soil Water

The amount of water potentially available for plants is the difference between amounts entering and stored in the soil and that lost by evaporation from soil and leaf surfaces. Water entering the soil is largely determined by the amounts of rainfall and irrigation but modified by such factors as perme-

TABLE 2.5. Practical Guide to Irrigation Timing Based on Critical Periods

Growth stage	Yield reduction if water stressed* %	Irrigate when soil moisture depleted reaches** %
Corn		
Vegetative	25	65
Silking and tasseling to soft dough	50	30
After soft dough	21	75
Cotton		
Prior to flowering	00	—
Early flowering	21	60
Peak flowering	32	35
Late flowering	20	60
Grain sorghum		
Vegetative (six- to eight-leaf stage)	25	65
Boot to heading	36	45
Heading to soft dough	45	35
After soft dough	25	65
Soybeans		
Vegetative	12	80
Early-to-peak flowering	24	45
Late flowering, early pod development	35	30
Late pod development to maturity	13	80

* CS x 100. See Table 2.2.
** Percent of water held between MHC and wilting point.

Source: Hiler, E. A. and T. A. Howell. 1983. Irrigating options to avoid critical stress: An overview. In H. M. Taylor, W. R. Jordan, and T. R. Sinclair (Eds.), *Limitations to Efficient Water Use in Crop Production.* American Society of Agronomy, Inc., Crop Science Society of America, Inc., and Soil Science Society of America, Inc., Madison, WI, p. 481.

ability of the soil, speed of water delivery, and soil slope. That lost by evaporation from the soil is increased by air temperature and wind velocity but lowered by soil cover and air humidity.

The maximum amount of water that a soil can hold against gravity is designated as field capacity and is a variable portion of its moisture holding capacity (MHC). Moisture held at the MHC, field capacity, and permanent wilting increase with increased amounts of OM and clay contents, establishing relationships between moisture held at field capacity and at the permanent wilting points to soil textural classes. Usually, available moisture increases with increasing clay content of the textural class, but amounts held at the permanent wilting points of some clays are so large that less moisture is available for plants (see Table 2.6).

The field capacity can be estimated from data for the different soil classes or calculated from actual measurements. Approximate field capacity data along with other important water facts for different soil textural classes are given in Table 2.7. Approximate amounts of water needed per acre of different soil textural classes and root zones to reach field capacity are also presented in Table 2.8.

Field capacity can be determined by measuring the moisture remaining in a sample of previously saturated soil that is held in a special apparatus at a tension of 5 psi or $^1/_3$ atmosphere. The wilting point is determined by measuring the amount of water remaining in the sample after it is held in the apparatus at 15 bars tension or 225 psi. A number of laboratories can provide field capacity and wilting point data from tension tests.

TABLE 2.6. Typical Amounts of Water Held and Available by Different Soil Textural Classes

Textural class	Inches of water per foot of soil depth		
	Field capacity	Wilting point	Available water
Sand	1.2	0.3	0.9
Fine sand	1.5	0.4	1.4
Sandy loam	1.9	0.6	1.3
Fine sandy loam	2.5	0.8	1.7
Loam	3.2	1.2	2.0
Silt loam	3.5	1.4	2.1
Sandy clay loam	3.7	1.6	2.1
Clay loam	3.8	1.8	2.0
Silty clay loam	3.8	2.1	1.7
Clay	3.9	2.4	1.5

Note: Silty clay loam and clay soils may have appreciably less available water than other soils despite high field capacity due to the much larger amounts of water held at the wilting point by these two textural classes.

Source: USDA. 1955. Water. In *Yearbook of Agriculture.* USDA, Washington, DC, p. 120.

TABLE 2.7. Moisture-Holding Capacities and Other Important Physical Properties of Different Textured Soils*

Soil texture	Infilt. perm. in/hr	Total pore space %	Field capacity %	Perm. wilting point %	Total available moisture Weight %	Volume %
Sandy	2 (1-10)	38 (32-42)	6 (6-12)	4 (2-6)	5 (4-6)	8 (6-10)
Sandy loam	1 (0.5-3.0)	43 (40-47)	14 (10-18)	6 (4-8)	8 (6-10)	12 (9-15)
Loam	0.52 (0.3-0.8)	47 (43-49)	22 (18-26)	10 (8-12)	12 (10-14)	17 (14-19)
Clay loam	0.32 (0.1-0.6)	49 (47-51)	27 (23-31)	13 (11-15)	14 (12-16)	19 (17-22)
Silty loam	0.10 (0.01-0.2)	51 (49-53)	31 (27-35)	15 (13-17)	16 (14-18)	21 (18-23)
Clay	0.2 (.004-.04)	53 (51-55)	35 (31-39)	17 (15-19)	18 (16-20)	23 (20-25)

* Normal values are shown in parentheses. The lower values in each category are more appropriate for soils low in OM and the higher values for high organic soils.

Source: Hansen, V. E., O. W. Israelson, and G. E. Stringham. 1980. *Irrigation Practices and Principles.* Copyright © John Wiley and Sons, Inc. Reprinted by permission of John Wiley and Sons, New York.

Infiltration and Percolation

Water entry into the soil, which affects the amount available to plants, depends on its infiltration rate. Infiltration, or the ease with which water enters the soil, and percolation, which describes its movement after the soil is wetted, affects the usefulness of irrigation or rainfall.

The infiltration rate depends upon soil slope, presence of mulch, speed of applied water, soil texture, soil structure, sequence of soil layers, depth of water table, moisture content, chemical content of the water, and length of time that water has been applied.

Infiltration of water can vary from less than 0.1 in/hr to more than 5 in/hr. Common infiltration rates are in the range of 0.3 to 1.0 in/hr. Values <0.2 in/hr are extremely slow and can lead to wasted water as runoff or ponded

TABLE 2.8. Approximate Amounts of Water Needed Per Acre for Different Soil Textural Classes and Root Zones to Reach Field Capacity*

Depth of root zone (in)	Sands		Loams		Clays	
	in	gal/ac	in	gal/ac	in	gal/ac
9	0.25-0.5	6,500-13,500	0.5-0.75	13,500-20,000	0.75-1.0	20,000-27,000
18	0.5-1.0	13,500-27,000	1.0-1.5	27,000-40,500	1.5-2.0	40,500-55,000
36	1.0-2.0	27,000-54,500	2.0-2.5	54,500-67,500	3.0-4.0	81,000-110,000

* Based on irrigation applied when one-half of available water in the effective root zone has been used. The lower values are for soils low in organic matter.

water that can cause serious problems of soil crusting and poor aeration. Rates of 2 to 5 in/hr are very rapid, signifying poor water retention and water being wasted by excessive percolation.

The best infiltration is from level or near-level land. If sloped, contour farming aids water infiltration. Application of water to soil with considerable slope or poor infiltration due to soil dispersion or crusting is still practical if water is added very slowly, as is possible with drip irrigation.

Percolation is primarily dependent on continuous pore spaces in the soil. The flow rate is greatly affected by the size of the continuous pores, increasing about four times for every doubling of the pore diameter. Because of this dependence, it is closely related to soil texture, increasing with coarse texture, but also increasing with good structure, increased amounts of organic matter, and absence of compaction. Other factors that increase its rate are increased depth of soil, decreased soil moisture (up to a point), and increased soil temperature.

Generally, sandy soils will have better percolation rates than the heavier soils, but fine-textured soils that have ample organic matter, good structure, and well developed stable aggregates will also infiltrate water rapidly—often more rapidly than the fine sands. (The relationship of infiltration rates and several soil classes is given in Table 2.9.)

The range for the coarse-textured soils is very wide, making such approximate values of little use for specific conditions. Even with loams, the infiltration rate can vary from 0.3 to 0.8 in/hr. It would be useful to have more specific data for individual fields, but determinations of infiltration in the laboratory or of undisturbed soil cores are laborious, expensive, and not always accurate. Establishing suitable infiltration rates for most growers is largely a trial and error procedure, although average data for the different textured soils can be used as a starting point (Tables 2.7 and 2.9) and a simple infiltrometer (Figure 2.1) can give a grower infiltration data that could be more useful than average values. Recording the time needed to empty the water from the jug into the soil supplies an approximate infiltration rate. Repeated trials will usually give slower values as the soil is wetted, but comparative values of new soils with previous runs can be useful in estimating infiltration.

Very low or high salt content of irrigation water impedes infiltration into soils with appreciable silt and clay. Waters very low in calcium or high in sodium or total salts tend to move very slowly into heavy soils. Remedies include adding calcium salts to irrigation waters low in calcium or total salts, and gypsum to soils for better penetration of waters high in salts.

Generally, infiltration rates also decrease as soil moisture rises, as happens with continued rainfall or irrigation, but some soils can become so dry that it is difficult to wet them.

Some of the slowdown due to continued watering is caused by the collapse of some large pores as the soil is wetted. Large pores formed by tillage are especially prone to this type of collapse. Raindrops or water drops from irrigation, as they impact on the soil surface, also tend to slow infiltration. The impaction tends to detach soil particles, producing a crust. Also, the detached particles move downward, plugging pores, and so reduce the rate. The harmful effects caused by large drops from overhead irrigation can be reduced by using different types of irrigation or changing

TABLE 2.9. Infiltration Rates as Affected by Soil Textural Classes

Final infiltration		
Rating	in/hr	Soil classes
High	1.2-3.2+	sand, sandy loam, loam, clay loam
Medium high	0.6-1.2	loam, silt loam
Medium low	0.2-0.6	clay loam, clay, silty clay loam
Low	0.08-0.2	clay, adobe clay

Source: FAO. 1971. *Irrigation Practice and Water Management.* Irrigation and drainage paper #1. Water Resources and Development Service, Land and Water Development Division. FAO, Rome.

FIGURE 2.1. A Simple Infiltrometer

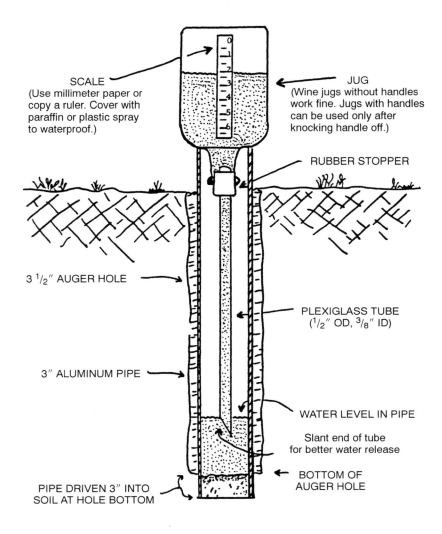

SCALE
(Use millimeter paper or copy a ruler. Cover with paraffin or plastic spray to waterproof.)

JUG
(Wine jugs without handles work fine. Jugs with handles can be used only after knocking handle off.)

RUBBER STOPPER

3 1/2" AUGER HOLE

PLEXIGLASS TUBE
(1/2" OD, 3/8" ID)

3" ALUMINUM PIPE

WATER LEVEL IN PIPE

Slant end of tube for better water release

BOTTOM OF AUGER HOLE

PIPE DRIVEN 3" INTO SOIL AT HOLE BOTTOM

Source: Neja, R. A., W. E. Wildman, and L. P. Christensen. 1982. *How to Appraise Soil Physical Factors for Irrigated Vineyards.* Leaflet 2946. Division of Agricultural Science, University of California, Berkeley, CA.

nozzles to supply smaller drops. Mulches, or large amounts of crop residues on the surface, are also helpful in maintaining good infiltration rates.

Water infiltration rates are slowed whenever there are contrasting layers, such as a clay layer with smaller pores underneath a sandier, more open layer, as is often found in the B horizon. Plow soles or other dense compact layers or even bedrock may also slow infiltration, increasing the likelihood of overly wet areas, even to the point of ponding.

In a few cases, infiltration is slowed by oily substances that can be present in some soils as the result of production of certain organic materials after prolonged growth of some plants or after some organic materials are burned, as is the case after brush fires.

Slow permeability in most cases can be increased by improving soil structure, using good quality water, adding composts or manures, and removing excess salts. Excessively rapid permeability is remedied by the addition of manures, organic matter, or barriers at strategic depths.

Water Storage

Amounts stored in the soil are the result of amounts infiltrated less losses from drainage and evaporation from soil or plant (transpiration). The amount retained by the soil against drainage is highly dependent on the amount of surface area exposed, making soils rich in clay and organic matter capable of storing large amounts of water.

Increased storage is an asset if there is sufficient aeration to utilize the water. Such water usually provides for greater crop yields, but in the final analysis, it is the extent and health of the root system that largely determines the amount of the stored water that is useful to the plant at any given moment.

Potentially Available Water

Only the water held between field capacity (maximum amount of water held by the soil immediately after wetting when excess is drained) and the wilting point (when plants wilt permanently) is potentially available to plants. All hygroscopic water and part of the capillary water held by the soil at tensions greater than 15 atmospheres is not available to most plants.

Not all soil water held between field capacity and the wilting point is equally available to plants because of its position or because it is held by attraction to other molecules of water or to the soil. Availability by position can be modified by large effective root systems and the presence of sufficient micropores that help move water to the roots. The force by which water is held in soils increases as the amount of water decreases, making it necessary to supply sufficient water for it to be effective.

This force has been quantified by atmospheres of pressure or atmospheres of suction (tension) or negative pressure necessary to remove the water. Water in a saturated soil is held at 0 atmospheres but increases to about $1/3$ atmosphere at field capacity (maximum amount of water held after drainage), to 15 atmospheres at permanent wilting, about 1000 atmospheres in an air-dry soil and 10,000 atmospheres in oven-dry soil. These amounts equal 0 bars suction at saturation, 0.338 bars at field capacity, 15.1 at the wilting point, 1013.3 for air-dried soil, and 10,133 for oven-dried soils.

Water tension is often expressed as bars, where one bar equals 0.987 atmospheres. For practical use, a bar is divided into 100 centibars, with one centibar equivalent to the suction of a water column 10 cm (4 inches) high. The tension or critical point at which several crops need to be irrigated is presented in Table 2.10.

TABLE 2.10. Soil Tensions at Which Several Crops Should Be Irrigated*

Crop	Tension (cbrs.)	Crop	Tension (cbrs.)
Avocado	50	Grape	40-80
Beet, sugar	60-85**	Lemon	40
Beet, table	40-60	Lettuce	40-60
Broccoli, early	45-55	Onion, dry	55-65
Broccoli, late	60-70	Onion, green	45-65
Carrot	55-65	Orange	20-80
Cabbage	60-70	Potato, Irish	30-50
Cantaloupe	35-40**	Potato, sweet	40-70
Cauliflower	60-70	Rhubarb	60-85
Celery	20-30	Sorghum	60-85
Deciduous fruit	50-80**	Watermelon	50-85
Endive	30-40		

* Where two readings are given, the lower one should be used for conditions of high evaporation. The higher values are for low evaporative conditions or, if marked with double asterisks (**), for the period when the crop approaches maturity.

Source: From a table presented by D. Goldberg, B. Gormat, and D. Rimon. 1976. *Drip Irrigation.* Drip Irrigation Scientific Publications, Kfar Schmaryanu, Israel, and modified by the author's data.

Although these forms of measurement are still used, there has been a movement in recent years to use soil water potential, or the ability of the water to do work as it is moved from its present state. The potential is measured in kilopascals (a standard metric unit), where 1 kilopascal (kPa) = 0.01 bar or 0.009869 atmospheres. The kPa are given a positive value as water is held in excess that will be drained or a negative value as it is held by the soil with various tensions. The potential or kPa is zero at saturation, -33 at full MHC and -1500 at the wilting point. The degree of work or energy needed to remove the water at the wilting point is about 45 times the amount needed at the MHC.

The availability of water also depends on the form in which it is present. The water may be present in three different forms with varying availability: (1) hygroscopic, which is tightly held by soil particles and of no value (-1500 to -3100 kPa); (2) capillary, held by surface tension in pore spaces between soil particles at -33 to -1500 kPa, and the principle source of water for the plant; and (3) gravitational or free water with kPa values of -33 to 0, which moves downward with gravity, and which has little or no value for plants. Free water after saturating the soil will percolate downward, leaving the soil at field capacity.

Available Water

The portion of potential water actually available depends on the amount of water held in the soil, its distribution in relation to the root system, and how tenaciously the water is held by soil particles. Only about 50 to 75 percent of the potential or total available moisture held between field capacity and the wilting point is readily available for most plants.

All the water held between field capacity and the wilting point is not equally available. Plants tend to readily absorb weakly held water near field capacity but find it more difficult to remove water as the amount is lessened, the difficulty increasing as the wilting point is approached. There is some water near or at the wilting point but plants generally cannot make use of it.

The availability of capillary water, which is the primary source of available water in soils, depends on soil micropores, which not only hold water but provide for its movement from lower layers. For effective water supply, there must be sufficient numbers of micropores to hold enough water against drainage forces and they must be located where they are readily reached by plant roots. Sufficient micropores properly handled allows for dryland farming. Soil with insufficient micropores tends to be droughty and unproductive unless ample rains or irrigation are available.

Movement of Soil Water

Saturated water, which has potentials less than − 33 kPa, moves downward primarily by gravitational forces. (In some sandy soils, appreciable water can be held against gravity in the − 10 to − 30 kPa range.) As potentials increase (become more negative), unsaturated water moves in any direction by matric and osmotic forces but from wetter to drier areas. The rate of movement increases as water potential gradient (difference between wet and dry areas) and size of the water-filled pores increase.

The point at which available water becomes critical as it falls below field capacity depends on the interception of new moist areas by developing roots but also on the upward movement of water in the capillaries. Movement downward due to gravity takes place only when water in the upper layers of soil is in excess of field capacity. Although such water can have an adverse effect on plant growth due to the leaching of nutrients, it has little effect in nurturing the plant. It is the capillary water moving upward in soils that helps maintain the plant.

The upward movement of water through the capillaries of coarse soils (sands) is very limited. The poor storage of water and poor upward movement in sands account for the great dependence of these soils on frequent rains or irrigation. The upward rise in fine-textured soils (clays), although much greater than in sands, is so slow that water at deeper layers often is of little value. Despite excellent storage facilities, it may be necessary to irrigate such soils if roots have not reached deeper layers where more water may be available. Generally, it is in the medium-textured soils (loams) that movement of water is rapid and extensive enough to provide a substantial amount of water from lower depths.

INSUFFICIENT WATER

Seldom does nature supply optimum amounts of water during the entire growing period. In many areas of the world, there is not enough water to grow most crops. In other areas, there may be excesses at times but more often there are periods when the water supply is insufficient.

The optimum amounts of water have to be considered in terms of the type of crops grown and also their yield possibilities. High-yield possibilities are often met by a combination of sufficient water and adequate fertilization. Fertilizer without sufficient water often is not profitable and at times can lower yields. But the judicious use of water with fertilizers increases crop yields and provides for more efficient use of water.

Water requirements to produce a unit weight of a crop can be reduced by adequate nutrition, at least up to a certain point. At low levels of fertility (yields reduced by 50 percent due to nutrient shortages), adding sufficient soil fertility can reduce water requirements to produce a unit of crop by one-half to two-thirds. Once soil fertility is improved so that yields are appreciably more than 50 percent of a full fertility program, the water requirement to produce a unit weight of crop tends to remain about the same.

Despite a constant water requirement to produce a unit of crop with fairly adequate fertility, water needs per acre to obtain the full potential of the crop may still rise with additional fertilizer application due to the increased yields.

Most crop production needs to be sustained with adequate fertilization to obtain economic yields. Generally, these will be at levels where the efficiency of water to produce a unit crop is relatively constant. Nevertheless, there will be shortages of water—at least for short periods—in many areas of the world. Some of these shortages can be met by (1) dryland farming or (2) irrigation.

Dryland Farming

By carefully husbanding rainfall, it is possible to grow drought-tolerant crops in many areas of the world that receive marginal amounts of rainfall. Most of these areas receive the limited amount of water in a concentrated rainy season. Proper management can utilize the marginal amount to grow satisfactory crops by (1) selecting crops that make most of their growth during the rainy season; (2) reducing unnecessary waste of water; and (3) storing rain from one season to the next by means of fallowing.

Dryland Crops

Grain crops grown during periods of the year when rains are expected have been used for most dryland farming. Wheat and barley are grown in the cooler regions, while sorghum and millet are better suited for the hotter regions. Barley is more suited than wheat for the cooler areas because (1) it needs a shorter growing period and (2) it does better on lighter soils, which warm up earlier in the spring. Sorghum can survive in very hot dry climates but good yields are possible only if it can have an adequate supply of water during the heading period. A good crop of winter wheat is possible only if there is sufficient water just before winter dormancy starts.

Amounts of Water Needed

A certain amount of water is needed to grow the crop before any yield is made. This varies in different locations. In Kansas, wheat is kept alive by about the first 8 to 10 inches of rain, but every inch above this minimum will yield about two additional bushels per acre. In the Great Plains, the depth of moist soil at the planting time of spring wheat decides the extent of crop failures. If less than 1 foot is present, a crop failure can be expected five out of six years but if the moist soil extends to 3 feet or more, crop failure is reduced to only one year out of four. In actual inches of available water, a crop failure of winter wheat is imminent if the soil contains less than 1.5 inches at the time of planting but success is assured if the amounts are greater than 3.0 inches.

Fallowing to Increase Soil Moisture

At times when there is insufficient moisture for planting, fallowing (soil preparation without seeding) can increase the amount of available water for the next season. Fallowing works only if rains fall during the fallow period in sufficient quantities to increase the moisture supply below the 4- to 8-inch depth. If it does not wet the soil to this depth, there is a good chance that most of it will be lost to evaporation. Other conditions include a soil that can store a significant amount of water down to a depth of 4 to 6 feet and be open enough so that plant roots can readily make use of it.

Soils must be maintained in an excellent tilth to effectively store useful water during the fallow period. Weeds need to be controlled, but left on the surface to reduce water loss. There is a good chance that considerable soil will be lost by erosion unless special precautions are taken. Cultivating the soil with ridging plows or listers in the direction of the contours will reduce erosion. It can also be reduced by alternating fallow strips with crops across the direction of the prevailing winds. Cultivation that allows the previous year's stubble to remain on the surface is superior to plowing. The practice of killing the crop with herbicides and leaving the organic matter on the surface as is done in reduced tillage is useful, providing that the surface soil can be prepared to readily accept rainfall. Equipment such as the Water Check (Marion, ND), which prepares numerous water-holding reservoirs, is helpful in trapping most of the rainfall. Cultivating the soil a few times to keep it open and free of weeds during the fallow season helps in storing water for the following crop.

Irrigation

Irrigation is being used with increasing frequency to supply needed water. In some of the areas being irrigated, there may be enough annual rainfall to grow a number of crops (at least those that have low water requirements), but irrigation is resorted to for two reasons: (1) rain does not fall in sufficient quantities to adequately maintain the crop at all times, and/or (2) it is profitable to grow crops with water requirements higher than nature will supply on a regular basis.

The benefits of irrigation, although appreciable in the short run, may not be long-lasting. History is full of examples of irrigation failing to maintain satisfactory crop production for many years. The principle reasons for the poor long-term performance are: (1) exhaustion and/or pollution of the water source; (2) use of poor-quality water that degrades the soil; and (3) excessive use of water that results in nutrient leaching or raised water tables with concomitant oxygen shortage and salinization. Many of these problems can be avoided by the use of high-quality water (see Chapter 15) and of adequate drains to remove excess water and the salts that may be introduced. Means to lessen these problems and make the benefits longer-lasting are discussed in various sections that follow.

The need for irrigation will largely depend upon the amounts of moisture held by a soil and how well plant roots are able to take up this moisture. The amounts of moisture stored in the soil depend on the efficiency of rainfall or irrigation to percolate into the soil and be held by it. The amounts taken up by plants depend on depths of root systems and the tenacity by which the water is held by the soil.

Critical Periods

Response to irrigation is likely at most stages of growth if the plant is under severe stress from lack of water, but response at the critical periods can be obtained if only mild stress is present. Conditions suitable for starting irrigation during these periods are presented in Tables 2.5 and 2.10.

Roots and Available Water

The amount of available water is finally determined by the extent and health of the root system. The extent of the root system varies with different species but is modified by such factors as soil compaction, height of water table, soil pests, soil pH, and nutrient content. Soil compaction generally restricts depth and extent of rooting as does a high water table. Low pH values, especially when accompanied by high aluminum and/or manganese concentrations, restrict depth and extent of rooting of many sensitive crops. Lack of nutrients, especially phosphorus and calcium, can also prevent the development of some plant roots in lower depths. Average rooting depths of several crops are given in Table 2.11.

Reductions in Crop Yields

Irrigation needs to be applied only if the soil contains inadequate amounts of available water. As was pointed out above, the point at which yields are adversely affected (critical point) varies with different crops

TABLE 2.11. Typical Rooting Depths of Various Crops

Shallow (12-24 in)	Moderately deep (25-48 in)	Deep (36-48+ in)
Broccoli	Barley	Alfalfa
Brussels sprouts	Bean, bush	Apple
Cabbage	Bean, pole	Artichoke
Cauliflower	Beet	Asparagus
Celery	Carrot	Bean, lima
Chinese cabbage	Chard	Cherry
Corn, sweet	Clover	Citrus
Endive	Cucumber	Corn, field
Garlic	Eggplant	Cotton
Leek	Muskmelon	Parsnip
Lettuce	Mustard	Peach
Onion	Pea	Pear
Parsley	Pepper	Pumpkin
Potato	Rutabaga	Sorghum
Radish	Squash, summer	Squash, winter
Spinach	Turnip	Sugarcane
Strawberry		Tomato
		Watermelon

Sources: Lorenz, O. A. 1988. *Knott's Handbook for Vegetable Growers,* Third Edition. John Wiley and Sons, New York; Hillel, D. 1988. *The Efficient Use of Water in Irrigation.* World Bank Technical Paper #64. The World Bank, Washington, DC; Neja, R. A., W. E. Wildman, and L. P. Christensen. 1982. *How to Appraise Soil Physical Factors for Irrigated Vineyards.* Leaflet 2946, Division of Agricultural Science, University of California, Berkeley, CA.

(see Table 2.3). Although it is common practice to irrigate when 50 percent of the available soil moisture (held between field capacity and wilting point) is depleted, there are some plants that suffer yield loss if the available water falls below 70 percent, and others are not affected until readily available water falls below 20 percent of the total. The different methods for evaluating the need for irrigation, the effects of water quality on yield, and the different irrigation systems are presented in Chapter 15.

EXCESS WATER

Damage from excess water is primarily due to the lack of oxygen as the excess water replaces the air in soil pores and limits the amount of gaseous exchange. Insufficient gaseous exchange causes variable harmful effects depending on the type of plant, its stage of development, duration of the flooding, and temperatures during the flooding period.

Susceptibility to Damaging Effects of Excess Water

Plants vary considerably as to the damage produced under a given set of conditions. Rice and willow can undergo long periods of flooding with no visible effects. Of the plants injured by excess, sensitivity of the small grains follows the order:

barley > winter wheat > rye

The order for warm-season turfgrasses is:

Bermuda grass > Bahia grass > Saint Augustine grass > zoysia grass > centipede grass

For fruits, the order is:

quince > pear > apple > citrus = plum > apricot = peach = almond > olive

Blueberry, cranberry, elderberry, gooseberry, and grape show some tolerance, but blackberry, black currant, kiwifruit, papaya, raspberry, and strawberry are intolerant. False aralia, *Carissa grandiflora*, and geraniums are extensively damaged by too much water. Pasture legumes generally are more sensitive to flooding than grasses. The grain legumes—French bean, cowpea, and soybean—are more sensitive than corn.

Tomato, tobacco, English pea, and papaya are readily harmed by short periods of flooding. If not overwhelmed initially, tomato can produce adventitious roots that help it survive long periods of high water tables. Tobacco, pea, and papaya, which cannot readily produce adventitious roots, are killed under conditions that tomato can survive.

Plant symptoms produced by excess water often are somewhat similar to those produced with shortages, e.g., rolling or wilting and the yellowing of leaves, although wilting is not common. Root hairs may become brown and root systems tend to be smaller than normal. Most roots will die with prolonged flooding. The high water table may accentuate several soil-borne diseases with resulting root and lower stem damage.

Excess water can also damage soil structure. Soil compaction tends to increase as wet soils are worked. Soil erosion is often aggravated with wet conditions.

Critical periods when relatively brief flooding can cause considerable harm are seed germination, rapid periods of root or shoot development, pollination, and the fruiting period. The meristematic tissue (rapidly dividing tissue) of roots is much more affected than older tissue.

The extent of damage for any plant or stage of development depends on the duration of flooding and the temperature. Greater harm will be done as temperature rises and the flooding period lengthens. If the excess water is drained away quickly, very little or no damage may be apparent, particularly if temperatures are low. If not quickly drained away, damage often can be lessened if plants are shaded during the flooding period.

Excess Water and Nutrition

Excess water can have devastating effects on plant nutrition. The water can leach away several important nutrient elements, particularly nitrogen, from many soils. Potassium, magnesium, and sulfate-sulfur can also be leached from light soils. The restricted amounts of oxygen limit microbial action, further reducing amounts of nitrogen produced by both symbiotic and nonsymbiotic organisms. The lowered oxygen content also reduces the activity of many other organisms, some of which are responsible for the release of nutrient elements from OM. Aggravating the effect of lower quantities of nutrient elements is that long periods of air shortages impede uptake of existing nutrients. The mechanism of restriction is not completely understood, but lack of oxygen impairs respiration, restricting the amount of energy needed for uptake of nutrients. Also, the incomplete breakdown of OM by microorganisms releases several volatile compounds that can inhibit root development or lead to root breakdown and decay. The poor function of roots is worsened by accumulations of carbon dioxide. (See Chapter 1 for additional effects of submerged conditions.)

SUMMARY

Water has a profound effect upon the other two sides of the fertile triangle. It usually is inversely related to soil air, so that a balance between water and aeration is essential for good crop production. Water also affects the nutrition side of the triangle, although the relationship of water to nutrition is not so closely related as that of air and water. Excess water can reduce the levels of many nutrients by leaching and that of nitrogen by denitrification. On the other hand, satisfactory levels of water increase nutrition levels by dissolving nutrients from minerals and fertilizers and by favoring the action of microorganisms.

Water is essential for crop production because it is essential for photosynthesis, respiration, growth, and reproduction. Lack of sufficient water can not only reduce crop yields and quality and increase susceptibility of crops to disease and insect attack, but ultimately lead to premature death of plants.

Crops vary in the amounts needed and in times when little stress can be tolerated. They also may require different amounts of water at different locations because of the effects of climate upon water use.

Dryland farming practices or irrigation can be used to grow successful crops when nature supplies inadequate amounts of water. Dryland farming utilizes crops requiring relatively small amounts of water and makes maximum use of existing water supplies by carefully husbanding that which falls by using systems of fallowing and soil preparation. Irrigation supplies extra water needed to complement inadequate amounts or that which fails to fall at critical times.

The need for irrigation and its effective use depends not only on the crop and its stage of development but also on the availability of soil water. Only the soil water held between field capacity (water remaining in soil after it has been flooded and water allowed to drain) and wilting point is available for plants. Because of the tension by which water is held in the soil, that water held close to the moisture holding capacity (MHC) is most available—with availability diminishing as the water level recedes from the MHC, becoming virtually unavailable at the wilting point.

The effectiveness of irrigation depends upon infiltration and percolation of water in the soil and on plant root systems capable of intercepting the added water. Infiltration is largely a matter of soil class, soil structure, slope, and rate of water supply. Poor infiltration can lead to flooding or erosion, but often can be improved by general improvement of structure, use of mulches, slow application of irrigation, or intermittent irrigation.

Excess water can be harmful to plants primarily due to lack of O_2 in overly wet soils. Crops vary in their tolerance to excess water and in critical stages when it is most damaging. Damage from excess water can be relieved by improving water infiltration through improvements in soil structure, providing irrigation slowly or intermittently or by introducing drainage.

Chapter 3

Nutrients

Nutrition, the third side of the fertile triangle, is closely related to the aeration and water sides, but unlike air and water, which are negatively correlated, nutrition and aeration, or nutrition and water, are usually positively correlated. Soil nutrients are of little value to the plant unless both sufficient air and water are present, but if air or water is excessive, the benefits of nutrition are reduced or eliminated.

Both air and water are closely related to the release of nutrients from OM, the production of nutrients by microorganisms, and the uptake of nutrients by the plant. It is only when both air and water are optimum that nutrient release and production as well as uptake by plants are at full capacity. Either low or high levels of air or water reduce availability of nutrients and their uptake. At low water levels, the solution of minerals, the movement of nutrients in soil solution, the production by microorganisms, and uptake itself are halted or slowed; at high water levels, nutrients may be lost to leaching or volatilization, and there may not be sufficient uptake due to lack of oxygen. Since aeration is usually inversely related to soil water content, nutrition also is largely dependent on the relationship between water and air.

In this chapter, we examine the various nutrient elements needed by the plant, how nutrients are moved from the soil to the plant, amounts removed by different crops, the timing of the removal, the effects of deficiencies or excesses, the amounts needed for optimum growth in soils and other media, influence of climatic factors on availability of nutrients, and the effects of several nonessential elements on plant growth.

The effects of soil factors on nutrients are examined in several other chapters: type of soil, aeration, and moisture in Chapter 4; organic matter (OM) in Chapter 5; pH in Chapter 6; cation exchange capacity (CEC) in Chapter 7; and salts in Chapter 8. The amounts of nutrients added by manures, composts, and sludges are covered in Chapter 5. Nutrients supplied by inorganic materials and the effect of different types of fertilizers as well as the manner in which they are applied are examined in Chapter 16.

IMPORTANT ELEMENTS

Plants are dependent on a minimum of 16 chemical elements for growth and reproduction. These elements and their chemical symbols are: carbon (C), oxygen (O), hydrogen (H), nitrogen (N), phosphorus (P), potassium (K), calcium (Ca), magnesium (Mg), sulfur (S), boron (B), chlorine (Cl), copper (Cu), iron (Fe), manganese (Mn), molybdenum (Mo), and zinc (Zn).

Several of these elements (C, H, O, N, and S) are needed for the production of organic matter: P for energy transfers; Fe and Mg for making chlorophyll; K, Ca, Mg, and Mn for activating enzymes (K, Ca, and Mg along with B, Na, and Cl also control osmotic potentials, membrane permeability, and conductance); Mo for the reduction of nitrate (NO_3-N); and some (Fe, Cu, and Zn) are constituents of enzymes that initiate and help carry out a number of chemical transformations at ambient temperatures and pressure.

Several other elements, such as aluminum (Al), cobalt (Co), chromium (Cr), nickel (Ni), silicon (Si), sodium (Na), titanium (Ti), and vanadium (V) appear to be useful for certain plants but have not been proven essential for all green plants. Cobalt is considered to be essential for N-fixing bacteria and the association of Rhizobium with legumes. Nickel has been beneficial for plants receiving most of their N from urea because of its effect on the function of urease, an enzyme that converts urea to ammonia (NH_3). Silicon serves as a strengthening agent for rice, small grains, and some grasses; it reduces susceptibility of these crops and of tomato and cucumber to certain fungal diseases, and stimulates the growth of tomatoes. Titanium may be stimulatory to several crops, possibly by aiding in the synthesis of chlorophyll.

Macro- and Micronutrients

The nutrients have been classified on the basis of amounts used by plants. In earlier periods, the primary nutrients added as fertilizers (N, P, and K) were listed as major nutrients. Other essential elements brought in by the ingredients supplying the N, P, and K or by liming materials were classified as secondary elements (Ca, Mg, S). As the importance of elements needed in very small amounts was determined, these were named trace or minor elements (B, Cu, Fe, Mn, Mo, Zn).

In recent years, the major and secondary nutrients along with C, O, and H have been referred to as macronutrients, while those formerly listed as minor or trace and others used by the plant in minute quantities are now considered micronutrients. Chlorine, an essential element required by the plant in very small amounts, is listed as a micronutrient, although it often is found within the plant in concentrations as great or greater than P, Mg, or S.

Other Elements in Plants

Nearly all elements found in soil or other substrates or in irrigation water can be found within the plant. Many of these elements are present in very small concentrations and appear to be of no significance. Sometimes, the presence of fluorine (F) may cause serious problems with plants sensitive to it and plants may contain enough chromium (Cr), lead (Pb), nickel (Ni), or selenium (Se) to be harmful to plants, animals, or humans.

Sensitive Elements

There are a number of elements, a few of which are essential for plant growth, to which various plants are very sensitive. The list includes aluminum (Al), arsenic (As), bromine (Br), boron (B), cadmium (Cd), chloride (Cl), copper (Cu), fluorine (F), lead (Pb), lithium (Li), and nickel (Ni). Aluminum is of concern because of its common presence in acid soils. Boron, an essential element, is also of concern because it is easily overdosed for certain crops, especially if grown on light soils. The heavy metals (Cd, Cu, Pb, and Ni) can be introduced in large amounts in sewage sludge and may be of greater concern in the future as the sludges are utilized more fully.

There is considerable variability in the sensitivity of various plants to several of these elements, making it possible to limit the damage to crops in their presence by avoiding crops that are sensitive. Some of the crops sensitive to several common elements are presented in Table 3.1.

Sources of the Elements

The nutrient side of the triangle is largely dependent on elements in the soil solution. Air supplies some nutrients (C and O) and water adds H and O. Some nutrients may enter the plant through the leaves or stems but for maximum crop production, most essential elements need to be present in the soil solution in available forms and in sufficient but not excessive quantities. Although O_2 is absorbed through the leaves, a certain amount of O_2 also must be present in the soil solution for most plant roots to function properly.

TABLE 3.1. Plants Sensitive to Low Levels of Several Elements

Aluminum	Boron	Chloride	Fluoride	Sodium
Alfalfa	Apple	Avocado	Asparagus,	Almond
Asparagus	Apricot	Apple	Sprengeri	Apple
Barley*	Artichoke,	Beans,	Apricot,	Avocado
Beets,	Jerusalem	Broad	Chinese	Banana
Sugar	Avocado	Dwarf	Blueberry	Beans,
Table	Beans,	Navy	Box Elder	Field
Blackberry	Snap	Runner	Coffee,	Kidney
Bluegrass,	Navy	Blackberry	Arabica	Lima
Kentucky	Blackberry,	Cherry	Cordyline,	Snap
Cabbage	Thornless	Gooseberry	Baby Doll	Celery
Carrot	Cherry	Lettuce	Dracena,	Cherry
Cantaloupe	Chrysanthe-	Onion	Striped	Citrus
Cauliflower	mum	Pea	Corn,	Clover,
Celery	Fig,	Plum	Sweet	Alsike
Clovers,	Kadota	Potato	Crocus	Crimson
Ladino	Grape	Raspberry	Dracena,	Dutch
Red	Grapefruit**	Strawberry	Striped	Ladino
Sweet	Lemon**		Elm,	Red
White	Peach		English	White
Lettuce	Pear		Fir,	Cucumber
Onion	Pecan		Douglas	Pea,
Parsnip	Persimmon		Freesia	English
Pea,	Plum		Gladiolus	Pear
English	Soybean		Cherry,	Plum
Soybean*	Tobacco		Jerusalem	Raspberry
Spinach	Walnut,		Larch,	Strawberry
Strawberry	English		Western	
Timothy			Lilies,	
Wheat*			Ace	
			Croft	
			Easter	
			Hybrid	
			Peach	
			Pines,	
			Eastern	
			White	
			Loblolly	
			Lodgepole	
			Mugo	
			Ponderosa	
			Scotch	
			Plum,	
			Bradshaw	
			Prune,	
			Italian	
			Sorghum	
			Spider plant	

* Some varieties.
** Greater injury from leaf-applied irrigation water.

Healthy seeds carry enough of these nutrients to get the plant started, but it is essential for the nutrient side of the triangle to supply sufficient quantities of all the essential elements before that in the seed is exhausted. Once the plant has a sufficient canopy, it is possible to supply all the needed micronutrients and a portion of the macronutrients through the leaves, although from a practical standpoint, applications by foliar sprays are usually reserved to supplement that which is obtained from the soil.

The ideal quantities of nutrients in the soil solution vary not only with different crops but also with different stages of the same crop. To complicate matters further, the amounts available in the soil solution are affected by such diverse factors as (1) soil characteristics (soil type, aeration, moisture content, pH, OM content, salt content, and CEC); (2) climatic conditions—particularly rainfall and temperature; (3) the amount of nutrients added to the soil in the way of fertilizers, manures, sludges, etc.; and (4) where and when the fertilizers are applied.

Element Forms and Their Utilization

Although a few elements may be taken up as simple organic molecules (N as urea), gases (O as O_2, C and O as CO_2), chelated complexes (Fe, Cu, Zn), or as water (H and O), most of the essential elements are not available for plant uptake unless they are in ionic form. Of those in ionic form, only eight of the elements are simple ions (K^+, Ca^{2+}, Mg^{2+}, Fe^{2+}, Mn^{2+}, Cu^{2+}, Zn^{2+}, Cl^-), while five of them are oxides (NO_3^-, $H2PO_4^-$, HPO_4^{2-}, BO_3^-, $B_4O_7^{2-}$) and one of them is a hydride (NH_4^+).

As ions, they can be present in the soil solution or held on the exchange complex. Both are readily available for plant use, with ions in the soil solution being the primary source and those in the exchange complex acting as a reserve, coming into the soil solution as they are replaced by other ions. Both cations and anions can be held in the exchange complex, but it is primarily the CEC that has much significance in affecting the nutrient leg of the triangle (see Chapter 7 for a more detailed description of cation exchange). Anion exchange of single charged anions (NO_3^- or Cl^-) is minimal and even that of divalent ions (SO_2^{2-}) is of minor importance except on acid soils with considerable Fe and Al oxides.

INSUFFICIENT NUTRIENTS

There are hundreds of reactions within the plant in which the essential elements take part. Because of the great need, failure to supply sufficient quantities of essential nutrients leads to a reduction in yield and/or quality. The reduction can take place without obvious symptoms of deficiency (hidden hunger). If the deficiency is severe enough, abnormal symptoms can appear on leaves, shoots, fruit, and roots. Under very serious shortages, death of plants can occur. Plant death due to shortages is seldom seen in commercial agriculture. Although more common, noticeable deficiencies are usually less prevalent than hidden hunger. Possible exceptions to the prevalence of noticeable deficiencies are those of Fe, Mn, and Zn found on plants growing on calcareous soils.

Loss of Elements

The presence of sufficient ions in the soil solution can be markedly reduced by leaching, volatilization, and crop removal. Fixation of several elements can remove some ions from solution—at least temporarily—and erosion can permanently remove elements from the soil profile.

Leaching

Leaching can remove large quantities of several elements, but amounts vary with rainfall, type of soil, and soil cover. Losses are most serious in humid climates, with little taking place in arid or semiarid regions unless they are excessively irrigated. The greatest losses are from light soils that have little cover or are in row crops, especially when plants are young and have limited root systems and canopy cover. Nitrogen is the element lost in greatest amounts, but K, Mg, and sulfate-S can also be lost from light soils. The nitrate form of N (NO_3-N) is much more subject to leaching, although considerable ammoniacal N (NH_4-N) can also be lost from light soils or those with little CEC.

Volatilization

Nitrogen is also the element lost in the greatest amount by volatilization. Overly wet soils permit denitrification, with formation of several volatile nitrogen compounds that are easily lost. Losses of N as ammonia can be severe on high-pH soils, especially if they have little holding capacity.

Fixation

Fixation is largely confined to P and K although NH_4-N and the micronutrients Cu and Zn can also be lost—at least temporarily—to fixation.

Fixation of NH_4-N in soils is not to be confused with the fixation of N_2 from the atmosphere by symbiotic or nonsymbiotic organisms to available soil nitrogen. The fixation of NH_4-N to unavailable forms involves a temporary immobilization of N by microorganisms primarily due to wide carbon/nitrogen (C/N) ratios of organic matter plus a more permanent fixation by clay minerals. Fixation of P takes place both in appreciably acid and alkaline soils; Zn is fixed in alkaline soils, while Cu is more readily fixed in organic soils.

Crop Removal

Plants differ in the amounts of nutrients removed, although these amounts vary depending upon nutrient availability as affected by soil moisture, pH, type, cation exchange, aeration, depth, and the climatic factors of rainfall, temperature, and solar radiation. The average amounts removed by good yields of some common crops are given in Table 3.2. Amounts of nutrients necessary to produce additional yield units for these crops are presented in Table 3.3.

Nutrient Removal with Time

The rate of removal is very small at first when the crop is young but increases rapidly as the plant ages. The uptake of N in the first third of an annual crop is usually less than 10 percent of the total. The uptake of P and K in this first period is usually greater than N but is seldom more than 20 percent of the total. As growth increases rapidly in the second third, the acre amounts of N, P, and K in the crop also increase greatly and can be 30 to 60 percent of the total uptake. The final third of growth can also result in 30 to 60 percent of the total uptake, the amounts varying with different crops and nutrient availability. The uptake of N, P, and K by several crops with time are presented in Figure 3.1.

The reason for limited uptake in the first few weeks of growth is not due to a slower growth rate, but rather to a small root system and limited production of dry matter. Although the total amounts of nutrients removed by most crops in their early growth is relatively low, soil concentrations of N, P, and K for fast-growing plants need to be high in this period. Cell division and expansion are at their fastest rate soon after the seedling emerges or leaf expansion in perennials is resumed. Concentrations of N, P, and K in leaves during this early period usually are at their highest, dropping as the plant ages. But root systems of annual crops are small, requiring high concentrations of N, P, and K near the root if the plant is to

TABLE 3.2. Average Amounts of Nutrients Removed per Acre by Good Yields of Some Common Crops (lb/ac)

Crop	Yield per acre	Primary nutrients			Secondary nutrients			Micronutrients					
		N	P₂O₅	K₂O	Ca	Mg	S	Cl	B	Cu	Fe	Mn	Zu
Alfalfa	10 ton	600*	120	600	248	53	51	70	1.60	0.30	1.50	1.20	1.00
Apples	250 cwt	100	46	180	—	24	—	—	—	—	—	—	—
Barley	80 bu	100	40	80	18	8	14	7	0.09	0.08	0.31	0.70	0.23
Beans	4 ton	138	33	163	—	17	—	—	—	—	—	—	—
Beets (sugar)	30 ton	255	40	530	90	80	40	—	—	0.14	1.10	0.22	—
Beets (table)	500 cwt	360	43	580	—	101	41	—	—	—	—	—	—
Bermuda (coastal)	10 ton	500	140	420	66	50	40	—	0.03	0.03	0.20	0.20	0.10
Bird's-foot trefoil	4 ton	192*	84	272	42	32	26	—	—	—	—	—	—
Bromegrass	5 ton	220	65	315	—	10	20	—	—	—	—	—	—
Cabbage	700 cwt	270	63	249	35	36	64	49	0.15	0.07	0.35	0.18	0.14
Cantaloupes	175 cwt	65	21	117	—	12	—	—	—	—	—	—	—
Celery	75 ton	280	165	750	150	45	90	—	0.12	0.12	0.60	0.25	0.12
Clover grass	6 ton	300*	90	360	—	30	30	—	—	—	—	—	—
Corn (grain)	200 bu	266	114	266	58	65	33	4	0.23	0.15	1.40	2.10	0.60
Corn (silage)	32 ton	266	114	266	65	33	—	—	—	—	—	—	—
Corn (sweet)	90 cwt	140	47	136	—	20	11	—	—	—	—	—	—
Cotton	3 bale	205	90	145	90	36	39	8	0.05	0.23	0.18	0.39	—
Fescue	3.5 ton	135	65	185	21	13	20	—	—	—	—	—	—
Flax	20 bu	54	22	45	—	—	6	—	—	—	—	—	—
Grapes	12 ton	102	35	156	36	18	26	—	0.13	—	—	—	—
Lettuce	450 cwt	90	30	185	—	9	—	—	0.08	—	1.30	0.34	—
Oats	100 bu	115	40	145	13	20	19	2	0.08	0.08	0.95	0.30	0.39
Onions	600 cwt	180	80	150	47	18	37	—	—	—	—	—	—
Oranges	540 cwt	265	55	330	76	38	28	—	—	—	—	—	—

Crop	Yield per acre		Primary nutrients			Secondary nutrients			Micronutrients					
		N	P_2O_5	KO	Ca	Mg	S	Cl	B	Cu	Fe	Mn	Zu
Orchard grass	6 ton	300	100	375	—	25	35	—	—	—	—	—	—
Pangola	11.8 ton	299	108	430	78	67	46	—	0.14	0.14	1.10	1.60	0.49
Peanuts	4000 lb	115	40	145	125	20	19	3	0.08	0.08	0.67	0.53	—
Peas (English)	25 cwt	164	35	105	175	18	10	8	0.04	0.06	0.60	0.40	0.02
Pensacola-bahia	7 ton	303	87	242	—	35	27	—	—	—	—	—	—
Potatoes (Irish)	500 cwt	269	90	546	20	50	22	—	—	—	—	—	—
Potatoes (sweet)	300 cwt	156	69	313	—	18	—	—	—	0.09	0.36	0.13	0.27
Rice	7000 lb	112	60	148	25	14	12	—	—	—	—	—	—
Ryegrass	5 ton	215	85	240	—	40	—	—	—	—	—	—	—
Sorghum (forage)	8 ton	198	67	286	—	35	18	—	—	0.04	—	—	—
Sorghum (grain)	4 ton	250	90	200	65	44	38	—	—	—	—	—	—
Sorghum (Sudan)	8 ton	319	122	467	64	47	32	—	—	—	—	—	—
Soybeans	60 bu	324*	64	142	85	27	25	21	0.09	0.06	1.50	0.56	0.35
Sugarcane	100 ton	360	156	610	120	100	86	—	—	—	—	—	—
Tobacco (burley)	4000 lb	290	37	321	—	33	24	—	—	—	—	—	—
Tobacco (flue-cured)	3000 lb	126	26	257	105	24	19	—	—	—	—	—	—
Tomatoes	30 ton	180	48	336	—	28	41	—	—	—	—	—	—
Wheat	80 bu	134	54	162	21	24	20	35	0.11	0.09	3.20	0.48	0.39

* Legumes normally obtain most of their nitrogen from Rhizobia.

Source: Wolf, B., J. Fleming, and J. Batchelor, 1985. *Fluid Fertilizer Manual*, pp. 4-11. National Fertilizer Solutions Association. Reproduced by permission of Agricultural Retailers Association, St. Louis, MO.

TABLE 3.3. Amounts of Nutrients Removed per Additional Increments of Some Crops (lb/ac)

Crop	Unit	N	P$_2$O$_5$	K$_2$O	S	Ca	Mg	Zn	Fe	Mn	Cu	B
Alfalfa	ton	60.00‡	73.60	82.80	21.00	56.00	13.00	0.32	0.39	0.28	0.038	0.08
Barley	bu	1.40	3.90	1.80	0.75	0.43	0.23	0.009	0.01	0.02	0.002	0.001
Beans	bu	3.50	6.40	1.60	0.88	0.28	0.04	0.006	0.03	0.003	0.002	0.003
Bermuda (16% protein)	ton	57.00	45.80	62.40	11.00	13.00	8.00	0.18	0.24	0.50	0.07	0.09
Clover grass	ton	52.00‡	48.50	60.00	11.00	42.00	10.00	0.16	0.30	0.20	0.05	0.04
Clover (various)	ton	30.00‡	112.20	87.60	10.00	50.00	11.00	0.17	0.26	0.14	0.05	0.05
Corn (grain)	bu	1.80	1.20	1.40	0.30	0.25	0.27	0.045	0.05	0.03	0.002	0.008
Corn (silage)	ton	52.00	50.40	43.20	10.00	12.00	16.00	0.30	0.50	0.25	0.04	0.06
Cotton (lint)	bale	62.00	178.60	80.40	17.00	34.00	12.00	0.32	0.46	0.16	0.06	0.50
Grass (various)	ton	57.00	52.70	63.40	13.00	12.00	6.00	0.14	0.36	0.28	0.05	0.02
Grass (Bermuda)	ton	67.00	73.30	75.60	15.00	13.00	7.00	0.10	0.36	0.30	0.05	0.02
Ladino clover	ton	93.00‡	103.10	129.60	12.00	58.00	20.00	0.28	0.96	0.42	0.03	0.03
Oats	bu	1.20	2.70	1.10	0.17	0.25	0.33	0.008	0.01	0.004	0.001	0.001
Peanuts	cwt	1.20‡	10.50	6.70	1.46	3.65	0.19	0.006	0.03	0.01	0.001	0.02
Potatoes (Irish)	bu	3.20	6.60	6.20	0.30	0.22	0.70	0.002	0.12	0.003	0.002	0.001
Potatoes (sweet)	bu	2.40	4.40	4.80	0.40	0.28	1.00	0.002	0.13	0.004	0.002	0.002
Rice	bbl	3.40	6.40	3.20	0.80	0.60	0.90	0.006	—	0.03	0.007	—
Rice	bu	9.40	17.60	9.00	2.20	1.70	2.50	0.017	—	0.08	0.019	—
Sorghum (milo)	cwt	3.00	1.40	1.40	0.33	0.30	0.16	0.01	0.01	0.03	0.006	0.001
Sorghum (silage)	ton	46.00	75.60	48.00	8.00	6.00	3.00	0.01	0.12	0.01	0.003	0.01

Crop	Unit	N	P_2O_5	K_2O	S	Ca	Mg	Zn	Fe	Mn	Cu	B
Soybeans	bu	1.40‡	6.60	3.80	0.63	2.30	1.50	0.014	0.13	0.03	0.005	0.006
Sudan	ton	50.00	52.70	48.00	8.00	15.00	19.00	0.10	0.40	0.20	0.03	0.02
Sugar cane	ton	3.60	13.50	16.60	3.40	1.80	2.00	—	—	—	—	—
Sugar beets	ton	4.50	11.00	6.10	4.00	4.30	3.20	0.15	0.03	0.10	0.004	—
Tomatoes	ton	8.40	15.60	13.40	1.30	0.30	0.50	0.02	0.10	0.01	0.009	0.01
Wheat	bu	1.80	5.50	1.70	0.36	0.32	0.56	0.007	0.02	0.06	0.002	0.001

† Nutrient requirements per increment of yield: expected efficiency of nutrient added as fertilizers from the most common fertilizer sources. Growing conditions may cause variations in absolute response from year to year while percentage responses remain almost constant.

‡ Legumes normally obtain most of their nitrogen from Rhizobia.

Source: Wolf, B., J. Fleming, and J. Batchelor. 1985. *Fluid Fertilizer Manual*, p. 4-12. National Fertilizer Solutions Association. Reproduced by permission of Agricultural Retailers Association, St. Louis, MO.

FIGURE 3.1. Dry Matter Production and Removal of Nitrogen, Phosphorus, and Potassium with Time by Corn

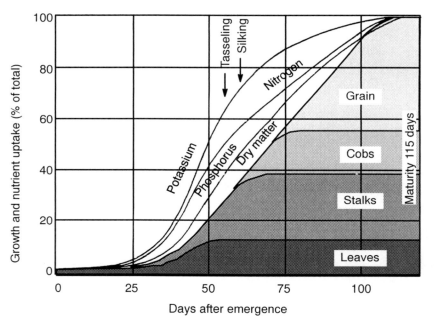

Source: Peck, N. H. 1975. Vegetable crop fertilization. *New York Food and Life Science Bulletin* 52.

take up enough to meet the high requirements of the very rapid growth so essential for high yields. Ample available P is essential for early development of the root system, and high levels of available N are needed to develop large efficient leaves. Large amounts of nutrients per acre are not needed at this time but the high concentrations close to the roots are essential. Amounts of fertilizer applied for this period can be small provided that high levels of nutrients in the limited root zone are achieved by careful placement of added nutrients.

The need for high soil concentrations during this early period is best exemplified by examination of N needs of a fast-growing plant such as muskmelons. Our best yields of this crop are obtained when available soil N in the root zone is about 50 to 75 ppm (total of NO_3-N and NH_4-N) soon after the plant germinates, but dropping closer to about 25 to 50 ppm at midterm, and less than about 15 ppm close to harvest. The low values as the plant matures are essential for production of high-quality fruit that ships well.

Replenishing the Soil Solution

The concentration of elements within the soil solution, which provide for good plant growth, can vary quite a bit—ranging somewhat in the order 60 to 300 ppm of NO_3^- (13-65 ppm of N), 0.05 to 0.5 ppm $H_2PO_4^{2-}$ (0.013-0.13 ppm P), 20 to 200 ppm K, 40 to 400 ppm Ca, 25 to 250 ppm Mg. The low levels are satisfactory and will give about the same yield as the upper levels if they are continuously replaced. Hydroponic culture of plants can obtain very high yields using low concentrations of nutrients because of the speed with which the solution can be renewed.

Nutrients in the soil solution removed by crops or lost by volatilization, leaching, or fixation must be renewed before levels drop low enough to affect yields or quality. Fertilizers, manures, composts, and liming materials are readily used for such replacement, but replacement can also come from one or more of the following: soil OM, CEC, dissolution of various soil minerals, irrigation water, and even small amounts brought down from the atmosphere as rain.

Just how low the nutrient concentration can get before it needs to be renewed obviously depends on the crop, its stage of development, and the rate of the soil solution renewal. The rate of renewal is affected by such factors as the nutrient release rate from the soil OM (dependent on soil moisture, aeration, and temperature), type of amendment used and its placement, the percentage saturation of nutrients in the CEC, amount and kind of minerals, and the amounts and kinds of nutrients supplied by irrigation water or rainfall.

Some of these considerations are covered below, but several of them are examined under the individual chapters devoted to OM (Chapter 5), CEC (Chapter 7), and adding nutrients (Chapter 17).

Quick replenishment of the soil solution usually is accomplished by fertilizers (soluble or readily available), nutrients held by the CEC (if saturation percentages are high enough), or by manures, composts, and other partially decomposed organic materials. If insufficient amounts of these materials are present, it is largely the nutrients in soil OM, the more resistant forms of manures or composts, soil minerals, and nutrients (mainly N) that are fixed either symbiotically or by free association that are utilized. If there are insufficient quantities of these, replenishment of the soil solution may be delayed with resulting harmful effects on crop yields and/or quality.

On many cultivated soils, there are not sufficient nutrients either in CEC or OM or easily dissolved minerals to quickly replenish the soil solution. The need is greater on sandy soils because of their lower CEC, OM, and mineral contents. Also, the lighter soils are more subject to leaching, making the need for renewal much greater than is necessary on the heavier soils.

Unfortunately, N, the element used in the largest amount, often has little or no backup in CEC or minerals and is primarily dependent on OM or the atmospheric N_2 fixed by microorganisms into forms available to plants. Frequently, the total present is insufficient and the addition of fertilizer nitrogen greatly increases yields.

Organic matter (soil OM, manures, composts, crop residues, organic fertilizers) is the major source of reserve N. Generally, plants cannot use organic N until it is broken down by microorganisms to the ammonium (NH_4) or nitrate (NO_3) forms. The organic sources are first broken down to NH_4-N which is then transformed to NO_3-N, but the production of the nitrate form is slowed if soil air is limited by excess moisture or compaction. About 40 pounds of N can be made available in a crop year for every percent OM in the soil, but amounts released depend on favorable microbial action.

Microbial action also affects the amounts of available soil N by influencing the amounts made available by fixation from the atmosphere. Generally, more N is made available from decomposition of organic sources and fixation if temperatures are favorable (60-95°F, 15-35°C), moisture is neither too low nor high, pH values are in the range of 6.0 to 7.0 and salts are not excessive (<0.2-1.5 mmhos/cm).

Microorganisms are also responsible for lowering available soil N by immobilizing available forms by incorporating it into their cells or by denitrifying available N in the soil to volatile forms that can escape. The immobilization of N is favored as organic materials with wide carbon/nitrogen (C/N) ratios are added to the soil. Generally, addition of materials testing less than 1.5 percent total N on a dry weight basis will tend to reduce available soil N, whereas decomposition of OM with higher N contents can be expected to increase it. Nitrogen loss by denitrification is accentuated by high soil moisture and temperatures, compaction, and the presence of NO_3-N.

The replacement of N, P, S, and a number of other elements is augmented if there is sufficient OM. That of N depends on adequate C/N ratios and that of P on proper carbon/phosphorus (C/P) ratios in OM. But the release of all elements from OM depends on microorganisms and their ability to function properly, which depends on temperature as well as the water and aeration sides of the triangle. (More about N release from OM in the section on nitrogen in this chapter and also in Chapter 5.)

The replacement of P in the soil solution is also dependent on occluded phosphates held by Al and Fe hydroxy compounds, by hydroxyphosphates of recent origin, and by various phosphate minerals. Phosphates can replace hydroxy groups in compounds such as aluminum hydroxide, $Al(OH)_3$ or they can react with Al, Fe, or Mn salts to form hydroxyphosphates. The hydroxyphosphates and the occluded phosphates become a secondary or "labile" source of P, which can help replace the quickly changing P levels in the soil solution. The mineral phosphates or "nonlabile" forms of P are only slowly available but are the backup for maintaining adequate soil solution levels of P.

Keeping an adequate level of phosphate, which is present in such small quantities in the soil solution, requires very frequent replenishing, sometimes as often as ten times a day, to meet the phosphate requirements of actively growing plants. To meet renewal needs, there needs to be enough occluded P, hydroxyphosphates, and readily converted organic P to quickly fill the breach and sufficient phosphate minerals and fertilizers as backup.

The cations (K^+, Ca^{2+}, and Mg^{2+}) frequently need not come from added nutrients as they can be quickly replaced if sufficient quantities are held in the exchange complex. The ammonium ion (NH_4^+) also can be held on the exchange complex, but it is usually quickly converted to the nitrate ion (NO_3^-) by microorganisms. The NO_3-N, weakly held by anion exchange, is easily leached from soil, making for very wide fluctuations of N levels in the soil solution, particularly in humid climates.

Symptoms of Deficiencies

Failure to replenish the soil or nutrient solution as nutrient levels fall below certain limits restricts growth of plants. Usually, early shortages are marked by a slowed growth rate, often undiscernible to most growers, that may be diagnosed by soil and plant analyses (Wolf, 1996). If shortages persist or become more acute, the slowed growth rate may be accompanied by various visual symptoms appearing on leaves, stems, and fruits. These symptoms may vary with different plants and stages of growth, and the severity of the deficiency.

Typical deficiency symptoms are outlined in Table 3.4 and Wolf (1996) but described or pictured in greater detail in a number of other books including Bennett (1993), Bould and Hewitt (1963), Glass (1989), Mengel and Kirby (1982), and Weir and Cresswell (1993, 1995). Descriptions and photos of deficiencies of tropical fruit and nut crops are given in Weir and

TABLE 3.4. Important Nutrient Elements, Symbols, Ionic Forms, Value to Plants, and Common Occurrences of Deficiencies and Excesses

Element chemical symbol and ionic form(s)	Value to plants	Deficiency symptoms and occurrences	Excess or toxicity symptoms and occurrences
Nitrogen (N) NH_4^+ & NO_3^-	Essential part of amino acids and proteins. Markedly affects growth perhaps more than any other element	Lightening of green color in early stages. Cessation of growth if deficiency becomes severe. Chlorosis (yellow-green color) and/or "firing" of leaf tips and margins being more severe in older leaves. Both yield and appearance of crop affected. Deficiency common on light mineral soils subject to leaching.	Deep green color. Lush plants with soft growth susceptible to insect and disease attack. Delayed maturation. Harvested product may have poor shelf life. Occurs more frequently on soils overfertilized with nitrogen or treated with large amounts of manure.
Phosphorus (P) $H_2PO_4^-$ HPO_4^{2-}	Essential for energy transfer system. Component of RNA and DNA, regulating genetic information. Also essential for phytin, an important component of seeds.	Stunted growth. Leaves may be dark green with tips dying or purple. Tillering in cereal is reduced. Maturity is delayed. Shoot and root growth and development or opening of buds is poor. Formation of fruits and seeds is affected adversely resulting in poor yields of low quality. Deficiency common on acid soils rich in aluminum or on soils with high pH.	Affects plant growth indirectly by reducing availability of iron, manganese, and zinc with the deficiency of zinc being most common. Excesses common in soil subject to heavy phosphorus fertilization particularly if aluminum is low.
Potassium (K) K^+	Essential for carbohydrate synthesis and transfer of sugars. Helps maintain proper balance of other ions. Regulates water status of plants. Activates several enzyme systems.	Weak stalks that lodge easily. Slowed growth. In advanced stages, withering or "burn" of leaf tips and margins beginning with older leaves. Subject to delayed maturity and disease. More common on light soils of low exchange capacity or soils with strong potassium fixation. High calcium levels or overliming also favor deficiency.	Excesses can lead to deficiencies of calcium and/or magnesium. Very high levels may cause salt damage (high conductivity). Excess is more common in soils of low exchange capacity.

Nutrient	Function	Deficiency Symptoms	Excess Symptoms
Calcium (Ca) Ca^{2+}	Essential for cell wall structure, membrane integrity and the formation of new cells. May also activate enzyme systems.	Distorted leaves. Withered or dying tips. Markedly branched roots which may be brown or discolored. Death of root tips. Leaves may have abnormally dark green color and stems may be weak. Buds and blossoms may shed prematurely. Various diseases are associated, such as blossom-end rot of tomatoes and peppers, black heart of celery, bitter pit of apples. Most common on light acid soils fertilized heavily with ammonium-N or potassium.	Excess may affect magnesium uptake and cause magnesium deficiency. Very high levels can also affect potassium adversely. Can raise pH values too high. The high pH may cause deficiencies of boron, iron, manganese and zinc with iron and manganese deficiencies being more common.
Magnesium (Mg) Mg^{2+}	Essential part of chlorophyll, which is necessary for photosynthesis. Activates a number of plant enzymes.	Yellow leaves with interveinal chlorosis, beginning on older leaves. Margins may yellow while midrib remains green. Most common on light soils with large amounts of potassium and/or calcium.	Not common but very high levels in comparison to calcium can lead to calcium deficiency.
Sulfur (S) SO_4^{2-} HSO_4^{-}	Essential for several amino acids and protein synthesis. Necessary for nodule formation in legumes.	Young leaves light green to yellow, although older leaves also may be affected. Plants tend to be spindly and small. Most common in Legumenosae and Cruciferae plants because of high sulfur requirements. More likely to occur in open soils located far from industrial sites.	Excesses not common but may be present on some saline soils. Reduction in growth rate. Leaves dark green. High levels of sulfur dioxide in air can cause necrotic spots on leaves.
Boron (B) $H_2BO_3^{-}$	Essential for development of the growing point of shoot and root. Probably involved in the formation of uracil, an essential component of RNA and a coenzyme for the formation of sugar.	Slowed and stunted growth followed by death of the terminal bud. "Witch's broom" effect. Leaves tend to thicken and curl. May be chlorotic. Stem tends to crack. Flower buds may fail to form or be misshapen. Poor pollination and fertilization. Fruits or tubers may be misshapen with soft, necrotic, or corky spots. Most common in high boron requirement plants, grown on open soils low in organic matter but with elevated pH.	Discoloration and death of leaf margins usually beginning with the leaf tip, yellowing and spreading between lateral veins and midrib. Most common in boron sensitive plants grown on light soils receiving excess from fertilizer or irrigation water.

TABLE 3.4 (continued)

Element chemical symbol and ionic form(s)	Value to plants	Deficiency symptoms and occurrences	Excess or toxicity symptoms and occurrences
Chlorine (Cl) Cl^-	Not much known about its function but believed to be a cofactor in photosynthesis. May regulate potassium and contribute to plant turgor.	Wilting of plant followed by chlorosis or bronzing of leaves. Growth may be slowed. Deficiencies are not common as more chlorine is carried by many waters and fertilizers. However, applications of chloride have given yield increases by suppressing diseases of certain plants.	"Burning" of leaf tips or margins. Bronzing with premature yellowing and dropping of leaves. More common than deficiencies. Occurs most often with chloride-sensitive plants grown with high-chloride water or on saline soils.
Copper (Cu) Cu^{2+}	Activates several enzymes and influences formation of a chloroplast protein important in photosynthesis.	Stunted growth with chlorosis of leaves that tend to wilt. Leaf tips may die. Shoots malformed. Dieback of tree terminals possible. Fruits may be small and malformed. Occurs more frequently on organic soils or light mineral soils of low exchange capacity.	Results in iron deficiency, also manganese and zinc deficiencies. Growth may be inhibited. Occurs more commonly on acid soils treated with fertilizers or sludges high in copper or receiving excess copper from copper ores or sprays.
Iron (Fe) Fe^{2+} Fe^{3+}	Important in chlorophyll formation and energy transfer functions involving photosynthesis, respiration, and nitrogen fixation.	Interveinal chlorosis and loss of chlorophyll of young leaves. Veins tend to remain green unless deficiency is severe. Twig dieback may occur and in severe cases, entire limbs or plant may die. Very common on high pH soils or soils having excess of heavy metals. Deficiencies more common with certain varieties.	Bronzing or leaf symptoms similar to manganese deficiency. Occurs commonly on rice soils subject to flooding or on acid soils rich in iron.
Manganese (Mn) Mn^{2+} Mn^{3+}	Activator for enzyme systems. Involved in oxidation-reduction processes. Affects structure of chloroplasts important in photosynthesis.	Interveinal chlorosis of young leaves. Unlike iron deficiency, there is no sharp distinction between veins and interveinal areas. Gray speck of oats, marsh spot of peas, interveinal white streak of wheat, and interveinal brown spot of barley are manganese deficiencies. Common on high pH soils particularly during cold wet periods. Also present on highly weathered sandy soils.	Brown spots and uneven distribution of chlorophyll particularly in older leaves. Younger leaves may appear to be iron deficient. Most common on manganese-sensitive plants grown on acid soil.

72

| Molybdenum (Mo) MoO_4^{2-} | Needed for utilization of nitrogen. Helps transform nitrate nitrogen to amino acids. Needed by legumes to fix nitrogen. Important part of some enzymes. | Early leaf symptoms similar to nitrogen deficiency but soon change as areas develop necrotic spots. Stunting and lack of vigor. Leaves are pale, tend to scorch, cup, or roll, and eventually wither leaving little but the midrib. "Whiptail" of cauliflower and yellow spotting of citrus result from molybdenum deficiency. Alfalfa, cabbage, soybeans, and clover also are commonly affected. Occurs more frequently on acid soils. | Toxicity of plants not a problem but levels toxic to animals have been found in forage grown on alkaline organic soils, young soils developed from volcanic ash, poorly drained neutral soils, or on alluvial plain soils. |
| Zinc (Zn) Zn^{2+} | Important in several enzyme systems and in the synthesis of indole acetic acid, a growth regulator. | Interveinal chlorosis of younger leaves. Striping or banding of corn leaves. Leaves may die and fall prematurely. Stem length is reduced and rosetting of fruit and nut trees is common. Bud and fruit formation is reduced. Bark is rough and brittle. Dieback may result with prolonged deficiency. Induced by excess of phosphorus or iron. More common on high pH soils but is also present on acid mineral soils of low exchange capacity. | Toxicity is not common, but can occur with very large additions of zinc or ores, sprays, and fertilizers on acid soils. Zinc-rich peats also can cause problems. |

Source: Wolf, B., J. Fleming, and J. Batchelor. 1985. *Fluid Fertilizer Manual.* National Fertilizer Solutions Association, Manchester, ND. Reprinted by permission of Agricultural Retailers Association, St. Louis, MO.

Cresswell (1993) and of temperate and subtropical fruit and nut crops in their 1995 book. Description and photos of many different kinds of plants are presented in Bennett (1993).

Although nutrient deficiencies can be diagnosed by comparison of symptoms as described or as viewed in photos, a much more accurate diagnosis, especially of initial stages, can be made by use of soil and plant analyses (Wolf, 1996).

EXCESS NUTRIENTS

An ideal range of essential nutrients is present in plants producing maximum yields and quality. The range will vary for different crops and even different stages of the same crop. Addition of nutrients beyond the optimum may first lead to "luxury consumption" with no noticeable effect on appearance or yield. But further additions can be followed by reduction in yield and/or quality. The reduction in yield may be accompanied by various abnormal or toxicity symptoms. Although rare, toxicity resulting from an excess of nutrients can be severe enough to cause death of the plant.

Essentially, there are four causes for crop reduction and damage from excess nutrients: (1) excess salts, (2) immobilization of certain essential nutrients, (3) imbalance between nutrients, and (4) sensitivity to the element.

Excess Salts

High salt concentration in the soil causes a decrease in root permeability to water, reducing the rate of water entry. Leaf cells have a limited capacity to accumulate salts. Salts moved by the transpiration stream collect in vessels and tracheids or outer spaces of the leaves. Salts accumulate as water evaporates and, unless the plant has the capacity to excrete salts, the resulting concentrations can rob cells of water, reducing their turgidity and in extreme cases causing plasmolysis (shrinkage of the cell contents from the cell wall) and death of cells. In some cases, the osmotic concentration increased by the presence of salts can be high enough to cause sufficient water to move out of the root to kill cells. Principle elements causing problems are Na, Cl, and K, but large amounts of Ca, Mg, and the ions ammonium (NH_4^+), nitrate (NO_3^-), sulfurous ion (HSO_4^-), and bicarbonate (HCO_3^-) can overload the system so that toxicity occurs. Sodium and HCO_3^- are particularly troublesome, since excesses can have serious harmful effects on soil porosity and water infiltration. More about high salts and their causes are covered in Chapter 8 and how they may be avoided or corrected in Chapter 12.

Immobilization

High levels of some elements can cause deficiencies because they precipitate or in some other fashion immobilize significant quantities of an essential element that may already be in short supply. Immobilization may take place at the root but also within the plant. High levels of P can have a negative effect on several micronutrients such as Cu, Fe, and Zn, with deficiencies of Fe and Zn being relatively common. It is suspected that immobilization of the micronutrients by phosphate may take place at the roots and also within the conducting tissues. Excess Cu and several heavy metals such as Cd, Pb, and Ni at the roots can have an adverse effect on the amount of Fe available, causing serious Fe deficiencies. High Cu levels can also bring about shortages of Mn. There is relatively little Cu-induced Fe or Mn deficiency in the heavier or organic soils because Cu is strongly bound to soil organic matter. The problem of soil-applied Cu inducing Fe deficiency has been much worse on low-pH soils because of the greatly increased Cu solubility at low pH levels.

Imbalances

Some problems of nutrient excesses are related to imbalances inducing deficiencies of essential elements. High levels of one element can interfere with the movement or utilization of another (antagonism), or it can cause greater demand for another element already in limited supply, or it can combine with other elements removing them from solution. Antagonisms between Ca^{2+} and K^+, NH_4^+ and K^+, and NO_3^- and Cl^- are well known. Elements such as N and K, which cause large increases in plant growth, can easily induce deficiencies of elements that move slowly (Ca) or were barely adequate before growth increased. The need for applying micronutrients was greatly increased as average corn yields increased from about 50 to about 150 bushels per acre with the use of better varieties and greater amounts of N. It is suspected that the deficiencies of Cu and Zn caused by excesses of P may be due in part to the precipitation of these elements by phosphates.

Several nutrient relationships seem to be necessary for plant health. A large unilateral increase of one appears to cause a deficiency of the other. These relationships may occur because two or more elements are needed for certain plant constituents. The element in short supply limits growth until it is added in sufficient quantity. The desired ratio of about 12 parts N to 1 part P appears to be related to the presence of these elements in certain energy-transforming compounds. The ratio of about 10 parts N to 1 part S probably is necessary because of the relative presence of these elements in certain plant proteins. The beneficial relationship of K to N, particularly NH_4-N, may be related to the presence of sufficient carbohydrates. Carbohydrate formation is often dominated by the supply of K, and an ample supply is necessary for new cell growth primarily initiated by N. Other important relationships are Ca and B, K and B, K and Fe, and Fe and Mn.

Sensitive Elements

A number of elements to which many plants are sensitive were listed above. The negative effect of high levels of Al, Cd, Cu, Ni, and Pb are lessened by the presence of sufficient Ca and the evaluation of heavy metal content must take Ca concentration into account. The actual mechanism of reducing the effects of the heavy metals by Ca is not completely understood but it may be due to the effects of Ca upon membrane permeability, precipitation of oxalic acid chelate that can keep the metals soluble, or the increase of pH, which can help reduce the solubility of the metals. The same increase in pH can also induce deficiencies of Fe, Mn, and Zn so that high levels of Ca from liming materials can be a mixed blessing.

Symptoms of Excesses

As with slight deficiencies, slight excesses of the various nutrients may be reflected only as a slowed growth rate, but if severe or prevailing for long periods, excesses can reveal themselves as various plant symptoms. A brief description of these and where they may occur are given in Table 3.4, but more details are given in the books cited for descriptions of plant nutrient deficiencies. It is also helpful to note that periodic soil and leaf analyses can be helpful in not only pinpointing causes of nutrient excesses, but preventing them—often long before marked symptoms are noticeable.

PROVIDING THE ESSENTIAL ELEMENTS

A detailed evaluation of elements important for crop production can lead to a better understanding of how they need to be handled to ensure maximum effectiveness.

Essential Macronutrients

Nitrogen

Nitrogen is required by most plants in very large amounts, usually exceeding the needs of any other single element (50 to 600 lb/ac). Some of the greatest removal is by legumes, which are not totally dependent on soil N but can fix a good portion of their needs from the air by symbiotic bacteria attached to their roots.

Concentrations of 150 to 200 ppm in solution cultures is about ideal, but as little as 50 ppm can maintain and 300 ppm can be tolerated by most crops. Concentrations of 50 to 75 ppm of NO_3-N and NH_4-N in soil or culture media extracts provide rapid early growth of nonleguminous plants, while as little as 15 to 25 ppm (primarily NO_3-N) appears to be satisfactory for later growth.

Both NH_4-N and NO_3-N are readily available to most plants. The NH_4 form is more readily absorbed from a neutral or elevated pH medium while NO_3-N is more readily taken up from acid medium. While uptake of both forms is depressed by low soil temperature, uptake of NH_4-N is affected less at low temperatures. The NH_4 form generally will supply more rapid growth. The slower growth with NO_3-N helps produce better quality of a number of harvested commodities that will hold up for long periods after harvest (cuttings for rooting, cut flowers, fruits and vegetables that have to be shipped long distances).

Plants may vary in their response to NH_4-N and NO_3-N. Most plants do best with a mixture of NH_4-N and NO_3-N but the ideal relative amounts of each may vary with the crop and its age and whether it is grown in soil or by hydroponics.

Generally, younger plants do better with a higher proportion of NH_4-N and older plants with more NO_3-N. The desirability of high NH_4-N during early growth of plants started during cool periods is especially fortuitous because the NH_4^+ ion is preferably absorbed during such periods by a number of plants. Yields of many crops will be greater with narrow ratios of NO_3-N/NH_4-N, but quality of the harvested crop is often favored by wider ratios. Shipping and storage quality of many horticultural products can be greatly reduced if the ratio of NO_3-N to NH_4-N applied in the fruiting stage is appreciably less than 1:1. Fortunately for these crops, NH_4-N in most soils is soon converted to NO_3-N if nitrification inhibitors are not used, allowing for the use of cheaper ammoniacal sources providing they are applied early in the crop cycle. Because nitrification is limited in hydroponic solutions, higher ratios of about 3:1 NO_3-N/NH_4-N are more suitable for these conditions. These higher ratios are also suitable for soils in which nitrification is restricted. Shifting the ratio of NO_3-N/ NH_4-N wider than about 4:1 generally is counterproductive as growth is slowed excessively, reducing yields.

The uptake of most elements is affected by the additions of N but the source of N may have a bearing on whether the absorption of a particular element will be stimulated or antagonized. Nitrate-N tends to depress uptake of Cl. Ammonium-N usually improves the uptake of P from super-phosphate or phosphorus solutions, and can stimulate the Fe and Mn uptake from high pH soils by lowering pH. It reduces the uptake of Ca and Mg, evidently by competing with these ions. The reduction of Ca or Mg uptake can cause economic losses of many crops. That of Ca adversely affects the quality of many fruits and vegetables.

Nitrification can be limited because of insufficient O_2 (heavy soils with poor structure that are compacted or excessively wet), insufficient mois-ture, low pH values, presence of certain pesticides, or if soils are subjected to extremes of temperature. For sensitive crops grown on any of these problem media, it is desirable to include enough NO_3-N to maintain an NO_3/NH_4 ratio of at least 2:1, especially in later stages of growth.

Maximum yields and quality associated with wide NO_3/NH_4-N ratios appear to be related to (1) slower but more compact growth with NO_3-N, (2) lowering of medium pH values if NH_4-N is more than about 20 percent of the total, and (3) toxicity with relatively low concentrations of NH_4^+ ions. In solution cultures, ammonium toxicity can occur with relatively low concentrations. The amounts vary with pH of the solution. Toxicity starts with as little as 10 ppm at pH 7.5, but as much as 150 ppm NH_4-N can be tolerated at pH 6.5.

These NO_3-N/NH_4-N relationships are of less importance in well aer-ated soils having good buffer capacity and subject to satisfactory tempera-tures. Under such conditions, there is a rapid turnover of NH_4-N to NO_3-N. But the turnover in moist or compacted soils may be slow enough to warrant limiting amounts of ammonium and urea fertilizers to supply no more than about 50 percent of the total N, especially late in the season when there may not be enough time to complete the nitrification.

The pH of soils with little buffering capacity (sands, sandy or gravelly loams) or of solution cultures can be lowered by continuous use of high levels of NH_4-N. In solution cultures, using NO_3/NH_4 ratios narrower than 80:20 tends to lower medium pH rather quickly. Narrower ratios can be used in soils with much less effect. In addition to buffering effects of the soil, changes are slowed because of rapid nitrification of NH_4-N to NO_3-N.

Ammonium-N is much less subject to leaching losses than NO_3-N, because NH_4^+ is held by clay. Such differences are negligible in sandy or organic soils. Losses of applied ammonium salts by leaching also takes place in the heavier soils as NH_4-N is readily converted to NO_3-N in well-aerated soils having satisfactory temperatures, moistures, aeration, and pH values, unless nitrification inhibitors have been added.

Some loss of N takes place as ammonia (NH_3). A small amount of N volatilizes from plants as NH_3 and can be lost to the atmosphere. Other losses take place as manures or organic materials rich in N, some ammoniacal compounds, or urea are added to soils of high pH, particularly if these materials are left on the soil surface for some time. Nitrogen losses as NH_3 can also be considerable if anhydrous ammonia is shallowly injected (>6 inches) or injected in dry soil or soil with low CEC. Besides losses as a gas, the NH_3 can diffuse and rise with capillary water to be lost as the water leaves the surface. The chances of this occurring are greater with shallow injections, or injections in dry or sandy soil.

Losses by leaching, denitrification, and fixation complicate maintaining sufficient N for crops. The problem is accentuated in light soils with little OM, particularly in regions of heavy rainfall, but deficiencies of N can exist on a wide range of soils in many different parts of the world.

Phosphorus

Amounts of P removed by crops varies from about 15 to 75 lb P (33-165 P_2O_5) per acre. Crops grown in solution cultures do very well with 40 to 50 ppm P although about as little as 20 ppm can be sufficient and as much as 50 ppm tolerated by many crops. Total P in soils is in the range of 1000 to 11,000 lb/ac of which 20 to 100 lb are easily soluble and only 0.5 to 6 lb/ac actually in solution. Because P in the soil solution is very low it can be quickly exhausted unless amounts present in the labile pool (easily soluble) are close enough and in sufficient concentration to replace that which is absorbed. Replacement amounts can come from organic sources, soil minerals, and added phosphatic fertilizers.

Organic sources can add considerable P in some soils. As with N release, P release depends on microorganisms that likewise are affected by temperature, moisture, pH, and salt content. Whether P is fixed (made unavailable for a time) or released from added organic materials depends on its P content. Materials with more than 0.2 percent total P tend to increase available P, while the decomposition of materials having a total P content of less than 0.2 percent tends to lower available P.

Phosphorus fixation. Phosphates in the labile pool, which are readily available, can be converted or fixed into a form unavailable for immediate use. Fixation is enhanced by the presence of clay minerals and either low or high pH values. Fixation is of less importance in sandy or organic soils and at pH values close to 6.5. Liming acid soils helps maintain P availability. Fixation of applied phosphates is also reduced in the heavier soils by banding the P to reduce contact with the soil.

Organic matter is beneficial in reducing fixation of added phosphate and its decomposition by microorganisms releases available P. Another group of microorganisms (mycorrhizal fungi) infecting roots of a number of plants aid in the uptake of P. Soil sterilization may kill enough of the mycorrhizia so that P uptake is decreased to the point of causing P deficiency.

Phosphorus loss. There has been more concern about P fixation, but P can be lost from soils by leaching or erosion. Leaching losses are primarily confined to sandy soils but the amount lost by erosion is worse in the heavier soils. Wind and water erosion tend to remove a larger proportion of the fine particles (clay and OM), which are richer in P than coarse soil particles.

Potassium

Crop removal of K varies from about 35 to 625 lb/ac (42-750 K_2O), much of which will be in the vegetative portion of the plant. Ideal concentrations in solution cultures is about 200 to 500 ppm K, but as little as 50 ppm and as much as 800 ppm can be effectively used for some plants and situations. Soil levels of about 50 ppm K, in both the soluble and exchangeable forms, have provided adequate supply for many crops on soils of low CEC but the level may have to be five to ten times as high on soils of high CEC (see Figure 3.2). The 50 ppm may be difficult to maintain in soils of low CEC due to leaching losses but maintenance of the higher levels in high CEC soils can be a problem due to fixation of K.

Potassium fixation. Fixation of K to nonexchangeable forms can cause serious problems of K fertility—at least temporarily as the fixed K may convert to the more available form in time. The amount of fixation will vary with the type of clay mineral and the amount of moisture. Fixation rates follow the order of vermiculite > illite > smectite, and are greater under dry conditions. Fixation in soils containing considerable vermiculite can make it impractical to meet the K requirements only by soil applications. Since the hydrogen ion (H^+) can be exchanged for K^+, fixation is less of a problem in acid soils. Fixation can also be reduced by limiting the contact of soil with fertilizer by proper banding and use of organic materials.

Leaching of potassium. The K^+ ions are held rather strongly by most clay particles but only weakly by kaolinite or humus. Leaching of K, therefore, is primarily limited to sandy or organic soils or soils containing considerable kaolinite. Applications of K to these soils in humid regions

FIGURE 3.2. Desirable Potassium Concentration in the Soil Solution As Affected by Clay Content of the Soil

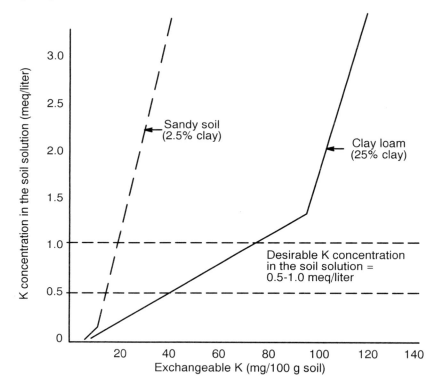

Note: 1 meq/liter (K) = 39.1 ppm.

Source: International Potash Institute. 1977. *Potassium Dynamics in the Soil.* IPI Extension Guide. International Potash Institute. Worblaufen-Bern, Switzerland.

should correspond closely to plant growth, limiting the amounts before planting and adding the remainder as growth is anticipated.

Calcium

The total amounts of Ca in soils range from about 10,000 to 100,000 lb/ac, of which about 1,000 to 10,000 are readily soluble and/or exchangeable. Only about 25 lb/ac are removed by small grains but many plants remove about 100 lb and alfalfa will remove about 200 lb. In solution cultures, the requirement is met by maintaining 150 to 300 ppm Ca. Satisfactory levels of Ca as measured by soil extracts range from 500 to 5,000 lb/ac depending on crop requirements and modes of extraction.

It appears that large amounts of Ca are needed in solution cultures to overcome various toxicities or to correct imbalances due to excesses of K, Mg, or NH_4. Failure to do so can lead to deficiencies of Ca in many plants.

Even greater amounts of available Ca are needed in the soil to grow good crops. Amounts varying from a few hundred lb/ac in light soils to well over 3,000 lb/ac in heavy soils, or enough to satisfy about 70 percent of CEC, come closer to answering the need for Ca. The large amounts needed for soil-grown crops probably are related to factors other than just supplying Ca. It has been surmised that these large amounts are needed for such related benefits as improvement of soil structure or reducing damage from high concentrations of salts or heavy metals.

Magnesium

The total amounts of Mg in some soils may be as much as Ca (10,000 lb/ac), but appreciably less (only about 2,500 lb/ac) in others. Heavily leached sandy soils (podzols) have the least; marsh soils and the brown earth silty soils the most. Satisfactory levels of available Mg (exchangeable and easily soluble) range from 100 to 1,000 lb/ac.

The amounts removed by crops range from about 25 lb/ac (small grains and grasses) to about 100 lb/ac for table beet or sugarcane. In solution cultures, crop requirements are met with about 50 ppm of Mg. In soils of low exchange capacity, an exchangeable Mg content of 25 ppm seems to be satisfactory for many crops but 50 ppm is more suitable for crops with high demand (beet, white potato, tomato, swedes, turnip, glasshouse crops). The amounts necessary for soils of high cation exchange capacity may well be 100 to 250 ppm.

Sulfur

The total amount of S in soils varies from about 50 to 4,000 lb/ac. Almost 100 percent of the S in peat soils may be present in organic forms, but a large part of the sulfur will exist as sulfates in dry regions. Large amounts of S are found in calcareous organic soils and peats but relatively small amounts in the acid podzolic soils. As with N, the S present in OM is not available to plants until the OM is decomposed by microorganisms. Sulfur mineralized in a year to sulfate form (SO_4) and available to plants ranges from about 2.5 to 25 lb/ac.

Amounts of S taken up by many plants are similar to that of P, generally in the range of 10 to 20 lb/ac. Crops grown in solution cultures appear to do well with S concentrations of about 50 ppm derived from sulfates. Plants rich in mustard oils (cabbage, rape) or in protein (alfalfa, clovers) or producing large quantities of dry matter (Bermuda grass, celery, corn, sugarcane) will remove larger quantities (about 50 lb/ac).

Essential Micronutrients

Boron

Total amounts of B in soils are in the range of 4 to 100 lb/ac. Available B, much of which is hot-water soluble, varies from 0.1 to 5 lb/ac. Small grains and grasses remove less than 0.2 lb/ac, but amounts removed by alfalfa, beets, broccoli, cauliflower, and celery are considerably larger. Alfalfa may remove more than 1 lb/ac.

Crop requirements can be ideally met in solution cultures with 0.25 to 0.5 ppm of boron, but a low concentration of 0.1 ppm or a high concentration of 1.5 ppm can be tolerated by some plants. In soils, adequate B for many crops is 0.35 ppm as measured by hot water extraction, but about twice this amount is desirable for alfalfa and other high-requirement crops.

Boron is readily leached from most soils. Its solubility is increased at low pH values, making shortages more common in acid, light-textured soils, or soils low in organic matter. Such soils are especially susceptible to B shortages as they are limed. Boron availability may be decreased in clay soils and in soils subjected to drying following a wet spring. Acid peats and mucks also may produce crops suffering from B shortages. Shortages are more common with crops having high demands for B (alfalfa, beets, cauliflower, celery, peanut, and turnips).

Boron excess. Excesses of B are more likely to occur in arid soils or soils derived from marine deposits. Damage commonly occurs if: (1) irrigation waters contain large quantities of B (from deep wells, those infiltrated by marine sources, or from natural waters in arid regions); (2) fertilizers or sludges contain excess B; or (3) crops sensitive to B follow too soon after B-tolerant crops were heavily fertilized with B. Beans, blueberry, fig, grape, grasses, and peach are very sensitive to B excess.

For many plants, there is a very narrow range between sufficiency and excess of B. Desirable amounts of added B vary from none or little (0.1 to 0.25 lb/ac) for sensitive crops such as grasses and cereals to a range of 1 to 3 lb/ac for crops requiring large amounts (alfalfa, beets, celery, turnips). If plants are grown in a rotation, it is often desirable to add B for the crop requiring the greatest amount and exclude it for the others.

Chlorine

Amounts of chloride present in soils range from a few pounds to about 1,000 lb/ac. It is easily leached from soils and so little is usually present in soils of humid regions. The amount of Cl required by most plants is very low, and can be met with an application of 4 to 10 lb/ac. This small amount is easily added by impurities in the air, irrigation, rain waters, or chlorides carried by fertilizer salts. As little as 100 lb/ac of muriate of potash (potassium chloride) supplies about 40 lb of Cl or about 20 ppm. Much larger amounts than 4 to 10 lb/ac is desirable for disease control of several cereals.

Chlorine deficiency. Although excesses of chlorine are more common than deficiencies, field deficiencies of certain crops do occur, primarily on light soils far from the oceans to which little or no chloride is added. The common muriate of potash or potassium chloride (KCl), which is the cheapest common potassium salt, is almost 50 percent chloride and its use usually avoids chloride deficiency.

Chlorine excess. Losses from excesses of Cl are common in dry areas when waters high in chlorides are used and there is insufficient rainfall to leach out chlorides. Excesses are frequently present in soils lacking permeability or receiving runoff from other areas. The poor permeability restricts leaching of Cl from many different sources. Excesses can also be present in coastal areas receiving flooding from salt waters. Damage can occur from excess chlorides added from fertilizers, but more commonly from irrigation waters. Overhead irrigation with water high in chlorine can be especially troublesome to several sensitive plants because Cl is readily absorbed by leaves of many plants.

Several plants (avocado, beans, some legumes, lemon, lettuce, peach, potato, stone fruits, and tobacco) are very sensitive and can easily be damaged by excess Cl. Reduction of yields of very sensitive crops (avocado, grapes, olive, and stone fruits) take place when leaves accumulate 0.3 to 0.5 percent Cl in dry matter, whereas tolerant plants (barley, corn, sugar beet, spinach, tomato) can contain as much as 4 percent of dry matter with no apparent damage. In addition to Cl moving up into the plant by the transpirational stream, Cl can be removed from sprinkler-applied irrigation. Frequent applications with intermittent drying allows greater amounts of Cl to accumulate in plant tissues with corresponding increases in damage noted.

Copper

Total Cu in soils ranges from about 2 to 200 lb/ac. Easily soluble amounts are much smaller (1 to 20 lb/ac).

The amount of Cu required by most plants is very small and varies from about 0.25 oz/ac (cereals) to about 2 oz/ac for some clovers. The requirement can be met by many soils. Those that are deficient in Cu usually belong to one of the following groups: (1) sandy podzolic soils that have been heavily leached; (2) calcareous soils; (3) certain clay soils, especially those with high pH; and (4) peaty and organic soils. Applications in the range of 0.25 to 1 lb/ac are usually sufficient for correcting deficiencies on sandy soils, while amounts in the range of 2 to 10 lb/ac are needed for the heavier mineral soils and up to about 20 lb/ac for the peaty and organic soils.

The large amounts needed for the heavier mineral soils and the peaty and organic soils are primarily due to fixation of copper. In hydroponic cultures, where large amounts of added Cu remain in solution, the Cu needs are met with extremely low concentrations in the 0.05 to 1.0 ppm range, with about 0.2 ppm being ideal for many plants.

Copper toxicity does not occur readily in most soils due to the strong bonding of Cu to clay and organic matter. Toxicities are more likely to occur in coarse soils that have little clay or organic matter. Coarse soils low in pH (<6.0) are especially susceptible to damage because solubility of the copper is greatly increased as pH levels fall. Toxicities that have been observed were the result of large application of copper as ores, fertilizer salts, or fungicidal sprays. Soil residues from longtime application of fungicidal sprays combined with applications from fertilizer dressings have been responsible for several cases of copper toxicity on Florida sands.

Damage from high concentration of fungicidal sprays applied to the plant appear as necrotic, elliptical lesions, some of which may be surrounded by chlorotic halos. Plants damaged by accumulation of copper in the soil from fungicidal residues, fertilizers, and ores tend to exhibit chlorosis, (resembling iron deficiency symptoms), root inhibition, reduced vigor, and dieback of tree twigs. If toxicity has not advanced too far, trees can be rejuvenated by increasing pH to the 6.0 to 6.5 range and judicious application of chelated iron to the soil.

Iron

Crops require very small amounts of iron, removing about 0.5 to 1.0 lb/ac. Only about 0.5 ppm soluble Fe is needed in the soil or 2 to 3 ppm in solution cultures to meet these demands. Many soils contain about 50 tons and very few soils contain less than 1 ton of total Fe per acre. Obviously, lack of Fe is primarily a problem of availability rather than sufficient amounts.

Availability of Fe in soil is largely controlled by pH, falling rapidly as the pH rises above 4.0. Very small amounts of soluble Fe are present as pH rises above 6.0, unless the Fe has been complexed or chelated (formation of a ring structure around a polyvalent metal, reducing its rate of precipitation). Complexes can form with exudates from certain plant roots, from constituents of OM, or from products produced by microorganisms or added as chelates.

Availability of Fe is also reduced by the presence of large quantities of bicarbonates, P, and heavy metals such as Cd, Cu, Mn, and Pb. Bicarbonates often present in large amounts in certain irrigation waters are particularly troublesome as they appear to interfere with Fe mobility in the plant. Bicarbonates accumulating in calcareous soil as calcium carbonate are dissolved in the presence of carbon dioxide also account for considerable Fe deficiency. This problem is minimized if soil aeration is sufficient to allow the carbon dioxide to escape readily from the soil.

Iron deficiency. Plants vary in susceptibility to Fe deficiency, especially at higher pH values. The calcifuges (azaleas, blueberry, camelias, ixora, rhododendron) are especially sensitive and usually need to be grown on soil with pH values <5.5 to avoid Fe deficiency. Other sensitive crops are citrus, deciduous fruit, and grapes, but these can be effectively grown at pH values to about 6.5.

Excess of iron. Excesses of soil-derived Fe are rather rare except in very acid or flooded soils. The rice crop, perhaps more than any other, is affected by iron toxicity. When Fe toxicity does occur, it is usually revealed as a deficiency of Mn. In rice, toxicity results from flooding and is known as "bronzing."

Damage from iron sprays, particularly under low light and high nutrient concentrations, is not uncommon. Symptoms appear as elliptical necrotic spots, which at times can be surrounded by a white chlorotic halo.

Manganese

The amount of Mn removed by crops is very small—usually about 0.5 lb/ac. In solution culture, this amount can readily be met by concentrations of 1.0 ppm in solution cultures. Much larger amounts exist in soils, which in some cases can exceed 200 lb/ac but more commonly are about 50 lb/ac. But these amounts may fail to supply sufficient Mn, because much of the Mn may not be available to crops. The availability is affected by soil pH, dropping markedly as pH rises. It has been estimated that soluble Mn decreases about one hundred-fold for each unit rise in pH.

Other factors that can decrease available Mn are reductions in soil moisture, increased microbial activity, and additions of organic matter, because these changes affect oxidation-reduction reactions, changing the more available Mn^{2+} to less available Mn^{3+} or Mn^{4+} forms.

Molybdenum (Mo)

Very little Mo (about 0.01 lb/ac) is removed by many crops. This amount can readily be provided by a concentration of 0.05 ppm in nutrient solutions or 0.2 ppm in soils. The total amount in soils lies between 0.5 to 3.5 ppm (1 to 7 lb/ac).

Molybdenum deficiency. Deficiencies occur because Mo may become less available in association with iron oxides or fixed in acid sandy soils. Soils containing ironstone accumulations are more commonly associated with Mo deficiency. Although OM decomposition usually releases available Mo, occasional shortages are noted on peaty soils as Mo is retained by humic acid in these soils.

Unlike most heavy metals, the availability of Mo is increased as pH rises, often making it possible to correct shortages by simply liming.

Zinc

About 0.2 lb/ac of Zn are removed by crops. A concentration of about 0.2 ppm in nutrient solutions is satisfactory for most plants, although as little as 0.05 ppm is sufficient and as much as 5 ppm can be tolerated by some plants. Soluble Zn in soils ranges from about 2 to 22 lb/ac.

Zinc deficiencies. Availability of Zn is decreased as pH rises and phosphates are increased. An increase of one pH unit decreases available Zn by about 50 percent. Deficiency occurs more frequently on sandy soils, soils low in organic matter, or soils that have been leveled. Peat and muck soils also have produced plants with Zn deficiency but most of these reports have come from calcareous soils. In most cases, Zn deficiency has been aggravated in the various soils by the presence of low soil temperatures.

A number of crops may suffer from Zn deficiency, the more common of which are beans, beets, citrus, corn, deciduous fruits, nuts, potato, sorghum, and small grains.

Zinc excesses. Most plants are tolerant of large amounts of Zn, although tolerance varies considerably with species and even cultivars. Damage to plants is largely limited to areas close to zinc spoils from mines or to fields dressed with large amounts of town refuse, sewage sludge, or composts made from these materials. Only occasionally is toxicity due to applications of zinc fertilizers.

NONESSENTIAL ELEMENTS OF IMPORTANCE

Aluminum

Aluminum is found in all plants, usually in concentrations of less than 0.02 percent of dry matter, although in a few plants (hickory, tea), concentrations may reach 2 percent. There have been reports of response to added Al and it is necessary for the blue coloration of hydrangeas. Despite the universal presence and some beneficial responses, Al is not considered an essential element. On the contrary, it is of much concern in many regions of the world because of its harmful effect on many plants.

About 8 percent of the earth's crust is Al, but much of this is insoluble. Its solubility is greatly increased as pH values fall below 5.5. In many acid soils, Al content can greatly restrict plant development, particularly of sensitive plants (alfalfa, barley, cotton, lettuce, timothy, sugar beet).

There is a great variability between plants and even between cultivars of the same species in sensitivity to Al. Most acid-loving plants, buckwheat, cranberry, mangold, mustard, perennial ryegrass, red clover, subterranean clover, turnip, and white clover are tolerant. Brussels sprouts, cabbage, oat, pea, radish, and rye are semitolerant.

Damage from excess Al is reduced in the presence of large amounts of OM, or by applications of phosphates or by increasing soil pH through the additions of liming materials.

An excess of Al, which has such a marked effect on roots, is less noticeable on plant tops. Where it is apparent, symptoms may resemble those of P or Ca deficiency. Generally, there is an overall stunting of the plant, occasionally with purpling of stems and leaves and a cupping or rolling of leaves, some of which may lose their normal green color and become dull, olive green, or even turn yellow and/or have necrotic areas. Red purpling of leaf bases can occur in barley and collapse of petioles with blackening of young leaves in celery.

Damage to roots is much more apparent. The root system is greatly restricted with limited amount of branching. Roots tend to be stubby with restricted root tips that may turn brown. Lateral roots may be thickened.

Bromine

Although Br$^-$ can substitute for at least part of a plant's requirements for Cl$-$, the importance of Br is primarily one of toxicity to plants, resulting from bromine fumigants. Amounts naturally present in soils are usually very low and can be disregarded. However, amounts of Br high enough to cause damage to plants can occur from applications of soil fumigants such as methyl bromide or ethyl dibromide. Generally, damage is confined to a short period of only a few days before Br may dissipate from the soil, but repeated applications can accumulate sufficient Br to be damaging to sensitive plants (bean, beet, cabbage, carnation, celery, chrysanthemum, citrus, garlic, melons, onion, pea, peanut, pepper, potato, snapdragon, spinach, sugar beet, sweet potato, and turnip). Flooding or heavy rainfall between fumigations can reduce levels to insignificant quantities.

Symptoms when toxic levels occur resemble those produced by excess salts, e.g., leaf tip and edge necrosis that expand with time; light or yellowish leaves in citrus; yellowing of bean leaves; and poor seed germination of several plants.

Fluorine

No deficiencies of F have been noted in the field but toxicities from excess have affected a number of plants. Fluorine toxicities usually do not arise from native soil contents, which are small (traces to 0.15 percent). Solubility is increased with low pH but uptake by soil roots is very small. Amounts present in the soil are increased by additions of superphosphates, ammonium sulfate, and diammonium phosphate. Several materials (perlite, fritted trace elements, resin-coated slow-release fertilizers), primarily used for potting media, have high levels of soluble F. Amounts in the soil are also increased by certain irrigation waters. Water with F concentrations greater than 0.25 ppm can injure plants such as "Baby Doll," which are very sensitive to F. Fortunately, soluble F in soil or growing medium can be markedly lowered by increasing pH to 6.5.

Some plants sensitive to soilborne F are "Baby Doll" cordyline plant, calatheas, dracenas, gladiolus, lilies, marantas, and Ti plant. Chlorosis followed by necrosis of tips and margins of elongating leaves are common symptoms. Leaf spotting near the tips is often followed by scorch, which may affect major portions of leaf tips.

Damage to plants also occurs as they are exposed to fumes of hydrofluoric acid emanating from several industrial operations, such as the production of aluminum and superphosphates. Fumes released by these plants can travel many miles and be brought to earth by rains to seriously damage sensitive plants (apricot, grape, gladiolus, iris, leek, peach, poinsettia) and crops even mildly sensitive such as corn, cherry, grape, lemon, maple, oat, orange, rice, rose, rye, walnut, and wheat. Symptoms include marginal chlorosis, interveinal chlorosis, marginal necrosis burnt tip, and necrotic areas from which tissue may be lost. Long exposure can cause death of the plant.

Sodium

A number of plants respond favorably to additions of Na. Many of these respond favorably if K is insufficient but some of them (beets, celery, spinach, and turnip) will give substantial increases to applied Na even when K is present in ample amounts.

Despite these responses, Na is not thought to be an essential element and is considered primarily because of adverse effects on soil and crop. Amounts of exchangeable Na greater than about 10 percent tend to disperse soil, increasing problems of water infiltration, aeration, and soil erosion. Soil Na increases soil pH values and adds to salt problems. Enough Na can accumulate in plant tissues to cause serious problems with plants sensitive to it.

Essentially, two types of damage result from presence of excess Na: (1) sufficient Na can accumulate in leaf tissue to cause damage; and (2) excess soil Na can raise pH values and salt content while adversely affecting soil physical properties, making it more difficult for crop plants to function normally.

High Na restricts amounts of water and air available to plants and so retards the uptake of many essential elements, reducing plant vigor and stunting growth of many plants. Plants vary in their sensitivity to Na with avocado, citrus, deciduous fruits, and nuts being very sensitive while alfalfa, barley, beets, tomatoes, the wheatgrasses, and Rhodes grass are quite tolerant. Symptoms of excess Na are somewhat similar to those caused by drought or root injury. Leaves tend to turn yellow, have damaged margins, and may show early autumn coloration.

The multiple effects of excess Na in the soil are revealed in several ways. High soil pH may lead to symptoms typical of high pH, e.g., Fe-, Mn-, and Zn-induced chlorosis. Increased soil salts can induce symptoms of excess salt damage (wilting, reduced growth, damaged roots, and symptoms typical of Ca deficiency). The resulting lowered soil aeration and permeability may be reflected in poor plant development, restricted root systems, and several nutrient deficiencies, primarily Fe.

Heavy Metals

Cadmium, Chromium, Cobalt, Lead, Nickel

These elements can be grouped together because their importance primarily relates to damage they do to plants, which usually results from additions of waste products to soils. They may be present in large quantities in sewage sludge or composts made up of waste materials. Before lead was removed from gasoline, plant damage from gasoline exhausts was not uncommon but usually was confined to small areas close to heavily traveled roads. Generally these metals are not naturally present in soils in sufficient quantities to cause damage, although quantities of Ni and Cr can be high enough in serpentine soils to be toxic. Toxic levels of these heavy metals in soils can be reduced by liming.

Generally, toxic symptoms are expressed as a chlorosis due to Fe deficiency commonly induced by most of these metals. In cereals, this may be represented by yellow stripes lengthwise through the leaf. Chromium toxicity may induce narrow brownish leaves with necrotic spots. Nickel can induce Mn deficiencies and Co can cause dark-green veins with interveinal chlorosis.

SUMMARY

Nutrition completes the fertile triangle. It has maximum effect on crop production when both aeration and water are in ideal ranges. The nutrient side of the triangle is dependent on 16 essential elements, most of which have to be present in the soil solution. The atmosphere supplies C and O_2, and water is the source of H and O, but O has also to be present in the soil solution along with the other 13 essential elements (N, P, K, Ca, Mg, S, B, Cl, Cu, Fe, Mn, Mo, and Zn).

For each of these elements, an ideal range of concentrations must be present in solution in soil or other media for optimum production. The amounts vary somewhat for different crops and at times for different stages of plant growth. For maximum use by the plant, nutrients in solution need to be neither too high nor too low in concentration. Lack of sufficient concentrations can cause reductions in crop yields or quality, some of which are accompanied by visual symptoms of deficiency. Excess concentrations can cause problems by raising salt levels too high, immobilizing certain nutrients, causing imbalance between elements, and/or damaging roots. Such excesses also may be accompanied by visual symptoms.

Unsatisfactory low levels of essential elements may exist in the soil solution as a result of plant uptake, leaching or volatilization from the soil or media, or fixation of elements in forms not readily available to the plant. In fertile soils, such low levels are quickly restored by either normal backup from the soil or the additions of suitable fertilizers or other amendments.

The principle backup for N, P, and S in some soils is OM, which also can be a good source of other elements. Rapid backup of these elements is possible from manures, composts, sludges, and other rapidly decomposable organic matter. Phosphorus can be replaced from labile sources of hydroxy and occluded phosphates. The CEC can be a source of rapid buildup of K, Ca, and Mg if it is properly saturated with these elements. Liming materials are good backup sources for Ca and Mg and in some cases P and Si as well. Soil minerals are the ultimate but relatively slow backup for all but N of the 13 essential elements normally taken up from the soil solution. But it is fertilizer, especially the water-soluble sources, that is usually used for rapid backup for any element in short supply.

Nutrient elements may be present as minerals, slow-release fertilizers, or organic materials but still not be able to recharge the soil solution because they are in an unavailable form. There may be insufficient moisture, high pH, or low temperatures which restrict solution or in the case of organic materials, low temperature, poor pH, and lack of soil aeration which limit activities of microorganisms.

Ranges of the individual nutrient elements are given as totals present in soils and as soluble levels in hydroponic cultures. The levels present in soil solution are most important, although totals that include amounts held by CEC, nutrients in organic materials, and soil minerals are helpful as backup for maintaining satisfactory concentrations in the solution.

A satisfactory range of elements in the soil solution does not guarantee adequate nutrition because roots may not be positioned to take advantage of the solution. Even if positioned properly, roots may still not be able to take up enough nutrients from the solution because they are not functioning properly (O_2 shortage, water shortage, excess salts, excess of certain nonessential elements, damage from pests).

Although most annual plants take up a major portion of their nutrients during the middle third of their growth, the highest nutrient concentrations for N, P, and K in the soil solution are required during their early stages of growth, when root systems are small. Because of these high requirements and small root systems, proper positioning of added nutrients at this early stage is highly essential for good crop production.

Besides requiring an ideal range of essential nutrients in the soil solution, high yields of many crops cannot be obtained unless several nonessential elements, such as Al, As, Br, Cd, F, Pb, Li, and Ni are restricted to low levels. Several essential elements (B, Cu, and Mn) also must be kept at low concentrations in the soil solution to obtain satisfactory yields of many crops.

SECTION II:
CHARACTERISTICS OF SOIL OR OTHER MEDIA AFFECTING THE FERTILE TRIANGLE

This section deals with the influence of several characteristics of soil or other media on the availability of air, water, and nutrients. Usually, more than one side is affected, and in some cases, all three sides are modified by properties of the soil or other medium.

It is not uncommon for both air and water to be affected simultaneously because of the usual inverse relationship of air and water. In most soils and other media, the volume of air increases as the volume of water decreases and vice versa. It is only in the drier heavy soils that the volume of air increases more than the volume of water decreases. This is due to the formation of cracks as the soil dries, permitting large quantities of air to enter the soil.

Air or water also is often closely related to nutrients, but directly rather than inversely. Some of the same factors that benefit soil structure (OM, soil flocculation, porosity, absence of compaction) have positive effects on air and nutrients.

Several physical properties of media that have an effect on air, water, and nutrients are presented in Chapter 4. Organic matter which must be considered from the standpoint of both physical and chemical properties, is covered in Chapter 5. The others, which primarily affect chemical properties, are covered as follows: pH in Chapter 6, cation exchange capacity (CEC) in Chapter 7, and salts in Chapter 8.

Chapter 4

Physical Properties

INFLUENCE OF PHYSICAL PROPERTIES ON THE FERTILE TRIANGLE

The several physical characteristics of soil or medium, which are covered below, have a primary bearing on the air and water sides of the fertile triangle. Most of these characteristics affect the amount of air or water that the medium can hold and how quickly one can be exchanged for the other. The nutrient side of the triangle is also affected because these same characteristics affect the amount of nutrients produced by microorganisms, the movement of all nutrients, and how easily roots can grow to nutrients and use them.

Texture

Soil texture describes the size distribution of the soil particles, and is a reflection of the percentages by weight of the fractions of clay, silt, and sand. Clay, because of the huge surface area exposed by the numerous small particles, has the greatest effects on basic soil properties. Silt, which is intermediate in size, has much less effect. Sand, the largest particle, has very little or no influence on retaining moisture but its presence in large amounts greatly increases the potential for water and air exchange and also water and nutrient losses.

The dimensions characterizing sand and silt vary depending whether the International Society of Soil Science (ISSS) or the United States Department of Agriculture (USDA) schemes of classification are used (see Table 4.1). The largest particles (gravel and stones), more than 2 mm in diameter, have little effect on basic soil properties and are excluded from the soil textural determination. They do affect management and use of soils. Their presence in large numbers is noted by adding the terms "gravelly" or "stony" to textural class names.

TABLE 4.1. Size Ranges of Different Soil Fractions

USDA Scheme		ISSS Scheme	
	Size (mm)		Size (mm)
Very coarse sand	2.0-1.0	Coarse sand	2.0-0.2
Coarse sand	1.0-0.5	Fine sand	0.2-0.02
Medium sand	0.5-0.25	Silt	0.02-0.002
Fine sand	0.25-0.10	Clay	<0.002
Very fine sand	0.10-0.05		
Silt	0.05-0.002		
Clay	<0.002		

Exposed Surface Area

The different fractions vary in exposed surface area, which greatly influences their properties. Most soil particles are not spherical, but some idea of the relative area of the different particles can be gained by comparing the surface area of spherical particles of about the same density (about 2.6) as most mineral soil. Although surface area increases tenfold as the size of particles decreases tenfold, relatively little surface area is present until particles are about the size of clay, or about 0.001 mm or smaller (see Table 4.2).

TABLE 4.2. Surface Area of Spherical Particles As Affected by Their Radii*

Particle radius (mm)	Specific surface area (m^2/g)
1	0.00115
0.1	0.0115
0.01	0.115
0.001	1.15
0.0001	11.5

* A density of 2.6 is assumed, which is characteristic of soil mineral particles.

Source: Flegmann, A. W. and R. A. T. George. 1977. *Soils and Other Growth Media.* Avi Publishing Company, Westport, CT.

Importance of Texture

Texture has great effects on infiltration and retention of water and the ability of the soil to hold and exchange air, and to hold certain nutrients against leaching. In addition to the major importance of texture in air and water exchange, it is desirable to note that texture also affects (1) CEC or the soil's ability to hold nutrients; (2) nutrient losses by fixation, leaching, erosion, and volatilization; (3) buffering capacity, which influences the maximum safe limits and ideal application methods of fertilizers, lime, and other amendments; (4) heat losses and gains; and (5) consistency properties, which affect power requirements to work the soil and ease of root penetration.

Textural Particles

Sands. Sands can be classified as very coarse (2.00-1.00 mm), coarse (1.00-0.50 mm), fine (0.50-0.25 mm), and very fine (0.10-0.05 mm). Large amounts of the very small sand sizes can reduce porosity, infiltration, and the amount of air held by soil. The excellent drainage and abundance of air associated with sands are properties of very coarse and coarse sands and may be absent if the soil consists primarily of fine sands.

Silts. Particles classified as silt range in diameter between large particles (0.002 mm) and the smallest fine sand (0.05 mm). For a soil to be classified as silt, it needs to have at least 80 percent silt and no more than 12 percent clay. Such soils, although superior to the clays as far as aeration is concerned, can still have problems with aeration and need to be handled very carefully.

Clays. Clay particles, which are smaller than 0.002 mm, add greatly to a soil's ability to hold water and nutrients but tend to limit the amount of air exchange and water infiltration while increasing the difficulty of working the soil. The rapid exchange of air and water necessary for most crop production is almost impossible in soils containing large amounts of clay or silt unless much of the clay or silt is no longer present as individual particles but has been combined with each other and/or organic matter to form larger aggregates or peds. The formation of aggregates provides large pores capable of moving air and water to provide for a healthy atmosphere at the roots.

Not only the amount of clay but its type has a bearing on soil properties. The type of clay is a product of weathering and the rocks from which it was formed. The principal types of clay are; montmorillonite, hydrous mica, kaolinite, and the oxides and hydroxides of iron and aluminum. Much of the properties of the different clays are due to differences in their surface area. The kaolinite-type clays have only external surface areas, which although very large (in the range of 5 to 10 m^2/g of soil or 1,400 to 2,800 ft^2/oz) are small compared to the external and internal surfaces of about 800 square meters per gram (22,400 ft^2/oz) for montmorillonite and vermiculite.

Soils derived from montmorillonite are found in regions of good base supply resulting from limited rain (grasslands) or, if there is ample rain, insufficient drainage, which helps preserve enough bases. The climate needs to be warm (semitropical or warm temperate) to ensure enough weathering. Montmorillonite clay holds nutrients well but soils with considerable montmorillonite clay present problems as they are wetted and dried. When wet, they are extremely sticky, difficult to work, and have low bearing strength. The low bearing strength makes roads almost impassable at times, allows building foundations to crack, and fenceposts to be dislodged. As the soil dries, it becomes very hard, and huge cracks can appear.

Hydrous mica clays are found in cool climates that have enough precipitation to remove soluble salts. These clays are less subject to swelling and shrinking and provide greater bearing strength when wet than the montmorillonite clays. Although potassium is an integral part of the clay, much of it is only very slowly available as the clay weathers. Also, this type of clay does not hold nutrients well against leaching.

The kaolinite-type clays have undergone more weathering and have very low capacity to hold nutrients or water. They are found in regions of considerable rainfall and warm temperatures. Their ability to hold nutrients is appreciably less than the hydrous micas.

The oxides and hydroxides of iron and aluminum represent the ultimate in weathering. Such clays are found in tropic and semitropic soils formed from parent materials rich in iron and aluminum and subject to considerable leaching. They have very poor holding capacity for nutrients.

Textural Class

In considering the attributes of the different particle sizes, one must not lose sight of the fact that most soils or other media consist of mixtures of the different sizes and their properties are affected by the combination of materials. The clays or other colloidal materials, by the presence of electric charges on their surfaces, are largely responsible for retention of nutrient ions and much of the physical properties. But suitable structure is not possible without the presence of particles or aggregates larger than colloidal size, which increase the pore volume for good air and water movement. Therefore, the more productive soils or media are combinations of coarse and fine particles, or individual fine particles and aggregates of these fine particles, providing a mixture of pore sizes.

Soils are classified as sands, silts, or clays, depending on the dominant particle size. Loams are mixtures of the different-sized particles that show the properties of the different fractions in about equal proportions. Because of the strong expression of properties by the clay particles, smaller amounts of clay than silt or sand are present in loams. Twelve basic soil textural classes are revealed in Figure 4.1, a schematic diagram illustrating the various amounts of sand (0.05-2.0 mm), silt (0.002-0.05 mm), and clay (<0.002 mm) defining the different textural classes. In addition, the adjectives very coarse, coarse, fine, and very fine are used to describe dominant sand size in the textural class of sand.

FIGURE 4.1. Textural Classification Based on Percentages of Clay, Sand, and Silt

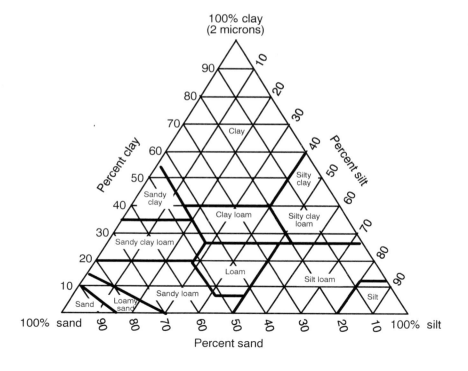

Importance of Textural Classification

Classification by texture can be helpful in choosing farm practices that maximize use of soil, fertilizer, and water. The efficiency of these practices is primarily related to the surface area exposed by the different fractions.

Generally, coarse or light soils, rich in sand, tend to warm and drain rapidly, and have good aeration, but do not hold water or nutrients well against leaching. The fine-textured or heavy soils, rich in clay and silt, are slow to warm but hold large amounts of water and nutrients. They tend to drain poorly and can be more compact, presenting greater problems with aeration. They are often much more difficult to work. The medium-textured soils, with mixtures of sand, silt and clay (loams, sandy loams, and silt loams) combine the best features of coarse and fine particles and tend to give higher productivity because of good holding capacities for water and nutrients while still allowing for sufficient air and water exchange.

There are gradations in the physical characteristics of medium-textured soils, depending upon the relative concentrations of sand and clay. Those having considerably more sand (sandy loams) have a greater ability to warm up in the spring; drain more rapidly and hold less water after rains; are more subject to leaching; are less prone to compaction and poor structure if worked wet; and have lower CEC and hold less nutrients than those with less sand. The air and water sides of these soils also can be improved by additions of organic matter, deep tillage, and certain chemical additives, although chemical additions are less useful for the sandier soils.

Structure

Structure describes the arrangement of the soil particles (sands, silts, and clays) into stable or cemented aggregates. The soil may be structureless, which usually occurs in sands where the individual grains are retained, or it may have cohesive masses in some loams or clays. If structured, the structure may be weak, moderate, or strong. The weakly structured soil peds or combined units are barely visible in moist soils. In moderately structured soil, the peds are visible in undisturbed soil and some of them can be handled without breaking. The strongly structured soil has peds that are clearly visible in moist soils and most of them can be handled without breaking.

The structure of the soil is greatly modified by: (1) the nature of OM and how it is handled; (2) soil tillage; and (3) chemical additions from water, fertilizers, liming materials, gypsum, and synthetic long-chained organic materials. Some of the effects of OM are outlined in this chapter with additional observations outlined in Chapter 5. Tillage is covered in

Chapter 9. Some of the impacts of chemicals on soil structure are given below but additional fertilizer effects are presented in Chapter 14; some of the water effects are outlined below with many more details in Chapter 17.

The Importance of Structure

Structure largely affects aeration and water (infiltration, drainage, and the amount of water held against gravitational forces) because of the spaces between aggregates that make up the pores or voids of the soil. The pores, largely dependent on the nature of the aggregates and how they are built into the structure, are also aided by channels or tunnels left by roots, rodents, insects, and earthworms and by the alternate shrinking and swelling of the clays. (The effect of pores on aeration has been covered in Chapter 1 and their effect on water in Chapter 2.)

Structure largely affects soil air and water by influencing the amount of water infiltrating the soil, the rapidity with which the excess is drained, and the amount retained by the soil for future plant use. Infiltration is influenced by surface and internal soil structure. Surface crusts or capping can greatly restrict water entry. Compacted soil or hardpans within the soil slow water infiltration and drainage. If drainage is slowed, there will be insufficient gaseous exchange of air for excess water in time to avoid damage to the crop.

Many soil characteristics besides good aeration and water-holding capacity are influenced by its structure. Good tilth, which allows for easy seedbed preparation and planting and increases the chances of rapid seedling and plant development, is primarily dependent on satisfactory soil structure. Good structure also provides rapid heat transfer, prevents moist soil from being excessively sticky, and allows it to retain many of its characteristics as it is worked while exerting a minimum of mechanical impedance. The crumb or granular structure comes closest to fulfilling these requirements, especially if the crumbs in the surface soil are large enough so that they do not blow away but are small enough to allow for good contact with seed to ensure good germination. The effectiveness of this type of structure is largely due to the balance between macro- and micropores that it is able to maintain.

Formation of Structure

Structure forms as a result of swelling, shrinkage, and cracking. The soil swells as it is wetted or frozen, and shrinks when it dries or thaws. Cracks develop because of stress along the lines of weakness, leaving the cemented soil masses between the cracks as structural units.

Although sand may remain as single soil particles, silt and clay particles tend to combine with humus and other particles to form larger units or peds, which provides for different soil structures. The larger units, or peds, are held together by cementing agents consisting of organic subtonics, iron oxides, carbonates, or silica. It is the peds that allow for much better moisture holding capacity, water infiltration and drainage, and greater exchange of air and water, while reducing the dangers of compaction.

Soils may be structureless because (1) the units retain their individual characteristics (primarily sandy materials), (2) particles cling together as a single mass (typical of clay), or (3) structure has been destroyed if the soil has been puddled or worked when too wet. But most soils do exhibit structure, the type of which, described below, affects the fertility of the soil by influencing the amounts of water, air, and nutrients that are available to plants.

Types of Soil Structure

Five major structural types largely define the basic characteristics of soil structure:

1. The platelike aggregates, which are arranged around a horizontal plane, tend to crack horizontally. They can be present in any part of the profile but puddling or ponding causes this type of structure on the surface. Continuous plowing at the same depth tends to develop the platelike plow sole or pan at the bottom of the furrow.
2. Blocky structure, which has units that are about as broad as they are long, is common in humid region subsoils and will exhibit both vertical and horizontal cracking.
3. Prismatic structure, usually found in subsoils, especially in dry regions, produces vertical rather than horizontal cracking and prism-like units that are much longer (two to five times) than broad.
4. Columnar structure also has units appreciably longer than broad, but the shape is approximately cylindrical, similar to a column. It is usually found in subsoils with considerable sodium.
5. Crumb or granular structure, represented by units that have no sharp edges, is present in surface soils rich in OM and is considered the most suitable for producing crops.

Factors Affecting Soil Structure

Structure in undisturbed soils is usually satisfactory. The results of weathering with alternate freezing and thawing or wetting and drying normally produce soils of good structure. Structure is enhanced by the presence of plants as roots penetrate soils, opening up channels, and by the decay of organic materials, which bind soil particles to provide better porosity. Poor structure in cultivated soils, however, is rather common, especially in high-tech monoculture. The single crop often limits the amount of organic materials returned to the soil; continuous cultivation usually associated with it helps deplete soil OM; and the heavy machinery usually used helps to unduly compact soils.

Nature is not always benign. Although slow freezing of wet heavy soils benefits structure, rapid freezing, especially of sandy soils, may be detrimental. The improvement in structure from wetting and drying is related to the proportion and type of clay. Soils that contain large proportions of montmorillonite clay tend to change volumes as the soil is wetted and dried due to swelling and shrinking. Cracks formed as the soil dries will remain for some time even though the soil swells as it is rewetted and compaction tends to be a minor problem on such soils. Clods are reduced as the soil is gradually wetted, but rapid wetting or flooding can destroy structure. Heavy rainfall in large drops increases compaction and can cause capping (compaction at the surface) of soils not protected by mulches.

Structure can be affected by several other factors, namely: (1) organic matter, (2) soil organisms, (3) percentage saturation of the CEC by various cations, (4) how water is applied, and (5) by chemical additions from irrigation water, fertilizers, liming materials, gypsum, and conditioners.

Effect of Organic Matter

The addition of organic matter tends to improve structure of all soils except those very rich in organic matter (peaty or muck soils). Organic matter tends to benefit structure in several ways. It limits the amount of compaction, thereby providing better water infiltration and gaseous exchange. Pieces of organic material between soil particles prevents sealing and provides air spaces as the soil is compacted. Growing crops produce channels as old roots decay. The presence of organic materials on the surface breaks the destructive force of large rain or irrigation drops with less resultant capping. It provides for better retention of moisture and a gradual beneficial wetting that helps maintain structure and prevent erosion. It also increases water infiltration and drainage, reducing ponding and excessively wet conditions in the soil.

Decomposition of organic matter produces substances that act as cements, binding the individual clay and silt particles into crumbs. The larger particles or aggregates formed from individual soil particles tend to give the heavier soils characteristics more like the sands in respect to rapid exchange of air and water. These aggregates, now 100 to 1,000 times the size of the clay and silt particles and about equal to that of fine gravel or coarse sand, tend to be porous and hold large quantities of air and water.

The various other benefits of OM, and the cultural practices that increase both the amounts and the effectiveness of it, will be considered in some detail in Chapter 5.

Soil Organisms

Various soil organisms affect structure. The microorganisms tend to bind the individual particles into peds, which aid structure. Filamentous fungi bind soil particles to form water-stable clods. Actinomycetes also can be beneficial but supposedly less so than the fungi. Bacteria, by producing cementing agents capable of aggregating various soil particles, probably change structure more than most other organisms.

Besides microorganisms, various other soil inhabitants affect structure, mostly in a beneficial way. Small animals (marmots, prairie dogs, ground squirrels, moles, shrews, and mice), by burrowing in the soil and moving it, tend to open soils and allow for greater drainage. Ants, mites, springtails, and termites also move considerable organic materials into the soil and, by helping to decompose them, improve structure. Unfortunately, the huge mounds built by termites in tropical regions or dislocations caused by some animals can impede cultivation. Earthworms improve structure by their movement in the soil and by their casts, which are rich in humic materials, Ca, and salts.

Usually, such changes are beneficial and conditions that favor bacterial development (suitable pH, ample OM, and satisfactory moisture, temperature, and oxygen content) tend to provide satisfactory structure. However, in very fine-textured soils with a predominance of micropores, the cements produced can plug some of the finer micropores, reducing the ability of the soil to infiltrate water or drain well.

Saturation of Cation Exchange Capacity
with Different Cations

Soil structure is affected by the ions held by colloidal matter (humus and clay) and the presence of certain salts. The ions and salts can come from water as indicated above or from various soil amendments (liming materials, fertilizers, gypsum, manures, etc.).

Soils containing considerable Ca and small quantities of free salts yield small clods when dry, and these clods retain their shape on wetting. Soils with considerable Na and few salts tend to produce large hard clods when dry. These clods will become liquid mud when wet. Soils with considerable K or Mg produce clods somewhat intermediate between the extremes of Na or Ca. Although not as hard or large as the Na clods, they tend to be unstable in the presence of water.

Acid soils rich in Al tend to produce clods similar to those of Ca-rich soils. The addition of Al has helped improve permeability of some clays but such additions can lower soil pH. The lowered soil pH and increase in Al can limit root growth of many plants, making the addition of Al a chancy procedure. The addition of Ca with Al reduces potential risk. Liming soils so that some exchangeable Al remains (to pH 6.5 instead of 7.0) could be a viable alternative.

Flocculation and Deflocculation

Saturation with different cations also affects flocculation or defloccula-tion of the soil colloids. In flocculated clays, particles tend to cling together, providing for better pore space. Deflocculated clay particles, on the other hand, tend to retain their individual character with negative effects on distribution of pore sizes. Flocculated clays can be deflocculated by replacing the electrolyte solution surrounding the clay with one that has lower valences or is lower in concentration. Clays saturated with solutions that have considerable sodium (Na^+) or lithium (Li^+) tend to disperse into individual particles or deflocculate when shaken in water. As they settle out, the particles will form a compact layer that is difficult to redisperse. If salts are added to the suspension, the particles will tend to flocculate or form loose aggregates that settle out as loose sediments, capable of being easily dispersed. The amount of salts necessary to flocculate a clay varies depending on the nature and charge of the cation. The greatest amount is needed for Na^+ or Li^+; less for NH_4^+, Mg^{2+}, or K^+; much less for Ca^{2+} or Al^{3+} and appreciably less for the multivalent cations or complex ions.

The flocculation or deflocculation of clay has practical implications. Not only is the stability of aggregates affected as the clay is deflocculated, but the soil also becomes sticky, is more difficult to work, is highly moldable at low moisture, and tends to form large hard clods as the soil dries. Management of air and water become exceedingly difficult. The flocculated clay is less sticky or moldable and dries to crumbly, friable aggregates. In the field, this condition is known as mellowness. A mellow soil is easily worked, capable of being satisfactorily tilled with appreciably more moisture than the deflocculated soil without forming hard clods.

In arid climates, deflocculation can result naturally because of the high Na content, especially if low-salt water is used on the sodium-rich soils. It also can be induced by use of irrigation water rich in Na and low in Ca and Mg or rich in HCO₃ (bicarbonates—see the section on Irrigation) and the use of sodium carbonate or other salts that raise the pH.

Application of Water

In addition to the effect on structure of chemical components in the water, structure also can be affected by the manner in which water is added. Irrigation applied overhead as large droplets can seal and compact soil, or too rapid an application can adversely affect structure. Large droplets applied can create a cap on the surface, negatively affecting soil structure. In this respect, irrigation water produces adverse effects similar to those of rain that falls very rapidly and in large drops. The presence of organic materials, especially as mulches, or of a crop canopy lessens some of the adverse effects of large droplet size.

For surface-type applications, the speed of water delivery has to be closely correlated to length of run, slope, type of soil, and its infiltration capacity. Because of its slow delivery of water, drip irrigation systems are much less prone to cause problems.

Structure Affected by Chemical Additions

Chemical additions affect structure by modifying the CEC saturation. The chemicals are added in the form of irrigation, fertilizers, liming materials, gypsum, and soil conditioners.

Irrigation

Irrigation can affect structure because ions carried by the water as excesses of carbonates and bicarbonates and/or Na gradually changes the base saturation of the soil.

Carbonates (CO_3^{2+}) and bicarbonates (HCO_3^{-}). Water with high carbonate or bicarbonate tends to harm soil structure as the Na usually associated with these ions tends to replace soil Ca. The Ca will react to form relatively insoluble calcium carbonate or bicarbonate as the soil dries, leaving a Na-saturated soil. The problem of excess bicarbonate is more common and serious than that of carbonate because of the larger amounts of bicarbonate usually found in irrigation water. Although the carbonate content is usually less than 15 ppm, bicarbonate can easily exceed 250 ppm. Concentrations of 120 to 180 ppm are considered as moderately harmful, 180 to 600 ppm as severely harmful, and >600 ppm as very severely harmful. More about water quality is presented in Chapter 17.

Sodium. Sodium in irrigation water also can adversely affect structure but the effects are modified by the presence of Ca^{2+} and Mg^{2+} in the water and the texture of the soil. The harmful effects of Na are related to the ratio of Na to total cations rather than the amounts of Na^+. The relationship can be expressed as soluble sodium percentage (SSP), which is determined by the equation:

$$SSP = \frac{Na^+}{Ca^{2+} + Mg^{2+} + Na^+} \times 100$$

using meq/liter for the cations.

The relative activity of the Na^+ as it reacts with the clay is expressed by the sodium adsorption ratio (SAR) given below:

$$SAR = \frac{meq/L\ Na^+}{\dfrac{\sqrt{meq/L\ Ca^{2+} + meq/L\ Mg^{2+}}}{2}}$$

Values of 3 or less represent no problems but permeability and alkali hazard will increase slightly as the SAR is 3 to 5; increase substantially as it falls in the range 5 to 8 and become practically intolerable if >8.

Fertilizers

Structure is also affected by chemical additions from fertilizers but perhaps to a lesser extent than from irrigation waters. The primary effects appear to be related to the displacement of Ca^{2+} from the CEC and re-placement with Na^+, NH_4^+, H^+, or Al^{3+}.

The use of ammonium salts rapidly leads to depletion of Ca^{2+} and ultimate replacement with H^+. As calcium saturation falls below about 60 percent, noticeable adverse effects on structure may be apparent on the heavier soils. The loss of Ca and replacement with H^+ tends to acidify all soils but as acidification continues, some of the hydrogen may be replaced with AL^{3+}. Soil structure of the heavier soils may be improved by the addition of AL^{3+}, but because many plants are sensitive to elevated levels of aluminum, yields are often depressed.

The Ca^{2+} loss can be reversed by liming, the use of gypsum, or the additions of calcium fertilizers, such as calcium nitrate or calcium phos-phates, but is accentuated by sodium nitrate. Large applications of sodium nitrate can lead to considerable sodium saturation with depressing effects on yields resulting from deflocculation and structural damage. Fortunately, the use of sodium nitrate, which at one time was extensive, has been markedly curtailed.

Liming Materials

A large group of compounds used to raise soil pH supply calcium, which increases the saturation of Ca^{2+} and can be helpful in displacing Na^+. The changes in pH and structure resulting from liming materials can boost yields of many crops. A description of compounds used and modes of application are presented in some detail in Chapter 6.

Gypsum

Calcium in the form of gypsum as well as liming materials can be expected to provide better soil structure with improvement of air and water utilization in many heavy soils, although it usually is of little value in improving structure of sandy soils. Providing sufficient Ca to saturate 70 to 75 percent of the exchange complex tends to group dispersed soil particles into clumps, providing better stable aggregates with better porosity, water infiltration, and reduced compaction.

The need for gypsum is suggested when heavy soils are poorly saturated with calcium (>60 percent), but many heavy soils can be helped by additions of gypsum if they have low amounts of stable aggregates, low porosity, poor water infiltration, or compact layers. Usually, permeability of the soil will be greatly increased by the use of gypsum if the soil contains appreciable Na (>10 percent exchangeable). The advantages are less as the amount of Na declines but can be appreciable on soils rich in clay but with relatively low Ca saturation (<60 percent), or if the addition of gypsum is combined with other procedures such as subsoiling or chiseling.

The effect of gypsum in opening soils has been effectively used in the reclamation of sodic soils. The gypsum replaces Na^+ with Ca^{2+} in the CEC and the Na^+ is leached out. Large applications (10 tons or more per acre) may be needed before sufficent Na^+ is replaced and structure is restored so that suitable air and water exchange can take place.

On some sodic soils, large deposits of free calcium carbonate are present that can be converted to gypsum. Gypsum is much more soluble than the calcium carbonates, allowing it to be dissociated to form Ca^{2+} ions that replace the Na^+. Sulfuric acid, sulfur, or various sulfur compounds that can be oxidized to sulfuric acid are applied and worked into the soil. Oxidation, which may take some time, eventually produces sulfuric acid, which reacts with calcium carbonate to produce gypsum. Some of the materials used to form gypsum in soils containing calcium carbonate and their equivalent value in relation to gypsum are presented in Table 4.3.

TABLE 4.3. Materials Capable of Acidifying Calcium Carbonate to Form Gypsum and Relative Amounts Needed Compared to Sulfuric Acid

Material	% S	Formula	Relative amount needed compared to sulfuric acid
Sulfuric acid (98%)	31.4	H_2SO_4	100
Sulfur	100	S	31
Aluminum sulfate	20	$Al_2 (SO_4)_3 \cdot 18H_2O$	218
Ammonium bisulfate*	17	NH_4HSO_4	143
Ammonium thiosulfate*	26	$(NH_4)_2S_2O_3$	96
Ammonium polysulfide*	45	$(NH_4)S_x$	56
Ammonia sulfur*	10	$(NH_3 + S)$	63
Iron sulfate (ferric)	17	$FeSO_4 \cdot 9H_2O$	185
Iron sulfate (ferrous)	18	$FeSO_4 \cdot 7H_2O$	274
Iron pyrites (87%)	46.5	FeS_2	148
Sulfur dioxide	50	SO_2	63

* = solutions.

Note: Calculations of relative effectiveness are based on amounts of pure calcium carbonate needed to neutralize 1 lb ammonium-N (1.8 lb) and 1 lb sulfur (3.125 lb).

Source: Agricultural Retailers Association, St. Louis, MO.

Amounts of gypsum needed. The approximate amounts of gypsum needed for soils containing exchangeable sodium can be calculated if the exchange capacity and percent sodium saturation are known. Using the equation

$$\text{gypsum} = \text{CEC} \times \text{Na sat (E)} - \text{Na sat (N)} \times 1840$$

where gypsum = gypsum needed per plowed acre (2,000,000 lb), Na sat (E) = percent saturation of Na^+ in existing soil, Na sat (N) = percent saturation of Na^+ desired, and 1840 is a factor converting milligrams gypsum per 100 grams of soil to pounds gypsum per acre. For example, if the existing Na^+ saturation = 15 percent in a soil having 20 meq CEC, and a 5 percent saturation is desired, the pounds of gypsum required per acre will be $20 \times (0.15 - 0.05) \times 1840$ or 3680 lb.

The amounts needed to improve the Ca saturation also can be calculated if the CEC and percent saturation of Ca is known. Again, if the CEC is 20 meq/100 g soil and the percent Ca saturation is 55 and a 70 percent saturation is desired, the lb gypsum needed per acre = $20 \times (0.70 - 0.55) \times 1840 = 5520$.

Gypsum application. Maximum benefits are obtained if the gypsum is broadcast and worked deeply into the surface soil. Some benefits can be obtained by treating a smaller portion of the total volume if treatment is confined to the volume directly under the planted row. The author has obtained benefits with the application of about 1/4 normal rates of gypsum if it was applied directly under the bed or ridge in a band about 1 foot wide and worked in well before forming the bed or ridge.

Combination of these smaller amounts of gypsum with chiseling prior to the bed or ridge formation has enabled plants to start rapidly due to early formation of an extensive root system.

Large Molecule Conditioners

The addition of large molecules acting as polyelectrolytes to soils has a beneficial effect on soil structure, evidently by flocculating or binding soil particles into larger peds. The improvement in soil structure results in better water penetration, drainage, resistance to erosion, and reduced draft needed for working soils. The flocculation induced by polyelectrolyte, which provides these benefits, appears to be much less reversible than that due to simple electrolytes, such as CA^{2+}.

The large molecules can be organic but synthetic polyelectrolytes have proven to be highly effective, some of them in minute quantities. Krilium was one of the first of such materials to be commercially tried, but it was not readily accepted because of costs. Recently, materials effective at much smaller rates have appeared on the market. Compounds such as ammonium lauryl sulfate or the polymers polyacrylamide (PAN) or poly-maleic acid (PAM) appear to be effective at rates of ounces to a few pounds per acre. At these rates, several of these materials can easily be cost effective for many crops.

Use of water-soluble polymers can be expected to stabilize soil aggregates. In so doing, movement of soil by wind and water is reduced, water infiltration is increased, and crusting of soils is minimized, allowing for better seed emergence, increased soil aeration, and enhanced tillability. The improvement in water infiltration can help the removal of excess salts and reduce erosion. The net result is better yields since many of the poor soil physical properties have been eliminated or diminished. There is some evidence to indicate that use of these polymers with modest amounts of organic matter can be more beneficial than can be expected from the combination of benefits from either input alone.

The Complete Green Company (Los Angeles, CA) has been promoting the use of PAM and has obtained considerable improvement of water infiltration and seed emergence with it. The material is sprayed on the soil or applied through irrigation lines.

FMC Corporation (Coral Gables, FL) has utilized PAM under the trade name Sper Sal in Mexico and Central America to help correct salinity problems. The material, by combining small soil particles into larger stable aggregates, permits more rapid elimination of Na^+ from solonetz or solonchak soils.

Anionic water-soluble polymer of high molecular weight added to irrigation water at 30 ppm is recommended by the JRM Chemical, Inc. (Cleveland, OH) for the purpose of reducing soil loss and water runoff.

Permeability

Permeability, or the ease with which water and air enters and moves down through soil, is a soil quality closely associated with structure. It affects the usefulness of irrigation or rainfall and the rates of air renewal, root penetration, and nutrient leaching. In addition to structure, permeability is affected by such factors as texture, structure, bulk density, clay films, and soil cracking. Four different classes of permeability are recognized (Donahue, Follett, and Tulloch, 1995):

1. Rapid—Loose and open sandy subsoils allow for rapid water drainage and air renewal. Such soils may provide poor crop production due to lack of water unless irrigated.
2. Moderate—Highly granular, blocky, or columnar clay loam subsoils provide for adequate water drainage and air replacement. Such soils are highly productive over a wide range of rainfall.
3. Slow—Permeability tends to be slow in a number of soils with clayey subsoils that have medium, blocky, or platelike structures that tend to be crumbly under ideal conditions. Many of these soils also have a thick transition zone between a deep A horizon and the very clayey zone of the B horizon.
4. Very Slow—Permeability is very slow in a number of soils that have fine clay or clay-pan subsoils and a platelike, coarse blocky, or massive structure. Soils, such as sodic soils, tend to be plastic and sticky when wet, limiting growth and root movement to cracks between blocks and plates.

Porosity

Porosity, or the amount of pore space in soil and other media, has been previously covered in some detail because of its importance in air exchange (Chapter 1) and in water movement in soils (Chapter 2). Porosity also affects the penetration of roots in the soil, thereby affecting the nutrient leg of the triangle.

Roots tend to penetrate soils by growing through existing pore spaces or cracks. The size of the root tip varies with different plants, making it easier for some plants to penetrate soils with limited pore space. If the open space is not large enough, the root tips will tend to elongate by moving aside soil particles. Elongation of the root tip is affected by soil strength, which is increased as compaction or bulk density increases.

In addition to good porosity helping nutrient uptake by providing satisfactory air and water environments, it actually increases the amount of nutrients available for uptake by favoring the activities of a number of microorganisms that release nutrients from OM and those that fix nitrogen both symbiotically or directly from the air.

Overall effects of porosity on the fertile triangle are most beneficial when there is a mix of different-sized pores. In fact, the distribution of pore size may be as important as the percentage of pore space. The large pores are essential for rapid water movement and gaseous exchange, but the smaller pores are necessary for capillary movement of water and nutrients to plant roots.

Bulk Density

One of the benefits of good soil structure is that it lowers bulk density, which is the weight of soil solids per unit volume of soil. It is obtained by dividing dry soil weight by its original volume. Lower bulk density means not only better aeration but also less weight of soils that are moved and less draft power needed to work soils. Such reduced power use can lessen damage to structure as soils are worked.

Bulk density, usually expressed as grams per cubic centimeter (g/cm^3), can be given in any terms of weight and volume. The common bulk densities of mineral soils range between 1.0 and 1.6 g/cm_3, with the finer-textured soils having lower bulk densities because of their greater pore spaces. If there were no pore spaces, the bulk density of soil would be 2.65 g/cm^3 (the average density of soil mineral matter). Soils with low densities (1.0-1.4 g/cm^3) offer little impediment to the extension of roots and allow for rapid movement of air and water. The bulk densities of organic soils will be below 1.0 and those of the sphagnum moss peats will be close to 0.1 g/cm^3. Organic matter lowers bulk density because it is much lighter in weight than a similar volume of mineral matter. Organic matter also increases the aggregate stability of a soil, which tends to benefit pore space, lowering bulk density. Common bulk density values for several soil classes are given in Table 4.4.

Bulk density is increased as the soil is compacted but is improved by practices that increase aeration. Tillage usually lowers bulk density, at least in the short term. Subsoiling lowers bulk density but puddling the soil tends to increase it.

Generally, crop yields tend to improve as bulk densities are lowered. Absolute values needed for optimum growth are difficult to define because the optimum values may vary with different crops and soil types. For example, the tolerance to soil compaction is in the order alfalfa > corn > soybeans > sugar beets > dry edible beans. Although a bulk density value of <1.4 g/cm^3 is satisfactory for many crops on clay soils, a slightly higher value of <1.6 g/cm^3 can be tolerated on sands. There are indications that lower bulk densities may be desirable to prevent some plant diseases. Values greater than 1.2 g/cm^3 on loams of Washington state were associated with higher incidence of bean and pea diseases (Miller and Donahue, 1995).

Bulk density has an effect on the movement of nutrients by altering their diffusion rate. Normally, soil moisture moves in a convoluted path,

TABLE 4.4. Average Bulk Densities of Different Soil Textural Classes

Soil texture	Bulk density (g/cm^3)
Sand	1.58
Loamy sand	1.52
Sandy loam	1.47
Loam	1.39
Silt loam	1.36
Clay loam	1.31
Sandy clay loam	1.44
Silty clay loam	1.40
Sandy clay	1.33
Silty clay	1.26
Clay	1.23

Source: Information capsule #137, Midwest Laboratories, Inc., Omaha, NE.

avoiding solid particles in the soil. As a very open soil is slightly compacted, the volumetric soil moisture content increases and diffusion is increased by the continuity of the water in the capillaries and as the length of the diffusion path is shortened. But as the soil is further compacted, the solid particles come together, making the diffusion path more tortuous as moisture must move around the particles.

The bulk densities at which diffusion is optimum probably varies with different soils, but the upper limits appear to be in the range of about 1.3 to 1.5 g/cm^3.

Pore space of soil or other media can be calculated if bulk density is known by using the equation:

$$\text{Percent pore space} = 100\% - \left(\frac{\text{bulk density}}{\text{particle density}}\right) \times 100$$

The particle density of mineral soils = 2.65 g/cm^3. If the bulk density of a particular soil = 1.23, the percent pore space in the soil is 46 percent, obtained by dividing 1.23 by 2.65, multiplying it by 100 and subtracting the answer from 100 percent.

Compaction

Compaction can be present in various portions of the soil. Surface crusting results from the impact of rain or irrigation droplets as they strike the soil and the drying of a layer of dispersed particles. Topsoil compaction results from compression by foot traffic or machinery. Hardpans made up of densely packed sediments can exist in subsurface layers. Some can be cemented to the point of being rocklike (fragipans or ortstein). Claypans having a very high clay content that are resistant to penetration of air and water when wet are present as subsoil layers.

Compaction can result from both natural and artificial forces. Gravity as well as the force of raindrops can compact soils, but their effects are greatly modified by cultural practices. The presence of sods or mulches substantially reduce the effects of rainfall. Alternate freezing and thawing of soils tends to alleviate the ill effects of compaction.

Although some compaction occurs naturally (rainfall impact and subsoil compaction due to load of topsoil), much of it is a result of modern farming operations. The extent of crusting is increased as the soil is deprived of cover or if it contains little OM to stabilize the soil aggregates. Many farm practices involving machinery, such as soil preparation, fertilizer and lime spreading, cultivation, spraying, and harvesting, can seriously compact soils. Vehicular movement, particularly on wet heavy soils, primarily compresses topsoil, although some of the force may persist to the subsoil. These forces plus slicks caused by plows and disks often produce compact layers or soles that can seriously affect water drainage and restrict root penetration.

Compaction problems usually become more serious as the weight of machinery increases, particularly as it is used on heavy soils and if such soils are wet. Fine-textured soils with little montmorillonite clay are especially subject to compaction.

Various methods can be used to limit compaction, such as: (1) reducing weight of machinery, (2) limiting use of machinery when soils are wet, (3) confining spraying and harvesting machinery to special roadways (use of long booms limits amount of ground lost to production), and (4) spreading the weight of machinery over larger areas by use of floater tires, crawler tracks, and all-wheel drive. Damage from overhead irrigation is reduced by using nozzles that tend to provide small droplets. Rotations that increase OM reduce the harmful effects of heavy machinery, and leaving organic residues on the surface to break the force of rain or irrigation droplets can reduce surface compaction. (The importance of OM for rotations and its special place as mulch to reduce compaction are covered more extensively in Chapter 5.)

Surface compaction can also be alleviated to a large extent by increasing stability of surface aggregates with such farming practices as suitable rotations and use of limestone or gypsum. Improving internal compaction problems by use of subsoilers may be short term unless the subsoiling is combined with the addition of gypsum and/or organic matter.

Several approaches to reducing compaction from machinery are considered in Chapter 9.

SUMMARY

The physical properties of a soil or other medium have important effects on the fertile triangle. Effects on the aeration and water sides of the triangle predominate, but because air and water affect the production and uptake of nutrients, the nutritional side is also influenced by physical soil properties.

The principle soil properties affecting the fertile triangle are texture, structure, porosity, bulk density, and compaction. The manner in which these various characteristics affect the fertile triangle and their importance have been outlined. Several means of improving these characteristics to lengthen the sides of the fertile triangle have also been given.

Chapter 5

Organic Matter

INFLUENCE OF ORGANIC MATTER ON THE FERTILE TRIANGLE

No single constituent of the soil is as important as organic matter in changing a pile of decomposed rocks into a vibrant, dynamic, living entity. In so doing, it changes all three sides of the fertile triangle, affecting air, water, and nutrients in significant ways. It reduces surface compaction, allows for better water infiltration and storage, provides for better gaseous exchange and microbial activity, increases cation exchange capacity and greatly benefits soil structure. By providing a source of energy for microorganisms, it nourishes the many biological reactions that are vital to good soil structure, transformation of N, P, and S, the release of nutrients, and the suppression of soilborne diseases.

Humus

Organic matter is not consistent or stable. It consists of a wide range of materials—from the undecayed plant residues, animal and insect remains, and microorganisms to various stages of decomposed portions of these items and to a complex substance with transitory stability, known as humus.

Sugars, starches, simple proteins, and cellulose of most organic materials are quickly attacked by various soil organisms (actinomycetes, bacteria, and fungi) providing that soil temperature, Ca, and moisture are satisfactory. The more complex compounds of hemicelluloses, lignins, fats, and waxes are decomposed more slowly. The attacks by organisms reduce organic materials to a product that is in equilibrium with its existing environment of moisture, temperature, and oxygen. This semistable product or humus is much less variable than the different organic components; it has a C/N ratio of about 10:1 and a C/P and a C/S ratio of 100:1, compared to C/N ratios of unmodified OM that can range from 12:1 to about 700/1, the C/P and C/S ratios from 100:1 to 450:1.

Humus, usually yellow-brown to black in color, is acidic and amorphous with rather indefinite composition. Although studied at great length, its actual composition is unknown, probably because of its great variability. Its important constituents appear to be fulvic acids, humic acids, and humins, the amounts of which vary with different lots of humus. Since molecular weights and solubilities of these important constituents are different, it is readily apparent why it has been difficult to establish an exact composition for humus.

Only a portion of the original organic material is held in the humus state. The amount remaining as humus varies from a high of about 40 percent to only about 1 to 2 percent. Although humus is resistant to microorganism breakdown, its decomposition is increased by aeration. Its ready destruction by oxygen accounts for reduced amounts in cultivated or open soils and better levels in wet or compact soils. Relatively open sandy soils will have lower levels than heavier soils with smaller pores.

With other factors constant, the accumulation of humus is greater as temperature is lowered, soil moisture is increased beyond a certain point, and clay content is higher. The lower temperature reduces the activities of microorganisms and accounts for the generally higher levels of OM in northern than in southern states. Increased soil moisture tends to reduce soil oxygen, which also limits microbial activity and tends to accumulate OM. Finer-textured soils with less oxygen also tend to accumulate more OM than the more open sands. This is fortunate because larger amounts of humus are necessary to provide suitable macropores for aeration and drainage with these finer-textured soils.

Humus is considered extremely important for a soil's productivity. It is valuable in improving friability, tilth, aeration, and water infiltration. It also greatly improves nutritional status, which is aided by the release of plant nutrients as humus decomposes and of nutrients held by CEC. The CEC of a soil is greatly increased by the presence of humus, although the increase is less in very acid soils than in slightly acid or alkaline soils.

Importance of Organic Matter

All portions of OM play an important role in air and water utilization. The undecomposed and partially decomposed portions physically reduce capping at the surface, limit compaction of soils, increase porosity of the soil at all levels, and provide the energy for microorganisms performing several vital functions of improving soil structure, nitrification, and maintaining availability of several elements. The more decomposed portion that is stabilized as humus improves soil structure, aids water movement in soils, increases MHC, provides for better aeration, increases CEC, provides a steady supply of nutrients beneficial to plant growth, and is a source of energy for the development of many microorganisms important in plant health.

The multiple effects on infiltration of water and on soil structure make OM an important component in erosion control. Because of its effect on MHC and the large amount of surface area exposed, OM also tends to buffer the soil against marked pH and salt changes resulting from large applications of liming materials, acid-forming materials, and fertilizers.

Sandy soils benefit from OM because of considerable improvement of their MHC, nutrient level, and microbial activity. The heavy soils (silts and clays) profit from OM because of considerable improvement in their structure and resultant benefits in porosity, water infiltration, and air exchange.

Knowledge of sources of organic matter, their relative practicability for different situations, their transformations, and the ways in which organic matter can be handled for maximum returns can help a grower make decisions about preferred procedures for maintaining adequate levels of this important asset.

BENEFICIAL ATTRIBUTES OF ORGANIC MATTER

Energy Source

Organic materials (soil OM, manures, composts, sludges, organic fertilizers, crop residues), are a source of energy for microorganisms. Moreover, microorganisms not only aid the decomposition of organic materials, which slowly releases a great deal of nutrients for succeeding crops, but also are responsible for continuing various microbiological activities, ranging from the improvement of soil structure to the fixation of N. The fixation of N, either symbiotically or free form, allows plants to use atmospheric N, normally not available to them. Generally, more N is made available from decomposition of organic sources and fixation if temperatures are favorable (60-95°F, 15.5-35°C), moisture is neither too low nor high, pH values are in the range of 6.0 to 7.0 and salts are not excessive (0.2-1.5 mmhos/cm).

Nutrients Supplied by Organic Matter

Organic matter can be a good source of several nutrients, but the amounts made available to plants are highly variable depending upon the source of the OM, its stage of development, soil conditions of pH, salts, moisture, aeration, and temperature, and the length of the crop season. Soil OM is the primary source of reserve N in many soils, and various organic additions (plant residues, cover crops, sods, manures, composts, sludges) can add considerably more N plus other essential elements. The primary nutrients supplied by a ton of various organic amendments are given in Table 5.1.

Legumes, many of which are used for soil improvement, are good sources of added N. Although amounts of N produced by leguminous crops vary depending on degree of inoculation, soil pH and calcium content, and age of crop when it is incorporated, the average values as given in Table 5.2 indicate that the N contributions of these crops can be substantial.

Release of Nutrients from Soil OM

Because plants cannot utilize N or other elements (primarily P, S) in organic forms until they are broken down by microorganisms, the availability of these elements for plant use is not constant. The organic N is broken down to the ammonium (NH_4) and then to the nitrate (NO_3) forms, both of which are highly available for plants. The production of the nitrate form is slowed in the presence of nitrification inhibitors, or if soil air is limited by excess moisture or compaction.

Various predictions have been made about the amounts of N released from soil OM. Generally, these vary from about 20 to 40 lb N per acre per crop year for each percent OM but amounts released depend on favorable microbial action.

The Vegetable Research Trust in Warwick, England, has published an N (along with P and K) predictor that suggests various N additions based on OM. The predictor provides four indices describing different OM levels. These allow the grower to choose a level of fertilization capable of giving maximum yields or any acceptable portion thereof for 20 vegetables. The four N indices correspond to the following:

0. Very low OM soils previously cropped for at least two years with cereals
1. Normal soil cropped one year with cereals or storage root crops
2. Normal soil in arable or ley (temporarily under grass) systems
3. Soil with large N reserves or that has received heavy applications of farmyard or other organic manures

TABLE 5.1. The Composition of Some Common Organic Materials

Material	Moisture %	Approximate pounds per ton of material		
		N	P_2O_5	K_2O
Alfalfa hay	10	50	11	50
Alfalfa straw	7	28	7	36
Barley hay	9	23	11	33
Barley straw	10	12	5	32
Bean straw	11	20	6	25
Beggarweed hay	9	50	12	56
Bermuda grass hay				
Coastal	10	50	8	40
Common	10	32	9	43
Bluegrass hay	10	35	12	35
Buckwheat straw	11	14	2	48
Clover hay				
Alyce	11	35	—	—
Bur	8	60	21	70
Crimson	11	45	11	67
Ladino	12	60	13	67
Subterranean	10	70	20	56
Sweet	8	60	12	38
Corn stover, field	10	22	8	32
Corn stover, sweet	12	30	8	24
Cowpea hay	10	60	13	36
Cowpea straw	9	20	5	38
Crested wheatgrass	10	62	11	44
Fescue hay	10	42	14	47
Field pea hay	11	28	11	33
Field pea straw	10	20	5	26
Horse bean hay	9	43	—	—
Lespedeza hay	11	41	8	22
Lespedeza straw	10	21	—	—
Oat hay	12	26	9	20
Oat straw	10	13	5	33
Orchard grass hay	10	45	14	55
Peat, muck	30	45	9	15
Peat, sphagnum	—	11	2	2
Peanut vines	10	60	12	32
Rice straw	10	12	4	27
Ryegrass hay	11	26	11	25
Rye hay	9	21	8	25
Rye straw	7	11	4	22
Sorghum stover	13	18	4	37

TABLE 5.1 (*continued*)

Material	Moisture %	Approximate pounds per ton of material		
		N	P_2O_5	K_2O
Soybean hay	12	46	11	20
Soybean straw	11	13	6	15
Sudan grass hay	11	28	12	31
Velvet bean hay	7	50	11	53
Vetch hay, common	11	43	15	53
Vetch hay, hairy	12	62	15	47
Wheat hay	10	20	8	35
Wheat straw	8	12	3	19

Source: Lorenz, O. A. and D. N. Maynard. 1988. *Knott's Handbook for Vegetable Growers,* Third Edition. John Wiley and Sons, New York, p. 101. Copyright © 1988 John Wiley and Sons, Inc. Reprinted by permission of John Wiley and Sons, Inc. Modified by author's data.

TABLE 5.2. Approximate Amounts of Nitrogen Fixed by Various Legumes

Crop	lb/ac	Crop	lb/ac
Alfalfa	190	Hairy vetch	165
Beans	40	Kudzu	110
Clovers		Lespedeza (annual)	85
Crimson	160	Pea (English)	80
Ladino	175	Peanut	50
Red	115	Perennial peanut	210
Sweet	120	Soybean	110
White	100	Winter pea	60
Cowpea	90		

Sources: Follett, R. H., L. S. Murphy, and R. L. Donahue. 1981. *Fertilizers and Soil Amendments.* Prentice-Hall Inc., Englewood Cliffs, NJ; Pieters, A. J., and R. McKee. 1938. The use of cover and green-manure crops. In G. Hambridge (Ed.), *Soils and Men: Yearbook of Agriculture 1938.* United States Government Printing Office, Washington, DC; Tisdale, S. L., and W. L. Nelson, *Soil Fertility and Fertilizers,* Third Edition. Macmillan Publishing, New York.

Recommended applications of N in kg/hectare to obtain maximum yields of a highly responsive crop such as table beets on the different indices is: 150 for #0, 120 for #1, 90 for #2, and 70 for #3. For a less responsive crop such as broad beans, it is 20 for #0, 15 for #1, 10 for #2, and 10 for #3. (The NPK Predictor can be purchased by contacting the Liason Officer NVRS, Wellesbourne, Warwick CV35 9EF, England.)

In the United States, the soil OM and the introduction of manures or legumes is also considered in making N recommendations. The Cooperative Extension Service in Arkansas uses soil OM as a basis for recommending amounts of N to be used for cotton or for the extent of reduction from standard applications (see Tables 5.3 and 5.4).

TABLE 5.3. Nitrogen Recommendations for Cotton As Affected by Soil Organic Matter

Soil OM %	Northern Arkansas lb/ac	Southern Arkansas lb/ac
3+	40-50	45-50
2-3	50-55	55-60
1-2	60-65	60-65
<1	65-70	70-80

Source: Miley, W. N. *Fertilizing Cotton with Nitrogen.* University of Arkansas Coop. Ext. Serv. Leaflet 526.

TABLE 5.4. Reductions in Recommended Amounts of N for Cotton Grown in Different Sections of Arkansas on Soils of Different OM Contents

Nitrate-N in upper 18 in	Northern Arkansas % soil OM		Southern Arkansas % soil OM	
	0-1.9	2.0 +	0-1.9	2.0+
lb/ac	Reduce N lb/ac		Reduce N lb/ac	
0-15	0	0	0	0
15-30	10-15	10-20	0	0
30-45	20-25	25-30	10-15	25-25
45-60	25-30	30-40	20-25	30-35
60+	35-45	40-50	30-35	35-40

Source: Miley, W. N. *Fertilizing Cotton with Nitrogen.* University of Arkansas Coop. Ext. Serv. Leaflet 526.

Nutrients Supplied by Cover Crops, Hays, and Plant Residues

The amounts of nutrients supplied by various organic materials and the nutrients available for the succeeding crop are quite variable. As indicated in Table 5.1, the N content of various leguminous materials, such as alfalfa, clovers, cowpea, lespedeza, peanut, and soybean are generally much higher than that of nonlegumes. The high N content is an important characteristic indicating the availability of this element and the speed of organic matter decomposition.

Stage of Organic Material

The amounts of nutrients listed in Table 5.1 are for mature materials. The percentages of N, P, and K in organic materials are appreciably greater at a more juvenile stage. Because of the higher N content and lower lignin content in younger tissue, decomposition is much more rapid.

It might be more suitable to incorporate the nonleguminous materials before they are mature to obtain rapid release of the nutrients, although the total amount at that time will usually be appreciably lower. An example of the effect of age upon decomposition of rye plants is given in Table 5.5.

TABLE 5.5. The Influence of Maturity on the Composition and Decomposition of Rye*

Stage of maturity	Moisture content %	Nitrogen content %	Mineral N liberated (+) or absorbed (−) (mg)
10-14 in. high plants	80	2.5	+22.2
Just before heads form	79	1.8	+3.0
Just before flowering**	57	1.0	−7.5
Grain in milk stage**	15	0.24	−8.9

* Fresh material to supply 2 g of dry material were added to 100 g of soil and incubated at 25-28°C for 27 days.
** Leaves and stalks.

Source: Russell, E. J. 1966. *Soil Conditions and Plant Growth,* Ninth Edition, Fourth Impression. Longmans, Green, and Co., London. Reprinted by permission of Addison Wesley Longman, Harlow, Essex, United Kingdom.

In addition to nutrients supplied by these materials, cover crops can add extra nutrients by serving as catch crops, which recover available nutrients ordinarily lost by leaching and erosion. The recovery of such nutrients is greatest when they are interplanted with the cash crop before the cash crop is to be harvested.

Unfortunately, some of the nutrients normally present in these materials may not be released in time to be valuable for the succeeding crop. The amounts released for the crop depend on favorable soil temperature and moisture, adequate N and P in the incorporated material, and the stage of the organic crop as it is worked into the soil.

Soil temperatures usually are too low for rapid release of nutrients in many northern states when these materials are incorporated for early crops, making use of highly available fertilizers along with organic materials a necessity for starting many crops.

Adequate N and P

Mature nonleguminous cover crops may provide little N or may even rob N from the succeeding crop. These crops may have such wide C/N ratios that the decomposition of the material by microorganisms will tie up available N. Organisms decomposing organic materials require about 1.5 percent N and 0.2 percent P of dry weight in the material for their processes. If the dry matter contains less than these amounts, organisms can rob the soil of available nutrients and delay the release of other elements from the incorporated organic material. No initial release of N can be expected if the cover crop contains less than about 1.75 percent N; extra N needs to be added at planting time if the N content of the cover crop (or any other organic material) is less than 1.5 percent; extra P needs to be added if the material contains less than 0.2 percent P.

Relatively immature nonlegumes can be expected to release N and several other elements for the succeeding crop, especially if they followed a cash crop that had been fertilized heavily. The amounts of nutrients released by such nonlegume crops are difficult to estimate. Nutrient value of such crops per acre can be determined by multiplying the average dry weight of a square yard (average of at least four sq yd) of cover crop × percent of the elements on a dry weight basis × 4890. A release of 33 to 50 percent of the total nutrients can be expected in the first year.

The lack of N for the succeeding crop following the incorporation of organic matter is seldom a problem with legumes because of their higher N contents. With these crops it would be desirable to delay incorporation to increase the amount of N supplied. For example, a crop of vetch cut on April 19 supplied 137 lb N per acre but if cutting was delayed to May 8, it produced 203 lb N; the N content of crimson clover increased from 140 to 188 lb and that of sweet clover from 124 to 160 lb per acre as it advanced to maturity. On the other hand, the incorporation of nonlegumes at an advanced age may require the addition of readily available N to prevent the succeeding crop from being adversely affected.

Nutrients Supplied by Manures, Sludges, and Composts

Unlike soil OM and some of the natural organic materials cited above, manures, sludges, and composts are usually a quick source of nutrients with less chance for adverse effects on the succeeding crop. (Sludges may contain heavy metals that can cause crop damage or be harmful to humans or animals.) Their nutrient contents though are still fairly low, requiring large amounts of materials to supply a crop's needs and usually increasing costs excessively unless they are produced on or near the farm.

Manures. Manures are good sources of macro- and micronutrients, although the content of P and sometimes K is low compared to the content of N. The slow release of nutrients, particularly N, can be especially beneficial for soils subject to leaching. The nutrients present in manures can partially or totally substitute for fertilizers, depending upon the quantity and quality of the manure added. Nutrients supplied by manure varies considerably and is affected by the kind of animal, its feed, litter used, handling, and storage. Typical analyses of several different kinds of manure are given in Table 5.6. Some of the variation is illustrated by analysis of dairy and poultry manure from two states given in Table 5.7.

Micronutrients present in manures and the chelation of these elements by various manure fractions help to prevent many micronutrient deficiencies, but the addition of large amounts of manure to high-pH soils may be partially responsible for causing deficiencies. The large quantities of CO_2 released by decomposing manures tend to increase or help maintain the high soil pH, thereby decreasing the effectiveness of the micronutrients in the manure and even reducing what already may have been inadequate amounts present in these soils.

Manures can be an important source of nutrients in regions close to their source, but their relatively low analyses make them expensive if they have to be hauled appreciable distances. Much of manure's nutrient value, especially the N and K_2O, can be lost by poor handling and storage. Some of the N loss can be reduced by the addition of superphosphate to the manure. The

TABLE 5.6. Common Macronutrient Contents of Manures, Composts, and Sludges

Material	N %	N lb/ton	P_2O_5 %	P_2O_5 lb/ton	K_2O %	K_2O lb/ton
Bat guano	7.5	150	3.0	60	1.5	30
Beef, feedlot	0.71	14	0.64	13	0.9	18
Compost*	0.9	18	2.3	46	0.1	4
Dairy	0.56	11	0.23	5	0.6	12
Dairy, liquid	0.25	5	0.05	1	0.25	5
Duck	1.1	22	1.45	29	0.5	10
Goose	1.1	22	0.55	11	0.5	10
Horse	0.7	14	0.25	5	0.7	14
Poultry, no litter	1.55	31	0.9	18	0.4	8
Poultry, liquid	0.15	3	0.05	1	0.3	6
Sewage sludge	0.9	18	0.8	16	0.25	5
Sewage sludge, act.	6.0	120	2.5	50	0.2	4
Sheep	1.4	28	0.5	10	1.2	24
Swine	0.5	10	0.32	6	0.45	9
Swine, liquid	0.1	2	0.05	1	0.1	2

* Finished and screened aerated sludge compost.

TABLE 5.7. Variation in N, P_2O_5, and K_2O Analyses of Dairy and Poultry Manure Collected from Two States (lb/ton)

State	Dairy N	Poultry N	Dairy P_2O_5	Poultry P_2O_5	Dairy K_2O	Poultry K_2O
N. Carolina	4-12	15-48	2-9	14-30	2-13	7-16
Maryland	4-14	4-136	2-9	10-111	1-15	1-79

Source: Livestock Manure Folder. Potash and Phosphate Institute and Foundation for Agronomic Research. Norcross, GA, p. 8.

superphosphate mixed with the manure at the rate of 25 to 30 pounds per ton helps to fix the N in a nonvolatile form. It also has the added advantage of supplying P, an element that is usually in short supply in many manures. Additional N and much of the K_2O can be saved by (1) using bedding to reduce loss of liquid excrement, (2) packing the manure tightly in lined pits with no drainage, (3) storing it under cover, or (4) applying it to the field and working it into the soil as soon as possible.

Manures can supply appreciable amounts of N and several other elements (see Table 5.6) but the release of the nutrients is also affected by soil factors as cited for OM. Only about one third to one half of the N can be expected to be released for the first crop year. The University of Maryland assumes only about a 5 lb N release per ton of manure as it suggests a 50 lb reduction in N per acre if 10 tons of manure are applied for corn. Somewhat similar estimates have been made by Michigan University in their recommendation for N use with and without manure applications.

Sewage effluents and sludges. About five trillion gal of sewage effluents are produced annually in the United States that need to be disposed of. The effluents, consisting of both liquids and solids, contain important nutrients and organic matter that could be highly beneficial to soils. Unfortunately, several of their components are potential environmental hazards because: (1) the C, N, and P can eutrophy water in streams and lakes; (2) nitrates can be leached to ground waters; (3) salts can accumulate or be leached to ground waters; (4) pathogenic organisms can be a source of disease in humans; and (5) heavy metals (Cd, Cu, Ni, Pb, and Zn) can harm soils and foods. These wastes must be carefully treated and handled to be of value to agriculture.

The use of liquid wastes or biosolids is regulated by the EPA and individual state agencies. Generally, they must be treated to destroy disease-causing organisms and reduce the attraction of flies and other vectors of human disease. Composting or heating the biosolids greatly reduces the pathogens. Drying or digesting the biosolids aerobically or anaerobically tends to solve the vector problem. Alkali additions are useful in controlling the vectors and is effective against various pathogens (Obreza and Muchovej, 1997).

Current processes can remove much of the water and produce two products of potential use. Sewage sludge is a high-moisture product of low analysis that might be useful on land very close to the source, but application of the large quantities needed is laborious and expensive. In addition, this product carries so much moisture that it is often necessary to wait for some time after it is applied before the soil can be used. Activated sewage sludge (aerated after being inoculated with microorganisms) has a higher analysis, is much easier to handle, and is being used commercially primarily for landscaping, growing ornamentals, and lawns. Typical sludge contents of N, P_2O_5, and K_2O are given in Table 5.6. Unfortunately, the relatively low nutrient content of sewage sludge makes its distribution expensive. Also, the potentially high content of heavy metals in these materials may limit their use to nonfood crops unless these metals can be prevented from entering the waste system or nullified by treatments. Even for nonfood crops, addition of wastes with elevated metal content needs to be limited to soils with pH values above 7.0 to limit harmful effects of several heavy metals.

As with manures, considerable use of sludges will probably be confined to areas close to their production but unlike manures, will probably be limited to nonfood crops. Additional use of these products could be made if some of the costs of the materials were underwritten by the communities that need to dispose of the wastes. Also, separating the industrial components from human wastes would open up a greater market for the products.

Composts. The value of composts is limited to restricted areas and uses because of its relatively low analysis (Table 5.6), but its C/N ratio is about 15:1, indicating good N availability. Using it as a sole source of nutrients or for improvement of soil structure requires relatively large tonnage per acre. The recent trend to compost urban organic wastes and sewage sludge may make composts more available for use in commercial vegetable production, but most compost is now used for gardens and landscaping, or for the production of artificial soils designed for growing specialty crops in containers. Studies indicate that standard compost made from wastewater sludge and yard trimmings is superior to those that may also contain mixed waste paper, refuse-derived fuel, or refuse-derived fuel residuals (Roe, Stoffella, and Graetz, 1997).

Loamless composts. Composts may be made with or without soil (loamless). The trend has been to prepare composts without soil because of the difficulty of obtaining soil free of herbicides or other pesticides. The preparation of loamless composts with ingredients, many of which are low in N, yields a product that tends to be low in nutrients. The addition of fertilizers during the decomposition process will largely overcome the deficiency. The use of inorganic fertilizers is satisfactory, but careful attention must be paid to conductivity to avoid excess salts. Use of relatively high-analysis slow-release organic materials, such as hoof and horn meal or coated fertilizer, can make this control easier.

Storage of composts. Because the high-analysis organic materials can produce considerable ammonia, their use requires attention to length of storage. If used within the first week after preparation, loamless composts prepared with organic materials tend to produce about the same amount of growth as composts prepared with inorganic fertilizer. If the compost prepared with organic fertilizer is stored for several weeks before using, it may produce very poor crops due to the ammonia produced as microorganisms attack the organic fertilizer. The ammonia is toxic in very small quantities and the high pH it produces tends to make several micronutrients unavailable.

The ammonia produced will in time (6 to 8 weeks) be nitrified and the pH will drop, making satisfactory growth again possible. The exact time that this occurs varies with materials used, moisture, and temperature, but the process can be followed by measuring the pH of the compost at regular intervals. When the pH is close to its starting point, most of the ammonia should have been nitrified.

Storage of composts with several slow-release inorganic materials can also be a problem, due to the salts which can increase dramatically within a week or two. At elevated temperatures materials such as Osmocote can release enough available nutrients in a relatively short period to make the composts unsatisfactory. Ideally, these composts should be used shortly after being prepared; if they must be stored, it should be done at relatively low temperatures.

Peats

Peats form under conditions of impeded drainage or high ground water levels that reduce O_2 levels and slow the decomposition of organic materials. The accumulation of organic matter, rich in lignin and hemicellulose, increases with greater submergence and cool temperatures.

Five different types of peat are recognized in the trade, namely sphagnum moss peat, hypnum moss peat, reed-sedge peat, peat humus or muck, and other peat.

With the exception of peat humus or muck, all peats provide relatively little N, P_2O_5, or K_2O and are valued primarily for their soil conditioning properties. The sphagnum peat has an acid reaction and very high MHC (15 to 30 times its weight). Although the reed-sedge peat has much lower MHC and may not be acidic, it is highly prized for golf courses and landscape use partly because of its longer-lasting properties. Some uses for the different peats are suggested in Table 10.2.

Other Organic Materials

Applications of various animal waste or other byproduct crops can contribute to much of the N needs of the succeeding crop. Amounts of N, P_2O_2, and K_2O of some of the more common legumes and animal products are given in Tables 5.6, 5.7, and 5.8. But as with OM and manures, the major release of these nutrients is affected by microorganism activity and only about one third to one half of the N can be expected in the first crop year. Also, many of these products have disappeared from the fertilizer market due to their greater value as animal feeds.

TABLE 5.8. Major Nutrient Contents of Plant and Animal Byproducts

Product	N %	P$_2$O$_5$ %	K$_2$O %
Animal tankage	7	9	—
Ashes, cotton hull	0	4-7	22-30
Ashes, hardwood	0	1	5
Beet sugar residue	3-4	0	8-10
Bone meal	4	22	—
Bone, precipitated	—	35-46	—
Bone tankage	3-10	7-20	3-9
Castor pomace	5-6	2	1
Cocoa shell meal	3	2	1
Cocoa tankage	4	1	2
Cottonseed meal	6-9	2-3	1-2
Crab scrap	3	3	0
Distillery waste	1	0.5	14
Dried blood	8-14	0.5-1.5	0.5-0.8
Dried king crab	9-12	—	—
Fish, acid	6	6	—
Fish meal	5-10	5-13	—
Fish scraps, fresh	2-8	2-6	—
Garbage tankage	3	2	3
Hoof and horn meal	13	—	—
Linseed meal	5	1	1
Olive pomace	1	1	0.5
Rapeseed meal	5-6	—	—
Peanut hulls	1.5	0.2	1
Seaweed kelp	2	1	4-13
Shrimp scrap, dried	7	4	—
Soybean meal	6	1	2
Steamed bone meal	3	20	—
Tobacco dust and stems	—	2	5
Winery pomace	1.5	1.5	1
Wool wastes	7	—	—

Sources: Collings, G. H. 1947. *Commercial Fertilizers,* Fifth Edition. McGraw-Hill, New York; Follett, R. H., L. S. Murphy, and R. L. Donahue. 1981. *Fertilizers and Soil Amendments.* Prentice-Hall Inc., Englewood Cliffs, NJ; Lorenz, O. A., and D. N. Maynard. 1988. *Knott's Handbook for Vegetable Crops,* Third Edition. John Wiley and Sons, New York.

Physical Improvement of Soil by Organic Matter

Organic matter alters several soil properties, most of which enhance the air and water sides of the triangle, which in turn beneficially affect nutrients.

Binding of Soil Particles

Organic matter tends to bind with soil particles to form peds or larger soil units with corresponding increase of large pores. Recently deposited organic matter (cover crops, sods, or plant residues) seems to be more useful in stabilizing the macroaggregates (>250 μm), whereas materials that have undergone extensive decomposition (humus) are more useful in stabilizing the smaller aggregates or microaggregates.

Various soil organisms (actinomycetes, bacteria, and filamentous fungi) are responsible for aggregate formation. Bacteria that produce slimes appear to be a major factor in binding soil particles and as such are important in improving infiltration and MHC and increasing gas exchange, but reducing erosion and compaction.

In addition to what appears to be a physical binding of organic substances with soil particles, there is a chemical bonding of organic molecules and clay. The clay-organic complex forms most probably as result of cation exchange or hydrogen bonding, adds stability to aggregates and reduces losses of OM.

Improvement of Soil Moisture

Soil OM tends to improve soil moisture because it increases MHC and provides for better infiltration of water. Moisture holding capacity is increased with the additional pore space usually resulting from added organic matter. The very addition of OM increases the potential for holding additional water, as OM can hold water in excess of its own weight (about 1.5 percent for each percent OM). Its effect per unit of OM is greater in sandy than in heavier soils probably because some OM is tied up in the clay-organic complexes as noted above. In a sandy soil, the 1 percent OM will hold about 1.5 percent water or about 10 percent of total MHC of this soil.

The improved infiltration of water resulting from the effects of OM on improved pore space greatly increases the potential amount of water that can be made available for the crop. The savings in water as a result of better infiltration are greater on sloped soils and during heavy rains or prolonged overhead irrigation.

Chelation of Mineral Elements

Organic matter produces a number of chelates, substances that keep several elements available over wide ranges of pH. Chelation is of great benefit in the maintenance of several micronutrients (Fe, Cu, Zn). The chelation of Fe has great economic importance, particularly on high-pH soils.

Within the plant, common chelated metal atoms are associated with the haem group as iron porphyrin or as chlorophyll. Malic and oxalic acids supplied by OM are recognized as having chelate properties. Root exudates capable of forming iron complexes that stay available for long periods are very useful in this respect. The plant chemical (α)-ketogluconic acid has been designated as being highly useful for this purpose (Webley and Duff, 1965), but there are probably many more within the plant or formed as organic materials decompose.

The addition of organic manures to high-pH soils for the purpose of providing chelates to correct lime-induced chlorosis, however, has not been very useful. The failure of organic manures to correct the problem is attributed to formation of considerable CO_2 as the organic materials decompose. The CO_2 favors the formation of bicarbonate, which reduces Fe uptake and translocation by the plant.

Soil Buffering

The changes in pH, conductivity, or available nutrients of a soil or other medium as liming materials, acidic materials, or fertilizers are added are modified by the buffering capacity of the medium. The changes taking place in liquid hydroponics are usually vast and occur rapidly; those in soils with high CEC are relatively small and slow. Generally, slow changes as a result of good buffering capacity are beneficial as they prevent extremes in pH, conductivity, or nutrients that may injure or retard plant development.

The reasons for differences are due largely to the CEC as there is a continuous exchange between ions in the soil solution and the many more held by the exchange complex. Soils or other media high in CEC have a much greater pool of adsorbed ions that can be exchanged with those in solution and thereby modify the amounts remaining in the soil solution, which readily show up in soil measurements.

Organic matter has considerable buffering effect and so alters the changes that can be expected from chemical additions to the soil. The reduction of changes are not only due to the CEC of OM but to large amounts of water held by it, as the extra water tends to dilute the added chemicals.

PLACEMENT OF ORGANIC MATTER AS MULCH

The placement of organic materials largely on the surface as mulch often increases its usefulness. Long-term effects tend to improve soil structure in the upper soil layers thereby improving both air and water for the plants. Organic mulches, by intercepting large drops of rain or irrigation, reduce soil capping or compaction close to the surface. Decomposition releases substances capable of aggregating soil particles into larger peds that allow for better water intake. Water storage is also increased by reduced water evaporation that results as temperatures are lowered by the mulch. The increased amounts of water, the ready source of organic materials, and lower soil temperatures help support a large biological population capable of further improving soil structure. This better structure not only allows for improved water storage but also increases air storage and exchange. Far greater amounts of water are stored if organic materials are retained as mulches than if they are disked or plowed into the soil. Much of the extra storage of water appears to be due to reduced losses from evaporation, since basin listing, which largely prevents runoff, fails to store as much water as straw applied as a mulch.

Protection of the surface largely depends on the extent of coverage by materials on the surface, which can be quite different with the same weight of materials because of different densities and diameters. For example, the benefits of wheat straw for preserving infiltration are greater than those of German millet, which in turn are greater than those of sorghum residues.

MAINTENANCE OF SOIL ORGANIC MATTER

As indicated above, organic matter is in a constant state of flux. Organic materials of various descriptions—remains of plants or their parts, microorganisms, insects, and animals—undergo decomposition until they reach the relatively stable form of humus. The humus is still subject to decomposition, albeit much slower than the original materials, but is constantly renewed from new organic materials. An equilibrium between the amounts of humus lost by decomposition and that gained from new materials is ultimately reached for particular conditions of plant cover, soil moisture, temperature, and aeration.

The decomposition of organic matter is a mixed blessing. On one hand, decomposition is beneficial for at least three important reasons: (1) it provides energy for a vast group of microorganisms that carry out several processes that are essential for good crop production, (2) some of the products of decomposition aid soil structure and greatly increase porosity and infiltration, and (3) it makes available considerable nutrients for succeeding crops.

On the other hand, decomposition can go too far, if losses greatly exceed replacements. If OM drops too low, several important soil properties are compromised. When this occurs, we can expect degradation of soil structure, unacceptable levels of beneficial microbial processes, reduction in the amount of nutrients released for the growing crop, increased erosion losses, loss of some MHC, greater soil compaction, and poorer gas exchange.

Decomposition of OM is not the problem because some decomposition is necessary if we are to reap the benefits of OM. Rather, it is the uncontrolled loss beyond a certain point that creates a number of problems for proper growth of many plants. It is only when insufficient organic materials have been added to replace that being decomposed, resulting in the diminution of several soil properties, that decomposition of OM lowers crop productivity. Correction of the problem can be made by adding more organic materials, slowing its rate of decomposition by reduced cultivation and adding enough nitrogen, placing the organic material to maximize its usefulness, or by combinations of these procedures.

Losses of Organic Matter

The levels of OM rapidly change as virgin soils with long periods of OM accumulation are suddenly opened to cultivation. The change in equilibrium is primarily due to the increased aeration of these soils but also to considerable changes in the type and amounts of organic materials that are left in or deposited on the soil. In most cases, the net result is a marked lowering of OM, rapidly at first, but slowing gradually until a new equilibrium between amounts of organic materials produced or deposited equal that lost by decomposition. Losses are greatest in soils of good OM content and the rate of loss is decreased with time. The average losses in the corn belt amounted to about 25 percent in the first twenty years, about 10 percent in the next twenty years, and another 7 percent in the next twenty years.

Much of the benefit of organic matter for crop production—primarily the effects on structure and the release of nutrients—is obtained as the organic matter is decomposed by microorganisms. The major benefits of large amounts of organic matter accumulated over the years are obtained as conditions are changed, which allows for a more rapid decomposition of the organic matter. The soil may be drained or opened by plow, allowing more air into the system and greatly accelerating the decomposition of large amounts of accumulated organic matter. The decomposition allows for a great release of nutrients but also helps to maintain good structure by cementing small soil particles in forms that allow for maximum air and water storage.

The losses of OM can be quite severe over a period of time if little or no effort is made to replace a major portion of it. Losses are greater under monoculture cropping systems with wide row spacing that allow for considerable aeration by cultivation. As losses increase, the ability of soils to hold sufficient water and air is decreased while soil compaction and erosion tend to increase. Even in the best of soils, the amounts left at an equilibrium stage for many monoculture crops may be insufficient to provide for optimum crops. But the smaller amounts of OM that exist after considerable cultivation has taken place can still be highly useful providing sufficient organic materials are routinely added to provide continuous benefits of erosion control, aggregation for structure and infiltration, and sufficient carbon to maintain existing OM levels.

The rate of OM decomposition as it is added to soils is dependent on several factors, such as: (1) kind of OM; (2) its maturity or stage of growth; and (3) soil temperature, moisture, pH, oxygen content, nutrient content, and the activity as well as the kind of micro- and macroorganisms. The influence of the kind of OM and its maturity upon decomposition has been briefly discussed above; that of the other factors are covered more fully in the following sections.

Effects of Climate

Decomposition of OM is markedly affected by climate as decreasing temperatures slow the decomposition rate more than they affect plant growth. The rate of decomposition is about doubled for every $10°C$ ($18°F$), making the buildup of organic matter much more difficult in warm climates even though the growing season is much longer than in temperate or cool ones. Shading the soil in warm climates can aid in maintaining or increasing its OM, since shade greatly reduces the maximum soil temperature.

The accumulation of OM in cool climates is apparent from the normal amounts present in virgin soils and by the amounts that are needed to maintain existing levels. The levels of OM in virgin soils of the northern plains may be well over 4 percent whereas it may be 2 percent or less in the southern plains. It has been estimated that an annual return of as little as 2 tons per acre of organic matter in the form of crop residues or manures would be sufficient to maintain existing soil OM levels in temperate zones. The amount necessary for maintenance of OM in cool climates is appreciably less, being on the order of 1 to 1.5 tons annually per acre.

Increased rainfall in most situations tends to increase OM primarily because of its beneficial effect on plant growth with little effect on OM decomposition. But amounts of rainfall that allow for wet soils for long periods can aid OM accumulation by limiting amounts of air in the soil.

Effects of Textural Class

The amount of OM that will provide adequate water and air in the soil varies with different soil textural classes. A difference of about 1 percent of OM often separates sandy loam and loam soils with a poor rating from those that are satisfactory.

Effects of Fertilizer and Lime

As is shown in Table 10.4, plant residues can supply substantial amounts of organic materials and thus play an important role in OM maintenance. Vines, stalks, leaves, straw, and roots left over after harvest can contribute to replacement of OM lost by normal decomposition. Lime and fertilizer, by increasing the amounts of organic materials produced as residues, greatly aid in OM maintenance. Proper use of both materials not only produces maximum economic yields of crops, but also provides large amounts of plant residues, which provide for better maintenance of soil OM.

Sufficient N is the key in helping to maintain good levels of soil OM. Nitrogen not only contributes to the production of large amounts of residues, but its presence in sufficient quantities helps maintain levels by limiting the decomposition of existing OM. Soil OM has a C/N ratio of about 10:1. Many organic materials will have C/N ratios greater than 20:1 with that of some materials 300:1 (see Table 5.9). Microorganisms responsible for the breakdown of organic materials will temporarily rob available N from the soil solution to compensate for low N in materials with wide C/N ratios. If sufficient amounts of available soil N are not present, microorganisms hasten the breakdown of existing soil OM to obtain the needed N. By adding sufficient available N, decomposition of existing OM is essentially stopped for the time being, allowing greater accumulation from the added organic materials.

The mechanisms of both increased residues and less decomposition of existing soil OM are probably responsible for increases in soil OM despite

TABLE 5.9. Carbon/Nitrogen (C/N) Ratios of Some Organic Materials

Material	C/N ratio
Slaughterhouse wastes	2:1
Soil bacteria	5:1
Soil actinomycetes	6:1
Soil fungi	10:1
Night soil	6:1-10:1
Soil humus	10:1
Sewage sludge	10:1-12:1
Alfalfa	12:1
Sweet clover, young	12:1
Grass clippings, young	12:1
Chicken manure, droppings	12:1
Alfalfa hay	15:1
Municipal garbage	15:1
Hog manure	17:1
Sheep manure	17:1
Chicken manure, litter	18:1
Sweet clover, mature	20:1
Grass clippings, fresh mature	20:1
Rotted manure	20:1
Cattle manure	30:1
Bluegrass	30:1
Bean straw	37:1
Olive pomace	37:1
Horse manure	49:1
Leaves, fresh	40:1-80:1
Pine needles	45:1
Peat moss	58:1
Corn stalks	40:1-80:1
Highmoor peat	80:1
Straw, small grain	80:1
Timothy	80:1
Cotton gin trash	80:1
Corn cobs	104:1
Red alder sawdust	135:1
Seaweed (kelp)	225:1
Hardwood sawdust	250:1
Douglas fir, sawdust	295:1
Douglas fir, old bark	295:1
Wheat straw	375:1
Pine sawdust	729:1

Sources: Follett, R. H., L. S. Murphy, and R. L. Donahue. 1981. *Fertilizers and Soil Amendments.* Prentice-Hall Inc., Englewood Cliffs, NJ; Lorenz, O. A., and D. N. Maynard. 1988. *Knott's Handbook for Vegetable Crops,* Third Edition. John Wiley and Sons, New York.

monoculture wheat and barley grown for more than 150 years at the Rothamsted Experimental Farm in Great Britain. Similar mechanisms are probably responsible for the higher OM levels reported in time with greater use of fertilizer N for cotton production (see Table 5.10). Extra N also increased soil OM at different depths for both conventional and no-tillage systems used for continuous corn in Kentucky (see Table 5.11). Rotations of crops can help maintain OM but the effects are much more pronounced if rotating crops are combined with adequate fertilization.

TABLE 5.10. Influence of Nitrogen Rates on Soil Organic Matter for Continuous Cotton

N Rate	Percent organic matter*		
(lb/ac)	1973	1984	1989
0	0.66	0.69	0.90
25	0.69	0.75	1.00
50	0.66	0.89	1.10
75	0.69	0.89	1.10
100	0.67	0.91	1.1
125	0.68	1.02	1.2
150	0.68	0.93	1.2

* Organic matter content of upper 6 inches of a Loring soil.

Source: Maples, R. L. 1989-1990. Nitrogen can increase cotton yield and soil organic matter. *Better Crops with Plant Food.* Winter. Potash and Phosphate Institute, Norcross, GA, p. 16.

TABLE 5.11. Soil Organic Matter at Different Depths As Influenced by Nitrogen Rate for Continuous Corn with Both Conventional and No-Till Systems

Soil depth (in)	Original blue-grass sod	Nitrogen rate (lb/ac)							
		0		75		150		300	
		NT*	CT*	NT	CT	NT	CT	NT	CT
		% OM							
0-2	5.18	3.68	2.37	3.96	2.60	4.11	2.78	4.53	2.79
2-6	2.47	2.22	2.23	2.28	2.53	2.15	2.60	2.46	2.52
6-12	1.36	1.36	1.17	1.37	1.57	1.24	1.47	1.47	1.37

* NT = no-tillage; CT = conventional tillage (moldboard plow).

Source: Griffith, D. R., J. V. Mannering, and J. E. Box. 1996. Soil and moisture management with reduced tillage. In *No-Tillage and Surface Tillage.* M. Sprague and G. B. Triplet (Eds.). John Wiley and Sons, New York, p. 25. Copyright © 1996, reprinted by permission of John Wiley and Sons, Inc.

IMPROVING SOIL ORGANIC MATTER

Large amounts of OM are necessary to make substantial changes in percentages of soil OM. In mineral soils, the plow layer weighs about 2,000,000 lb/ac and one percent of this weight is 20,000 lb. Soil OM is the equilibrium product resulting from the initial decomposition of organic materials. Raising the soil OM by 1 percent would require the addition of organic materials of at least several times 20,000 lb. This is an enormous task that takes years to accomplish. Large amounts added over a short period, besides being difficult to accomplish, would probably overwhelm the soil's capacity to process the materials into the stable form of OM or humus.

Even over long periods, soil OM seldom increases unless radical departures are made in growing of crops. A major shift in types of crops or the manner in which cash crops are grown is necessary before OM will improve. This seldom happens unless long-term crops such as pastures, sods, or hayfields are substituted at least for some of the cultivated crops, or the rate of OM decomposition is slowed appreciably by adopting methods of reduced tillage. The growing of long-term crops provides large quantities of organic materials in the form of roots, fallen leaves, and stems while at the same time restricting the amounts of air entering the soil, which limits the rate of decomposition.

Although increases in soil OM contents are highly desirable, the additions of even relatively small amounts of organic materials without any appreciable change in OM can still be beneficial just by providing temporary benefits of soil aggregation, reduction of erosion, and better water infiltration. Some methods by which OM can be improved are discussed below and some approaches for better utilization of OM are presented in Chapter 10.

A satisfactory situation for most soils that have lost a major portion of their virgin OM is to institute practices that will reverse the declining trend by increasing the total amount of organic materials added, by making better use of added organic materials, or combinations of both. Extra organic materials can be added by bringing in manures, composts, sludges, etc. or growing them in place as sods, hays, cover crops, or residues. Good liming and fertilizer practices, combined with rotations, can be used to increase the amounts of organic matter grown in place. Placement of organic materials as mulches increases the effectiveness of added organic materials for many soils and crops. No-till and reduced-till methods place more of the added organic materials as mulches, while the reduction in cultivation from using these methods slows the decomposition of soil OM. Substituting small grain crops for crops grown in wide cultivated rows reduces cultivation, slowing the destruction of OM.

BENEFITS OF PLACING ORGANIC MATERIALS AS MULCHES

The placement of organic materials largely on the surface as mulches often increases their usefulness. Long-term effects tend to improve soil structure in the upper soil layers, thereby improving both air and water for the plants. Organic mulches, by intercepting large drops of rain or irrigation, reduce soil capping or compaction close to the surface. The decomposition of the organic matter releases substances capable of aggregating soil particles into larger peds, which allow for better water intake. Water storage is also increased by less water evaporation, which results as temperatures are reduced by the mulch. The increased amounts of water, the ready source of organic materials, and lower soil temperatures help support a large biological population capable of further improving soil structure. This better structure not only allows for improved water storage but also increases air storage and exchange. Far greater amounts of water are stored and such water tends to penetrate deeper if organic materials are retained as mulches than if these materials are disked or plowed into the soil.

HARMFUL EFFECTS OF ORGANIC MATTER ADDITIONS

It is not uncommon to have adverse crop responses to incorporation of organic materials. Some of these responses have been attributed to disease or pests accompanying or favored by organic matter additions, wide C/N ratios, or allelopathic substances added by organic matter.

It is also possible to have an excess of OM—or at least of raw organic materials. The excesses are associated more with raw organic materials than the more decomposed materials, although large additions of manure may exceed the soluble salt content and supply more nutrients than can be safely handled by the crop. Many of the excesses that have caused temporary setbacks in plant production have been related to (1) large amounts of materials with wide C/N ratios; (2) excesses of certain gases released by the decomposition of the organic matter; (3) a shortage of oxygen resulting from the decomposition; (4) poor contact of soil with seeds; (5) the disruption of capillary movement of water; (5) increased micronutrient deficiencies; and (6) the large amounts of allelopathic substances introduced.

Carbon/Nitrogen Ratios

As has been noted above, the C/N of humus is about 10:1 but that of raw organic material can be many times greater. To decompose the wide

C/N ratio materials, organisms require extra N, which is taken from available sources close by, often robbing the cash crop. Generally, the decomposition of materials with C/N ratios wider than 18:1 will require extra N for some months after adding these materials. Despite the very wide C/N ratios of sawdust, wood chips, or bark, there may be no need to add extra nitrogen with these materials, especially if they are applied as mulches. These materials are slowly attacked by microorganisms because of their high lignin content, resulting in minimal loss of available N.

The determination of C is a bit tedious and expensive and efforts have been made to measure the need of extra N from N analysis only. Usually, little or no N is released from organic materials that contain less than 1.3 percent N, but some release can be expected with N contents >1.75 percent. A value of 1.5 percent N or 30 pounds per ton of dry material appears to be a good average N content that just allows for the microorganisms' N needs. A dry cereal straw that contains only 0.8 percent N (16 lb/ton), therefore, would need an additional 14 pounds of N per ton of straw to avoid N deficiency of the following crop. Some representative C/N ratios of a number of materials are presented in Table 5.9.

Presence of Pests

The presence of various plant pests in the remains of crops has led to considerable losses of succeeding crops. A common practice to reduce the carryover of these pests is to burn remains of various crops. This is effective in reducing damage from pests (fungi, bacteria, viruses, insects, and weed seeds) but has several undesirable side effects. Valuable organic matter along with its N and S are lost. Burning organic residues often leads to wind and water erosion. It also may reduce infiltration of water in certain soils, making them more difficult to wet and increasing the chances of erosion. The process also increases air pollution, which is becoming more and more unacceptable. Alternative procedures for disposing of crop wastes may be preferable to burning to reduce damage from pests harboring in crop residues. One procedure is to work the materials into the soil as soon as possible. Since most soilborne pathogens are found in the upper 6 to 8 inches, inverting the soil with a plow and following with a disk turns under the infested layer and replaces it with a noninfested one. Spores and insects are buried and the decomposition of the organic materials is quickly started.

In North Carolina, a pest management program utilizing quick plow-down of residues with plowing and disking, known as the R-9-P (Reduce 9 Pests), has been very effective in reducing tobacco infections of mosaic, vein banding, brownspot, and nematodes while suppressing budworms, fleabeetles, and hornworms and helping to control weeds and grasses. The program, which includes another disking about three weeks after incorporating the residues and planting a cover crop, has worked so well because of the general acceptance by growers. They have incorporated tobacco residues almost immediately after finishing the crop, thereby limiting the amount of stalks and roots that can serve as maintenance and cover for many of these pests.

Ordinary plowing and disking may not be sufficient for certain organisms. Some organisms, such as those causing southern blight of peanuts and soybeans *(Sclerotium rolfsii)* exist at deeper layers of the soil and deep plowing is necessary to help control them.

It is important to point out that one of the reasons the North Carolina program has been so successful is that the early follow-up with a cover crop after the plowing and disking does not leave the soil exposed over the winter season, and this practice adds some additional organic matter.

Moisture and Aeration Problems

Some problems of moisture and air have been noted with the incorporation of large amounts of organic materials. If the materials have high levels of moisture, as they start to rot there may be temporary high soil moisture levels that limit O_2 availability. The shortage can be compounded as the wet materials tend to pack. The shortage of O_2 and overly wet conditions, as may be present with large applications of certain partially decomposed material, can lead to the production of methane and other undesirable gases that form under anaerobic conditions. At other times, the inclusion of large amounts of moist material greatly stimulate soil microorganisms, which can seriously diminish existing soil O_2 levels while greatly increasing the CO_2.

The addition of large quantities of dry materials may temporarily result in moisture shortages and prevent good contact of soil and seed. Large pieces of organic materials tend to aggravate this type of problem. In addition to the effects on seed germination, these large pieces of organic materials can interrupt capillary movement, aggravating the effects of diminished water availability resulting from the addition of large amounts of dry materials.

Deficiencies of Fe, Zn, and Cu

The additions of very large amounts of organic materials have led to deficiencies of Fe, Zn, and Cu. Organic matter can fix available Cu in less available forms and Cu deficiency is quite common on organic soils. The cause of increased Cu, Fe, and Zn deficiencies in soils with relatively lower OM contents is not fully understood but seems to be due to (1) the inability of the soil solution to keep up with the huge demand for nutrients of the increased microbial population, (2) reduced plant uptake of elements because of shortages of O_2 caused by the rapid decomposition of large quantities of OM, and (3) lower availability of these elements because of higher soil pH resulting from large amounts of CO_2 generated by decomposition of the organic matter. The added CO_2 increases the bicarbonate content of many high-pH soils, greatly increasing the frequency of Fe deficiencies.

Allelopathic Effects

Plants contain many substances that selectively inhibit growth of certain soil organisms and limit germination of many different seeds and restrict growth of many plants. These allelopathic substances, consisting of volatile terpenoids, quinones, thiocyanates, coumarins, tannins, flavanoids, and cyanogenic glucosides, are used by the plant as defensive or offensive mechanisms to limit competition and regulate density. Their presence in organic materials can limit the early growth of the succeeding crop.

Reducing Harmful Effects of Organic Matter Additions

Composting to Reduce Problems

Composting organic materials or allowing them to break down before the new crop is established seems to negate the effects of many allelopathic substances. Composting pine bark prior to its incorporation into the medium eliminates restrictions of crops grown in mixtures of bark, sand, and peat.

Delayed Planting

The harmful effects on the cash crop following incorporation of organic materials with insufficient N and/or P, or the incorporation of large amounts of organic materials, often can be avoided by delaying the planting of the cash crop.

Where the growing season is long, it usually pays to delay planting until initial decomposition of the added organic matter is well under way. The delay allows for better pest control and allows more of the materials to decompose before the new crop is started, reducing competition between soil microorganisns and the cash crop. The delay is especially beneficial when large amounts of organic matter are incorporated. This delay is worthwhile even if the nutrient needs of the microorganisms are met by natural decomposition of the organic matter or by additions of nutrients.

Although beneficial in most cases, the delay in planting can seriously affect the yield of the succeeding crops in short growing seasons. Adding the extra N or P needed to speed the decomposition of some organic crops often is a more practical approach for handling the problem of organic materials low in these elements.

Placement of Organic Matter

The deleterious effect of organic matter upon the succeeding crop may be absent or of less concern if the organic matter is left on the surface of the soil as mulches, as in reduced-tillage operations. Evidently, by not incorporating the material in the soil, the demand for N and O by microorganisms is slowed; the release of harmful substances is also slowed or the substances are temporarily removed from the active root zone. The placement of organic materials on the surface may not work as well if certain pests are problems (see "Presence of Pests.")

SUMMARY

Organic matter has a marked influence on all three sides of the fertile triangle. Its physical presence can reduce compaction, allowing for better aeration and water relationships; its decomposition provides cements that bind the small soil particles into aggregates that improve aeration and water. Decomposition of OM also contributes considerably to the nutrient side of the fertile triangle.

The benefits of organic matter are manyfold. It is the source of nutrients important for crop production, primarily N and S. It is an energy source for many different microorganisms that have beneficial effects on crop production, namely, organic matter decomposition, nitrification, nitrogen fixation, and the production of stable soil aggregates. Stable aggregates greatly aid soil porosity and water infiltration. Organic matter also tends to improve MHC not only because of better infiltration but also because of the large amount of water held by soil OM. In addition to all these benefits, the decomposition of OM releases several chelating materials that greatly aid in keeping several micronutrients in an available state.

Many of the benefits of OM accrue as it decomposes. The process is largely conducted by microorganisms, and OM is the energy source that drives the decomposition. Organic matter decomposition is not only necessary for orderly control of plant and animal remains but, within limits, it is beneficial for soils and crops. Microbial breakdown of organic materials is highly beneficial in such important soil functions as nitrification, nitrogen fixing, maintenance of soil structure, and the suppression of several soilborne diseases.

The beneficial effects of OM decomposition can be seriously compromised if decomposition greatly exceeds renewal and the level of OM drops too low, adversely affecting all three sides of the triangle. From the standpoint of modern farming operations, the impairment of the water and air sides of the triangle is more critical than that of the nutritional side since the latter can readily be corrected by the addition of fertilizers, while improvement of the other two is usually limited to long-term commitments.

Occasionally, too much organic material is added, adversely affecting crop production. Such harmful effects usually can be avoided by adding sufficient N to maintain adequate C/N balances, allowing crop residue or organic matter additions to undergo considerable decomposition before planting the succeeding crop, or keeping most of the organic material on the soil surface as a mulch.

Chapter 6

Soil pH

INFLUENCE OF SOIL pH ON THE FERTILE TRIANGLE

Soil pH is one of the more important criteria affecting crop production because of the sensitivity of most plants to pH. The causes for the relationship between media pH and crop productivity are many but most of them deal with the effect of pH on the nutrient side of the fertile triangle.

Some of the more important causes of the close relationship between crop productivity and pH are due to the fact that media pH affects (1) the solubility and availability of several nutrient elements; (2) the amounts of toxic elements (Al^{3+}, Mn^{2+}, heavy metals) in the soil solution; (3) soil structure due to relative amounts of Ca^{2+} and Na^+ in base saturation; (4) uptake of nutrients; and (5) the activities of several important soil microorganisms.

FUNDAMENTALS OF pH

The term pH is defined as the negative logarithm of the hydrogen ion (H^+) concentration and it expresses the relationship of hydrogen ions and hydroxyl ions (OH^-). The term, designated by the Swedish chemist Sorensen, is a simplified method for expressing the relationship of the two ions and is based on the ionization of water. At equilibrium, the ionization (K) can be expressed by the equation:

$$K = \frac{(H^+) \times (OH^-)}{(H_2O)}$$

and the ionization constant at $25°C = (H^+) \times (OH^-) = 1 \times 10^{-14}$.

From the ionization constant, increasing H^+ concentration lowers the OH^- concentration, while increasing the OH^- concentration lowers the H^+ ions. Theoretically, at a pH of 7.0, the $H^+ = OH^-$ with the H^+ increasing and the OH^- decreasing as the pH is lowered to 1.0 (lowest point on the scale), and the OH^- increasing with the H^+ decreasing as the pH is increased to 14 (highest point on the scale).

Since pH is represented by logarithms, the actual concentration of either H^+ or OH^- at any given pH is $10\times$ that of the preceding pH. For example, the hydrogen ion concentration at pH 5 is 10 times that at pH 6 and 100 times that at pH 7; The OH^- concentration, on the other hand, would be only 1/10th as great at pH 7 compared to that at pH 6 and 1/100th of that at pH 5. These relationships, along with the pH values of several substances and the availability of several ions, are depicted in Figure 6.1.

THE IMPORTANCE OF SOIL pH

The vast majority of plants on mineral soils do best if the soil is slightly acid (6.0 to 7.0), although this ideal range needs to be slightly higher (6.5 to 7.0) if the soil contains considerable clay. The higher value for the heavier soils appears to be related to amounts of soluble Al (an element toxic to many plants in low concentrations), which is released by clay in acid soils. Allowing the pH to fall below 5.0 will bring in large amounts of Al^{3+}, but liming to pH 5.5 ensures precipitation of Al in the top layer (Espinosa, 1996).

A small number of plants (azalea, camelia, ixora, rhododendron) respond to lower pH value (<5.5) and a few (alfalfa, sweet clover) do better in slightly alkaline soils (7.0 to 7.5). Some (blueberry, coffee, pineapple, and stylosanthes) do well at a pH of 5.0 to 5.5 but are tolerant of fairly wide ranges of pH. But most plants do poorly as mineral soils become strongly acid (<5.9) or increasingly alkaline (>7.5). The preferences of a number of plants are presented in Table 6.1.

Ideal pH Values of Histosols, Oxisols, and Ultisols

There is a close correlation between desirable pH and OM contents. The desirable pH level of 6.0 to 7.0 for most crops holds for mineral soils that contain less than about 5 percent OM, but the ideal pH will be about 5.8 if the soil contains more than 10 percent OM and closer to about 5.0 if the OM is greater than about 20 percent.

The availability of nutrients in histosols (organic soils containing 12 to 18 percent or more organic carbon) is greatest at pH values of 5.0 to 6.0 (Lucas and Davis, 1961), but in mineral soils with less than 5 percent OM it is in the 6.0 to 7.0 range. Figures 6.2 and 6.3 illustrate the comparative availability of nutrient elements at different pH values in organic and mineral soils.

FIGURE 6.1. The pH Scale with Relative Intensities of Alkalinity and Acidity Along with pH Values of Some Common Substances

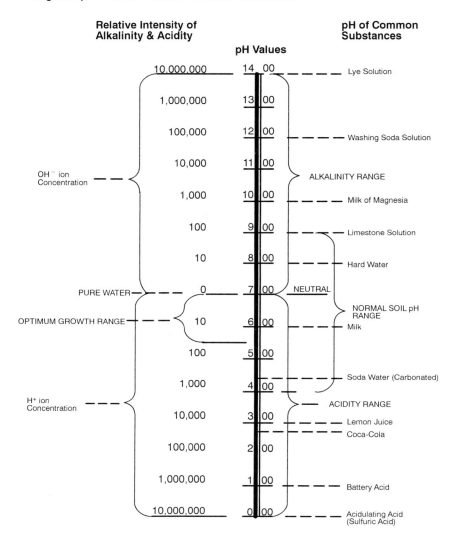

Source: National Fertilizer Solutions Association. 1967. Using pH values. *Liquid Fertilizer Manual,* National Fertilizer Solutions Association, Peoria, IL. Reprinted by permission of Agricultural Retailers Association, St. Louis, MO.

TABLE 6.1. Desirable pH Ranges for Some Agricultural Plants

Common name	Botanical name	pH range
Abelia	*Abelia grandiflora*	6.0-7.5
Alfalfa	*Medicago sativa*	6.5-7.5
Almond	*Amygdalus communis*	6.0-7.0
Alyssum	*Alyssum maritimum*	6.0-7.0
Amaryllis	*Amaryllis belladonna*	5.5-6.5
Apple	*Malus pumila*	5.5-6.5
Apricot, Siberian	*Prunus sibirica*	6.0-7.0
Arborvitae	*Thuja occidentalis*	6.0-7.0
Ash		
Amer. mountain	*Sorbus americana*	4.5-5.5
black	*Fraximus nigra*	6.0-7.0
white	*Fraximus americana*	6.0-7.0
Asparagus	*Asparagus officinalis*	6.0-7.5
Aster		
bigleaf	*Aster macrophyllus*	5.0-6.0
seaside	*A. spectabalis*	5.0-6.0
sky-drop	*A. patens*	5.0-6.0
stiff	*A. linarifolius*	5.0-6.0
wave	*A. undulatus*	6.0-7.0
Astilbe	*Astilbe* spp.	6.0-7.5
Avocado	*Persea americana*	6.0-8.0
Azalea, hiryu	*Rhododendron obtusum*	4.5-6.0
Azalea, pink	*R. periclymenoides*	4.5-5.2
Barley	*Hordeum sativum*	6.5-7.5
Bean		
kidney	*Phaseolus vulgaris*	6.0-7.0
lima	*Phaseolus limensis*	6.0-7.0
snap	*Phaseolus vulgaris*	6.0-7.0
velvet	*Stizolobium Deeringianum*	5.5-7.0
wax	*Phaseolus vulgaris*	6.0-7.0
Beech	*Phagus grandiflora*	5.0-6.5
Beet		
mangel wurzel	*Beta vulgaris, Crassa* group	6.0-7.5
sugar	*Beta vulgaris, Crassa* group	6.5-8.0
table	*Beta vulgaris, Crassa* group	6.0-7.0
Begonia	*Begonia meloir*	5.5-7.0

Common name	Botanical name	pH range
Birch		
Amer. white	*Betula papyrifera*	5.0-6.5
cherry	*Betula lenta*	4.5-7.0
European white	*Betula verrucosa*	4.5-6.0
Blackberry	*Rubus allegheniensis*	5.0-6.0
Blueberry		
high bush	*Vaccinium corymbosum*	4.0-5.0
low bush	*V. vacillans*	4.5-6.0
Boxwood	*Buxus sempervirens*	6.0-7.0
Broccoli	*Brassica oleracea, Italica* group	6.0-6.8
Brussels sprouts	*Brassica oleracea, Gemnifera* group	6.0-7.0
Buckwheat	*Fagopyrum esculentum*	5.5-7.0
Cabbage	*Brassica oleracea,* var. *capitata*	6.0-7.0
Cabbage, Chinese	*Brassica pekinensis*	6.0-7.0
Caladium	*Caladium esculentum*	6.0-7.0
Camellia	*Camellia japonica*	4.5-5.5
	Camelia sasanqua	4.5-6.5
Canna	*Canna* spp.	6.0-8.0
Cantaloupe	*Cucumis melo*	6.0-7.0
Carnation	*Dianthus caryophyllus*	6.0-7.0
Carrot	*Daucus carota*	6.0-7.0
Cauliflower	*Brassica oleracea, Botrytis* group	6.0-7.0
Celery	*Apium graveolens,* var. *dulce*	6.0-7.0
Chard, Swiss	*Beta vulgaris, Cicla* group	6.0-6.8
Cherry, sour	*Prunus cerasus*	6.0-7.0
Cherry, sweet	*P. avium*	6.0-7.5
Chicory	*Chicorium endiva*	5.0-6.8
Chive	*Allium schoenoprasum*	6.0-7.0
Chrysanthemum	*Chrysanthemum morifolium*	6.0-7.5
Clover		
alsike	*Trifolium hybridum*	5.5-7.0
bur	*Medicago denticulata*	5.0-6.5
crimson	*T. incarnatum*	5.5-7.0
hubam	*Melilotus alba,* var. *annua*	6.0-7.5
red	*T. pratense*	6.0-7.5
white	*T. repens*	6.5-7.5
white sweet	*M. alba*	6.5-7.5
yellow sweet	*M. officinalis*	6.5-7.5

TABLE 6.1 *(continued)*

Common name	Botanical name	pH range
Collards	*Brassica oleracea, var. acephala*	6.0-7.0
Corn		
common	*Zea mays*	5.5-7.0
pop	*Z. mays,* var. *verta*	6.0-7.0
sweet	*Z. mays,* var. *rugosa*	5.5-7.0
Cotoneaster	*Cotoneaster tomentous*	6.0-7.5
Cotton	*Gossypium hirsutum*	5.5-6.5
Cowpea	*Vigna sinensis*	5.0-6.5
Cucumber	*Cucumis sativus*	6.0-7.0
Currant, black	*Ribes nigra*	6.0-7.5
Currant, red	*Ribes sativum*	6.0-7.0
Cyclamen	*Cyclamen persicum*	6.0-7.0
Daffodil	*Narcissus bulbocodium*	6.0-6.5
Dahlia	*Dahlia variablis*	6.0-7.5
Delphinium	*Delphinium grandiflorum*	6.0-7.5
Dogwood, flowering	*Cornus florida*	5.0-6.5
Eggplant	*Solanum melongena,* var. *esculentum*	5.5-6.5
Elm	*Ulmus americana*	6.0-7.5
Endive	*Chicorum endiva*	5.0-6.8
Fennel	*Foeniculum vulgare*	5.0-6.8
Fern		
Boston	*Nephrolepsis exalta,* var. *Bostoniensis*	5.5-6.5
maidenhair	*Adiantum pedatum*	6.0-7.5
Forsythia	*Forsythia* spp.	6.0-8.0
Gardenia	*Gardenia jasminoides*	5.0-6.0
Garlic	*Allium sativum*	5.5-6.8
Grape	*Vitusca labrusca*	5.5-7.0
Grape	*V. vinifera*	5.5-6.5
Grapefruit	*Citrus maxima*	6.0-7.0
Grass		
bent, colonial	*Agrostis capillaris*	5.5-6.5
bent, Rhode Island	*Agrostis tenuis*	4.5-5.5

Common name	Botanical name	pH range
Grass *(continued)*		
bent, seaside	*Agrostis palustris*	5.5-6.5
bent, velvet	*A. canina*	5.5-6.5
Bermuda	*Cynodon dactylon*	6.0-7.0
blue, annual	*Poa annua*	5.5-6.5
blue, Kentucky	*P. pratensis*	5.5-6.5
brome, awnless	*Bromus inermis*	6.0-7.5
brome, fringed	*B. ciliatus*	6.0-7.5
fescue, Chewings	*Festuca rubra,* var. *fallax*	5.5-6.5
fescue, meadow	*F. pratensis*	4.5-7.0
fescue, red	*F. rubra* var. *genuina*	5.5-6.5
orchard	*Dactylis glomerata*	6.0-7.0
Rhodes	*Chloris gayana*	6.0-7.5
rye, perennial	*Lolium perenne*	6.0-7.0
St. Augustine	*Stenotaphrum secundatum*	6.0-7.5
timothy	*Phleum pratense*	5.5-7.5
Hibiscus	*Hibiscus rosa-sinensis*	6.0-7.5
Holly		
American	*I. opaca*	5.0-6.0
English	*Ilex aquifolium*	4.0-5.5
Japanese	*I. crenata*	5.0-6.5
mountain	*Nemopanthus mucronata*	5.0-6.0
Horseradish	*Armoracia rusticana*	6.0-7.0
Hyacinth	*Hyacinthus candicans*	6.5-7.5
Hydrangea		
blue-flowered	*Hydrangea macrophylla*	4.0-5.0
pink-flowered	*Hydrangea macrophylla*	6.0-7.0
Iris		
blue flag	*Iris versicolor*	5.0-7.0
Carolina	*I. caroliniana*	5.5-7.0
Japanese	*I. kaempferi*	5.5-6.5
vernal	*I. verna*	4.0-5.0
Ivy, Boston	*Parthenocissus tricuspidata,* var. *veitchii*	6.0-7.5
Ivy, English	*Hedera helix*	6.0-7.5

TABLE 6.1 *(continued)*

Common name	Botanical name	pH range
Juniper		
common	*Juniperus communis*	6.0-7.0
creeping	*J. horizontalis*	5.0-6.0
mountain	*J. communis montana*	5.0-6.0
Kale	*Brassica oleracea, Acephala* group	5.5-6.8
Kohlrabi	*Brassica oleracea, Gongylodes* group	5.5-6.8
Leek	*Allium ampeloprasum, Porum* group	6.0–6.8
Lettuce	*Lactuca sativa*	6.0-7.0
Lilac, common	*Syringa vulgaris*	6.0-7.5
Lilac, Persian	*S. persica*	6.0-7.5
Lily		
easter	*Lilium longiflorum*	6.0-7.0
Japanese	*L. speciosum*	6.0-7.0
regal	*L. regale*	6.0-7.0
tiger	*L. tigrinum*	6.0-7.0
Lily-of-the-valley	*Convallaria majalis*	4.5-6.0
Linden	*Tilia* spp.	6.0-8.0
Lupine	*Lupinus mutabalis or nanus*	5.0-6.0
blue	*L. hirsutus*	5.0-6.0
rose	*L. Hartwegi*	6.0-7.0
white	*L. albus*	5.5-7.0
yellow	*L. luteus*	5.0-6.0
Lychee	*Litchi chinensis*	6.0-7.0
Magnolia		
saucer	*Magnolia soulangiana*	5.0-5.8
southern	*M. grandiflora*	5.0-5.8
star	*M. stellata*	5.0-5.8
Maple, sugar	*Acer saccharum*	6.0-7.2
Millet	*Setaria italica*	5.0-6.5
Milo, dwarf yellow	*Sorghum vulgare*	5.5-7.5
Mint	*Mentha arvensis*	6.8-7.5
Mountain laurel	*Kalmia latifolia*	4.5-5.5
Muskmelon	*Cucumis melo*	6.0-7.0
Mustard	*Brassica juncea*	5.5-6.8
Myrtle crepe	*Lagerstroemia indica*	5.0-6.0

Common name	Botanical name	pH range
Oak		
black	*Quercus velutina*	6.0-7.0
English	*Q. robur*	6.0-7.5
live	*Q. virginiana*	5.0-6.0
pin	*Q. palustris*	5.0-6.5
red	*Q. borealis*	4.5-6.0
white	*Q. alba*	5.0-6.5
Oats	*Avena sativa*	5.0-7.5
Okra	*Hibiscus esculentus*	6.0-7.5
Onion	*Allium cepa, Cepa* group	6.0-7.0
Orange, sweet	*Citrus sinensis*	6.0-7.0
Pachysandra, Japanese	*Pachysandra terminalis*	5.0-8.0
Palm		
Canary Island	*Phoenix canariensis*	6.0-7.5
cocus	*Cocus australis*	6.0-7.5
Washington	*Washingtonia filifera*	6.0-7.5
Pansy	*Viola tricolor*	5.5-6.5
Parsley	*Petroselinum hortense*	5.0-7.0
Parsnip	*Pastinaca sativa*	5.5-7.0
Pea		
early dwarf	*Pisum sativum*	6.0-7.0
field	*P. arvense*	6.0-7.0
sweet	*Lathyrus odoratus*	6.0-7.5
Peach	*Amygdalis persica*	6.0-7.0
Peanut	*Arachis hypogaea*	5.5-6.5
Pear	*Pyrus communis*	5.5-6.5
Pecan	*Carya pecan*	6.5-7.5
Peony	*Paeonia albiflora*	6.0-7.5
Pepper, sweet	*Capsicum frutescens,* var. *grossum*	5.5-7.0
Pine		
jack	*Pinus Banksiana*	4.5-5.5
loblolly	*P. taeda*	5.0-6.0
longleaf	*P. palustris*	4.5-5.0
red	*P. resinosa*	5.0-6.0
white	*P. strobus*	4.5-6.0
yellow	*P. echinata*	5.0-6.0

TABLE 6.1 *(continued)*

Common name	Botanical name	pH range
Pineapple	*Ananus sativus*	5.0-6.0
Plum	*Prunus americana*	6.0-7.0
Poinsettia	*Euphorbia pulcherrima*	6.0-7.0
Poplar, silver	*Populus alba*	6.0-7.5
Poppy	*Papaver alpinum* or *aurantiacum*	6.0-7.5
blue	*Meconopsis Bailera*	6.0-7.0
California	*Eschscholitzia californa*	6.0-7.5
Iceland	*P. nudicaule*	6.0-7.5
oriental	*P. orientale*	6.0-7.5
Potato, Irish	*Solanum tuberosum*	4.8-5.5
Potato, sweet	*Ipomaea batalus*	5.2-5.8
Pumpkin	*Cucurbita pepo*	5.5-7.5
Radish	*Raphanus raphanistrum*	5.5-6.5
Raspberry, black	*Rubus occidentalis*	5.0-6.5
Raspberry, red	*R. idaeus,* var. *strigosus*	5.5-7.0
Rhododendron	*Rhododendron albiflorum, carolinianum, caucasicum, lapponicum, mucronulatum*	4.5-6.0
	R. ferrugineum	4.0-7.0
	R. hirsutum	6.0-7.5
Rhubarb	*Rheum rhaponticum*	5.5-7.0
Rice	*Oryza sativa*	5.0-6.5
Rose		
climbing	*R. cathayensis*	6.0-7.0
hybrid tea	*Rosa* sp.	5.5-7.0
ideal	*R. polyantha*	5.0-6.5
rugosa	*R. rugosa*	6.0-7.0
Rutabaga	*Brassica napobrassica*	5.5-7.0
Rye	*Secale cereale*	5.0-7.0
Salsify	*Tragopopon porrifolius*	6.0-6.8
Shallot	*Allium cepa, Aggregatum* group	5.0-6.8
Sorrel	*Rumex scutatus*	5.0-6.8
Soybean	*Glycine Max*	6.0-7.0
Spinach	*Spinacia oleracea*	6.0-7.5
Spruce		
black	*Picea mariana*	4.0-5.0
Colorado	*P. Pungens*	6.0-7.0
Norway	*P. abies*	5.0-6.0

Common name	Botanical name	pH range
Spruce *(continued)*		
red	*P. rubra*	4.5-5.0
Sitka	*P. sitchensis*	5.0-6.0
white	*P. glauca*	5.0-6.0
Squash, crookneck	*Curcubita moschata*	6.0-7.0
Squash, hubbard	*C. maxima*	5.5-7.0
Strawberry	*Fragaria collina*	5.0-6.5
Sugarcane	*Saccharum officinarum*	6.0-7.5
Sunflower	*Helianthus augustifolius* and *annuus*	6.0-7.5
Tobacco	*Nicotiana tabacum*	5.5-7.5
Tomato	*Lycopersicum esculentum*	5.5-7.5
Trefoil, bird's-foot	*Lotus corniculatus*	5.5-7.0
Trefoil, yellow	*Medicago lupulina*	6.0-7.5
Tulip	*Tulipa gesneriana*	6.0-7.0
Turnip	*Brassica rapa, Rapifera* group	5.5-6.8
Vetch, hairy	*Vicia villosa*	5.5-7.0
Viburnum	*Viburnum tomentosum*	6.5-7.5
Viburnum	*V. acerifolium*	4.0-5.0
Violet	*Viola canina*	6.0-7.5
Violet, blue common	*V. paplionacea*	5.0-7.5
Walnut, black	*Juglans nigra*	6.0-7.5
Watercress	*Nasturtium aquaticum*	6.0-7.5
Watermelon	*Citrullus vulgaris*	5.0-5.5
Wheat	*Triticum aestivum*	5.5-7.5
Wisteria, Japanese	*Wisteria floribunda*	6.5-7.5
Yew, Canada	*Taxus canadensis*	5.0-6.0
Yew, Japanese	*T. cuspidata*	6.0-7.0
Yucca	*Yucca* spp.	6.0-8.0
Zinnia	*Zinnia elegans*	5.5-7.5

Sources: Lamotte Chemical Products Co. 1950. *Lamotte Soil Handbook.* Lamotte Chemical Products Co., Towson, MD; Lorenz, O. A. and D. M. Maynard. 1988. *Knott's Handbook for Vegetable Growers,* Third Edition. John Wiley and Sons, New York; Spurway, C.H. 1941. *Soil Reaction (pH) Preferences of Plants.* Special Bull. 306, Michigan State College, East Lansing, MI.

FIGURE 6.2. The Influence of pH on the Availability of Nutrients in Organic Soils*

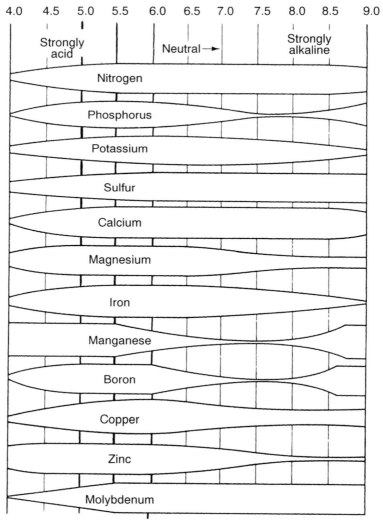

* Availability of the nutrient is indicated by the width of the outlined areas.

Source: Lucas, R. E. and J. F. Davis. 1961. Relationship between pH values of organic soils and availability of 12 nutrients. *Soil Science* 92: 178. Copyright © 1961 Williams and Wilkins. Reprinted by permission of Waverly, Baltimore, MD.

FIGURE 6.3. The Influence of pH on the Availability of Nutrients in an Inorganic Soil*

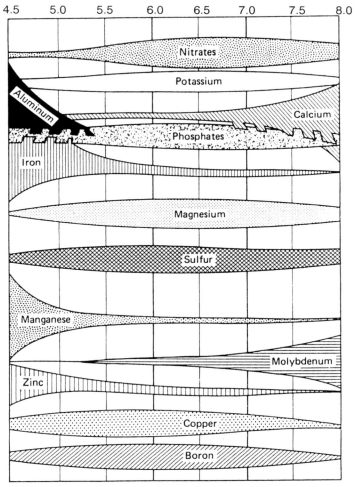

* Availability of the nutrients is indicated by the width of the outlined areas. Aluminum, although not an essential element, may become toxic at low pH values.

Source: Anonymous. 1978. *Liming Acid Soils.* University of Kentucky Agricultural Extension Service Leaflet AGR-19.

Ideal pH values of 5.0 to 6.0 have also been suggested for the tropical or semitropical oxisols (highly weathered low-CEC soils with clay in oxide or kaolinite forms) and ultisols (highly weathered, low-CEC soils with clay horizons and kaolinite as the dominant clay). The tolerance of appreciably lower pH values by plants grown on histosols can be accounted for in part by the buffering effect of large amounts of organic matter so that less toxic Al or Mn is present to adversely affect growth. The exact reasons why lower pH values of the oxisols or ultisols are ideal are not completely understood, but may be due to one or more of the following causes: (1) the commonly low amounts of several micronutrients (B, Mn, Zn) in these weathered soils are lowered even further at pH 6.0+ so as to become deficient; (2) soluble phosphates are converted to insoluble apatite as pH rises in these soils; and (3) the Al and Fe are present in mineral clays (oxides and kaolinite) that are stable down to almost pH 5.0.

Cultivar and Root Stocks

Not only are there considerable differences between species as to tolerances of both low and high pH values, but differences can be quite marked between cultivars of the same species, or on different rootstocks. Tolerance to low pH values appears to be associated primarily with the ability to grow despite high levels of Al, but the ability to tolerate other ions that may be in high concentrations at low pH values ($B(OH)_4^-$, Cu^{2+}, Fe^{2+}, Mn^{2+}, Zn^{2+}, and NH^{4+}), or to function with low levels of Ca, may also be contributing factors. Tolerance of high pH values has been associated with the ability of some plant roots to lower pH in the rhizosphere (area adjacent to root hairs) sufficiently so that certain elements usually unavailable in high pH soils are solubilized (Fe, Mn, Cu, Zn).

Some of these factors that increase tolerance at unfavorable pH values appear to be genetically related and there has been considerable screening of different cultivars to select ones that are tolerant for current use and also for breeding purposes.

Solubility and Availability of Elements

The pH has an important effect on plant growth because soil or medium pH has an important bearing on available nutrients due to its effect on the solubility of soil minerals and activities of microorganisms. The solubility of carbonates, phosphates, and sulfates is increased as pH drops. Low pH also favors the weathering of minerals and release of K^+, Ca^{2+}, Mg^{2+}, Al^{3+}, Cu^{2+}, and Mn^{2+}. The solution of phosphates and sulfates and the increase in the basic cations generally increase crop production by increasing the nutrient, but in humid climates, these elements may be leached from the soil, leaving it with low levels of important nutrients.

Release of considerable Al^{3+}, Cu^{2+}, Mn^{2+}, and several other heavy metals in acid soils can be disastrous to a number of different crops. Excess amounts of Al^{3+} are common in acid soils; excess Mn^{2+} is less so but constitutes a serious problem in many acid soils. Excess Cu^{2+} is rather uncommon but has caused problems because of undue amounts applied to certain crops (grapes, citrus) as fungicidal sprays or because of accumulation from mine wastes or use of sewage sludge or wastewaters.

PRIMARY NUTRIENTS

Nitrogen

The form of N existing in the soil and its availability to plants is largely influenced by microorganisms and enzymes, both of which are greatly affected by soil pH. Plants cannot utilize N from the atmosphere or from organic matter until it is transformed by microorganisms.

The gaseous N in the atmosphere is made available to plants after it is fixed into nitrogenous compounds by either free-living or symbiotic organisms. The free-living organisms (primarily *Azotobacter*) do best in soils with pH values 6.0 to 8.0, their activities being greatly restricted as pH values fall below 5.7. The symbiotic organisms, most commonly *Rhizobia* and associated with legumes, function best in rather narrow pH ranges. Several species, such as those associated with sweet clover and alfalfa, grow best at a pH value close to 7.0, but a few, such as those that colonize subterranean clover or *Stylosanthus* are very efficient at pH values close to 4.5. Even those species that prefer the higher values can function at much lower pH values if there is sufficient Mo in the soil.

The limiting low pH values for nitrifying bacteria (organisms that oxidize ammonium to nitrites and then to nitrates) are 4.0 to 4.4, with the optimum reaction range being 6.8 to 7.3.

The process that makes the N of OM available to plants takes place in a series of steps which can be diagrammed as follows:

```
              Fungi and      Nitrosomonas   Nitrobacter
               bacteria        bacteria       bacteria
Organic N------> ammonium N------>nitrite N ---->nitrate N
                 NH4-N     -------> NO2-N   -----> NO2-N
```

Decomposition takes place over a wide range of pH values with fungi dominating the first step if pH values are low. Fungal decomposition is slower and not as complete as that of bacteria. The conversion of NH_4-N to NO_3-N is primarily accomplished by bacteria and, while it can occur in acid soils, is slowed appreciably if the pH falls below 6.0.

Immobilization or Loss of N

Satisfactory pH values, by improving microbial activity, usually favorably affect the availability of N but can also temporarily reduce available soil N. The reduction of available N takes place if the N released from the decomposition of OM is insufficient to meet the N requirements of the organisms. This temporary immobilization of N occurs if large amounts of materials such as sawdust or the residues of corn, wheat, oats, sorghum, etc. with wide C/N ratios are added to the soil. Low pH values achieved by slowing the decomposition of OM can at times benefit crops that are adapted to them. Some facts about C/N ratios and methods of avoiding crop losses due to wide C/N ratios were given in Chapter 5.

The losses induced both by microorganisms and chemical processes take place as ammonia (NH_3), di-nitrogen (N_2), and nitrogen oxides (NO_2, N_2O, and NO) volatilize as gases or if NO_3-N is leached from the soil.

The NH_4^+ does not volatilize but it can be converted as indicated in the diagram below to highly volatile ammonia gas (NH_3).

Ammonium ion		hyroxyl ion		ammonia gas		water
NH_4^+	$+$	OH^-	$------->$	NH_3	$+$	H_2O

A high pH (above 7.0) or addition of materials such as lime that increase the pH (increased OH^-) shift the reaction to the right with greater production of volatile NH_3 from the addition of added NH_4 salts or urea.

The loss of N from urea is influenced by the presence of the enzyme urease, the amount of which is related to the action of microorganisms but also to the preceding crop. As urea is added to the soil, it undergoes hydrolysis, raising the pH. The loss of N takes place as NH_3 is formed from NH_4^+, similarly to the reaction indicated above, but the increase of pH by hydrolysis makes urea quite susceptible to N losses if not handled properly. The transformation of the urea to NH_4^+ takes place in a series of steps that can be diagrammed as follows:

	Urea		water	+ urease or other enzymes	= ammonium carbonate
(1)	$CO(NH_2)_2$	$+$	H_2O	$----------------->$	$(NH_4)_2CO_3$
	Ammonium carbonate		water	= ammonium ion	+ water + carbon dioxide
(2)	$(NH_4)_2CO_3$	$+$	H_2O	$---> NH_4^+$	$+ H_2O + CO_2$

Denitrification losses of N as NO_2, N_2O, NO, and N_2 usually take place under anaerobic conditions under a wide range of pH values (4.0 to 10.0) but losses are most serious at pH values of 7.0 to 7.5 because of the increased microbial activity at these values. The losses can be minimized by improving soil aeration and/or by using nitrification inhibitors. Where increase of aeration is not possible, as in flooded rice soils, denitrification losses are limited by the avoidance of NO_3-N. Nitrates also are less efficient during rainy seasons when used on compacted soils containing crop residues.

Losses of N as NO_3 are indirectly affected by pH as the pH influences the transformation of NH_4-N to NO_3-N (see above). This loss is also greater in soils of low CEC due to the poor ability of these soils to hold NH_4-N. The losses can be reduced by the use of nitrifying inhibitors or by limiting the application of N at any one time to amounts that can be adsorbed by the soil or removed by plants.

The losses of N from different materials are related in part to the pH changes resulting from the use of the material. Much of the loss is due to the relatively high pH resulting from the addition of ammonia in or generated by materials such as anhydrous ammonia, liquids containing free ammonia, diammonium phosphate, and urea.

Phosphorus

The availability of P is markedly influenced by pH, with maximum availability in the range 6.5 to 7.0 and decreasing as the pH falls below 6.5 or rises above 7.0. In acid soils, the reduction in solubility is related to increased amounts of Al and Mn, both of which can combine with P. Also, these elements in high concentrations restrict root growth of many plants and thereby lessen absorption. At pH values above 7.0, P availability is reduced by the formation of relatively insoluble calcium phosphates.

There are instances when increasing the pH of acid soils by the addition of liming materials does not increase availability of P and in certain cases may actually reduce availability. Generally, soils high in Fe and Al oxides respond to liming with increased availability of P but soils low in exchangeable Al^{3+} do not. It is thought that in the absence of exchangeable Al^{3+}, liming increases the precipitation of the added P as apatite, the solubility of which decreases with increasing pH.

Availability of Other Macronutrients

Although K, Ca, Mg, and S are usually available over much wider pH ranges, their utilization is still affected by pH. The slightly better uptake of these elements in the pH range of 6.0 to 7.0 appears to be due indirectly to better root development at these pH values. But liming also can make appreciable changes in the availability of these elements.

The pH, increased by liming many southern acid soils, does decrease the availability of K^+. Much of the CEC of these soils is pH dependent and liming causes an important shift of solution K^+ to the exchange complex. The limed soil also retains more of applied K^+ against leaching because of the greater shift of K^+ to the exchange complex.

Liming increases Ca^{2+} and Mg^{2+}, the amounts depending on the amounts of liming materials applied and their composition. Use of dolomitic liming materials increases the Mg^{2+} but usually lessens the amount of Ca^{2+} compared to high-calcium materials.

Sulfur, similar to N, is released from OM by microbial decomposition, which in turn is affected by pH. Liming acid soils generally tends to hasten the decomposition of OM. Availability of S is further affected by pH because at low pH values (<6.0), S is removed from the soil solution by surface adsorption and precipitation.

Micronutrients

Lowering the pH below 7.0 tends to increase availability of the essential micronutrients B, Cu, Fe, Mn, and Zn, and the nonessential Al, whereas raising the pH above 7.0 tends to decrease their availability but increases that of Mo.

The solubility of Mn can increase so rapidly as the pH falls below 5.0 that it may be toxic to plants, although most toxicities at very low pH values are more likely to be due to Al.

The amounts of available Fe, Mn, and Zn can decrease rapidly as the pH rises above 7.0, often to the point of inducing deficiencies. The need to add B is also greater on soils with pH values 7.0 to 8.5, but the solubility of it and also Zn increases slightly as the pH rises above 8.5.

The efficiency of adding common sources of these elements directly to high-pH soils is low. Availability of Fe is greatly increased if the Fe is in the chelate form. Bypassing the soil by adding these elements directly to the plant as foliar sprays often is a cheaper means of correcting deficiencies of these elements on high-pH soils.

TOXIC ELEMENTS

Aluminum

The poor performance of many crops at low pH values appears to be due primarily to the availability of Al^{3+}, which is greatly increased in many soils as pH values fall below 5.5. The exact causes of aluminum toxicity are not known but it is believed that it may be due in large part to alteration of root membranes, affecting uptake of nutrients and water use. It is known that high levels of AL^{3+} seriously affect root growth. The poor development of roots in many subsoils is believed to be due to lower pH values and higher Al contents in the subsoil. The toxicity appears to be related to amounts of Al in the soil solution rather than that which is exchangeable.

Aluminum toxicity is known to seriously affect availability of P, probably by affecting solubility of several phosphates. Large amounts of applied phosphates can mitigate the effects of excess aluminum, which is a practical means of growing Irish potatoes at a low pH. Applying large amounts of superphosphate has been a means of getting good crops of potatoes, which are grown at low pH values to avoid scab development on the tubers.

Aluminum also interferes with the uptake and utilization of Ca, Mg, K, and Fe. Al toxicity of some plants can be reduced by increasing concentrations of Ca, Mg, K, or Na.

Not all plants are affected by excess levels of Al^{3+}. In fact, there is often considerable variation in plant response to excess Al^{3+} among different cultivars, some of them making satisfactory growth at Al^{3+} levels detrimental to others.

There does not seem to be a single reason for the tolerance of some plants to large amounts of Al in solution. Some of the tolerance appears to be due to the ability of some plants to raise the pH near their roots, while sensitive plants fail to raise pH and may even lower it. The higher pH immobilizes much of the Al^{3+}. A number of plants are able to exclude Al at the roots, while others that accumulate it in the tops appear to have high internal tolerance to the element. In some plants, excess Al^{3+} induces iron deficiency, but tolerant cultivars are able to make good growth because they can chelate the Fe and keep it available. Other tolerant plants are able to take up greater amounts of K, Mg, or Si in the presence of excess Al^{3+}.

Manganese

Available Mn, which is increased as pH values fall below 6.5, can at very low pH values rise to levels that are highly toxic to a number of plants. Toxicity usually occurs at pH values below 5.5 but primarily on soils with high total Mn contents. Poor aeration, which favors the production of Mn^{2+}, increases the chances of toxicity and in such cases may cause problems in slightly acid soils. Fumigation with steam or methyl bromide can induce or intensify manganese toxicity, but usually the addition of OM lessens the toxicity.

Some of the common disorders associated with manganese toxicity are gray leaf of tobacco, crinkle leaf of cotton, stem streak necrosis of potato, and internal bark necrosis of apple trees. Common symptoms include black or dark brown flecks on several tropical grasses, overall yellowing of some plants, puckering of leaf surfaces, interveinal chlorosis, and in case of severe toxicities, necrotic tips on emerging leaves.

Heavy Metals

The solubility of the heavy metals (cadmium, copper, lead, nickel, and zinc) is increased as pH values fall below 6.5. Use of large amounts of sewage sludge and wastewaters as fertilizers can readily result in excesses of these elements for plants and potential harm to humans or animals consuming these products. Liming soils to a pH close to 7.0 greatly reduces the elements' availability. This still may not make it safe to use these products for edible plants but can lessen the possibility of phytotoxicity of landscape and ornamental plants.

Membrane Effects

The effect of excess elements is accentuated because pH affects root systems of many crops, influencing amounts of nutrients absorbed. Altering the pH or Ca level can significantly modify uptake of several elements. Much of the poor uptake of nutrients at low pH has been attributed to excess Al^{3+}, but may be due to the low Ca^{2+} concentrations usually present. At least, increasing Ca^{2+} often enhances the uptake of inorganic ions, particularly K^+ and other monovalent ions. The lack of Ca^{2+} increases the leakiness of cell walls, allowing losses of K^+ and $H_2PO_4^-$, and lessens intake of these ions. Much of the ill effects of heavy metals can be also be overcome by the addition of Ca^{2+}.

Hydrogen Effects

It is difficult to determine the effects of pH per se on plants grown in soil because of its effect on several elements and microbial growth. For example, as the pH is lowered in most soils to about 5.5, there is usually enough Al^{3+} or Mn^2 to damage root systems of many crops. Also, low pH values limit microbial action, with deleterious effects on nitrification and decomposition of OM. But bypassing the soil and using nutrient solutions for growing plants shows that much lower pH values can be tolerated by most plants.

The effects of low pH therefore are seldom due to excess H$^+$ but rather to its effects on various elements that affect plant growth. Some of the more important influences of pH on availability of the different nutrients were outlined previously. Some other effects of pH on growth of plants are listed below.

OTHER EFFECTS OF pH

Besides the indirect influence of pH on N and several other elements through its effect on microorganisms, it also plays a major but indirect role in the development of several diseases and the effectiveness of certain herbicides. It may also affect the activities of certain nematodes.

Some diseases controlled by regulating pH are potato scab (*Streptomyces scabies*), chocolate spot of bean (*Botrytis* spp.), *Pythium* damping-off of seedlings, fusarium wilt in tomato and club root of brassicas. A low pH (<5.5) reduces damage from potato scab and black shank, whereas high pH (6.8+) is associated with lessened damage from fusarium in tomato, club root in brassicas, and botrytis infection, but causing chocolate spot in bean and nematode damage of tobacco.

There is some evidence that pH affects diseases and nematodes through its effect on certain nutrients. Manganese, which tends to increase as the pH decreases, has a positive effect in reducing potato scab. It is interesting to note that spring wheat, inoculated with the organisms *Gaeumannomyces graminis* var. *tritica* suffered less damage if Cu, Fe, Mn, and Zn are added with the seed at planting. But there is an indication that *Fusarium oxysporum* virulence tends to increase with increased concentrations of these micronutrients, while virulence of *F. mariti,* causing root rot of English pea, is lessened with increased Al, Cu, and Fe. These elements become more available as pH is lowered. Also, the botrytis organisms causing chocolate spot in bean is increased by deficiencies of Ca, Mg, P, and K. Availability of Ca, Mg, and P in many soils is increased as the pH rises from 4.5 to 7.0.

The effect of pH on disease severity may be due in large part to its effect on the conversion of NH$_4$-N to NO$_3$-N. A very high ratio of NH$_4$-N/NO$_3$-N lowers severity of several diseases (take-all of wheat, corn stalk rot [*Diplodia maydis* and *Gibberella zeae*], foot rot of small grains [*Pseudo-Cercosperela herpotrichoidii*] and *Xanthomonas* leaf spot of schefflera). The severity of take-all is negatively correlated with the length of time in which the ratio of NH$_4$-N/NO$_3$-N is greater than 3:1. The lower pH values tend to limit the production of NO_3-N, but so does the application of chloride or Nitrapyrin, both of which effectively reduce the damage of take-all. Delaying the application of NH_4-N to the spring so that less of it is converted to NO_3-N also decreases take-all.

The effect of pH on herbicides is related not only to the degradation of herbicides by microorganism but also to changes in adsorption of the herbicide on clay and OM. Degradation tends to be greatest at maximum microbiological activity, which takes place at an approximate pH of 7.0.

Critical Values and Crop Growth

As pointed out above, the influence of pH upon nutrient availability affects ideal ranges of pH in which plants grow best. These ideal ranges are not absolutes as they may differ on some soils. The infertility of acid soils appears to be due to toxicities of Al, Mn, and H and deficiencies of Ca, Mo, and occasionally Mg; that of alkaline soils to deficiencies of Fe, Mn, B, Zn, and perhaps P. Therefore, it is not uncommon that soils with low contents of Al and ample levels of Ca often can support good crops at pH values much lower than the ideal range indicated; soils with good levels of micronutrients can yield good crops at higher pH values.

The values in Table 6.1 were compiled on the assumption that there is a critical pH value below which plants will not produce satisfactory yields and plant growth will respond to liming. As indicated, these ideal ranges are not absolutes as they may differ on some soils. Some of the causes, other than nutrient availability, affecting these ideal ranges are listed briefly below.

Soluble Salts

The pH of a soil will usually drop as fertilizers are added to a soil. In poorly buffered soils with low clay and organic matter contents, the drop may be 0.5 to 1.0 pH units. The drop appears to be due, at least in part, to displacement of H^+ and Al^{3+} by fertilizer cations from the exchange complex to the soil solution. The pH will rise again as nutrients are removed by plants or as fertilizers are leached from the soil.

Not all salts will lower the pH for Na, and Ca salts can have an opposite effect. Additions of compounds of these elements tend to raise the pH of the soil or medium, especially if there is a differential absorption favoring the anion, as is the case with calcium or sodium nitrates.

Time of Year

The critical pH value is also affected by the time of year in which the pH measurement is made. Generally, pH values tend to fall during the summer months or active growing periods. This has been attributed to increased fertilization and nitrification, which generally occurs during these periods.

Soil Moisture

The changes in fertilizer salts and nitrates in particular do not appear to completely explain the variation of pH with time of year since the drop in pH is greater in some years than others. The anomalies, however, may be partially explained by the effect of different moisture regimes on pH.

Years of ample rainfall may lower salts and nitrates by leaching. Also, the moisture content of a soil directly influences the pH reading, as dry soils tend to lower pH values. Air drying a soil will lower the pH value of most soils 0.2 to 0.3 units.

ACID SOILS

Types of Acidity

Soil acidity is the result of active acidity, which is reflected by H^+ in the soil solution and the potential acidity or H^+ adsorbed on the exchange complex. It is generally believed that soil acidity is due to the loss of the basic cations Ca^{2+}, Mg^{2+}, and K^+ and the presence of exchangeable aluminum (Al^{3+}). The lost cations are replaced with H^+ ions. The Al^{3+} ions displaced from clay minerals hydrolyze and are readsorbed by clay minerals, inducing additional hydrolysis, which in turn releases H^+ ions on the exchange complex. A simplified categorization of the hydrolysis and production of H^+, as presented by Tisdale and Nelson (1966) is given below:

$$Al^{+++} \quad + \quad H_2O ----> Al(OH)^{++} \quad + H^+$$
$$Al(OH)^{++} \quad + \quad H_2O ----> Al(OH)^{++} \quad + H^+$$
$$Al(OH)^{++} \quad + \quad H_2O ----> Al(OH)^{+++} \quad + H^+$$

Causes of Acid Soils

Acid Parent Materials

The parent material from which a soil originates has an influence on its pH. Acid parent materials, such as granites, sandstones, and shales tend to give rise to soils low in pH. Soils derived from such materials are usually acid because the parent materials have so few basic cations (Ca^{2+}, Mg^{2+}, K^+, and Na^+).

The Effect of Climate

Climate is a dominant force in determining soil pH, with the parent material exerting strong influences. Climate affects pH directly by leaching cations and indirectly by altering the dominant vegetation. Removing the cations allows for replacement with H^+ and Al^{3+}, while altering the vegetation influences the cations that are maintained in the system.

The parent material or original geologic formation also has an effect on the type of vegetation that is naturally maintained. But the parent material is also modified by climate, which in turn affects the vegetation. Some of the effects of climate, vegetation, and modified parent material are reflected in the average pH values of several soils, given in Table 6.2.

Leaching

Leaching is the most important of all the climatic factors in producing acid soils. In humid regions, where the loss of soil water by leaching exceeds that lost by evapotranspiration for at least part of the year, sufficient bases are lost so that the colloidal complex becomes saturated with H^+ and Al^{3+}. Acid soils predominate in the eastern part of the mainland United States and in a narrow belt along the Pacific Ocean where rainfall is greater than 20 inches per year. The leaching effect is enhanced by the

TABLE 6.2. Average pH Values of Several Soils

Soil	pH range
Drained organic soils containing sulfur	1.0-3.0
Conifer forest soils in humid regions	4.0-5.0
Hardwood forest soils in humid regions	5.0-6.5
Grassland soils in subhumid regions	5.0-7.0
Grassland soils in arid regions	6.5-8.0
Soils with excess calcium	7.5-8.5
Soils with excess sodium	8.0-10.0
Highly productive soils	6.0-7.0

presence of carbonic acid, formed from the combination of water and carbon dioxide exuded from the roots. The effect also appears to be aggravated by acid rain (rainwater containing dissolved emissions of sulfur dioxide and nitrous oxides from highly populated industrialized areas).

Whereas the effect of acid rain is probably insignificant on soils with considerable bases, pH can be lowered on soils derived from parent materials that contain few bases, e.g., granite, shale, and sandstone. Acid rain can further deplete the few bases remaining so that acidity is increased.

Differential Absorption of Cations

Plants exchange H^+ for cations and bicarbonate ions (HCO_3^-) for anions of the soil solution. There is a preferential absorption of cations over anions, particularly with legumes, bringing much more H^+ into the soil solution to contribute to acidity.

Chemical Additions

Humans modify soil pH by making chemical additions to the soil via fertilizers, manures, liming materials, sprays, and irrigation waters. Also, the soil pH is altered by changing the cations retained or lost through such practices as crop selection, organic matter additions, and cultural methods that affect erosion. The use of acid-forming fertilizers is one of the more dominant ways in which soil pH is lowered.

Changes in pH Affected by NO_3-N:NH_4-N Ratios

The ratio of nitrate-N (NO_3-N) to ammoniacal-N (NH_4-N) has a bearing on the pH of the medium. Ratios greater than about 9:1 tend to increase the pH, whereas ratios less than 9:1 tend to lower it.

The effect on pH by varying NO_3:NH_4 ratios is much more important in media having little buffering capacity (hydroponic cultures or light soils) than in the better-buffered soils (loams, silts, clays, or those with considerable OM). The lowering of pH by use of NH_4-N is also less pronounced in well-aerated soils, since NH_4-N is usually quickly converted to NO_3-N, unless soils have been treated with nitrifying inhibitors.

Acid-Forming Fertilizers

Acid-forming fertilizers are responsible for lowering the pH of many soils. The ammonium sources of N (ammonium sulfate, nitrate, chloride, and monoammonium phosphate) as well as urea, sulfur, and its compounds form acid solutions. It has been estimated that 100 lb of NH_4-N will require 357 lb of $CaCO_3$ (pure calcium carbonate) to neutralize its acidity. Long-term use of nitrogenous fertilizers has shown the degree of acidification to be in this order: ammonium sulfate = monoammonium phosphate = ammonium phosphate sulfate > ammonium chloride > ammonium nitrate = anhydrous ammonia = aqua ammonia = diammonium phosphate = urea > ureaform. On the other hand, carbonates, nitrates, or hydroxides combined with Ca, Mg, Na, and K do not lower and may raise the pH, although 100 lb of N leached as NO_3-N carries with it the equivalent of 357 lb of basic cations. Potential acidity or basicity of a number of fertilizer materials are given in Table 6.3.

The effects of adding liming materials will be covered in Chapter 11 and those of water on pH values in Chapter 15.

Crop Removal

The soil pH is also altered by cultural practices that change the cations retained or lost through such practices as crop selection, organic matter additions, and cultural methods that affect erosion.

The reduction in soil pH values is intensified by the removal of large quantities of crops that contain high concentrations of basic cations. When a major portion of the crop is removed, as with hays, silage, and pastures, the process is accelerated. The legumes, alfalfa, peas, and high-yielding non-legumes (corn silage, sugar cane) remove large quantities of bases, and can be expected to lower pH values, while the small grains remove very small amounts (particularly if the straw is returned to the soil) with negligible effects on pH (see Table 6.3).

ALKALINE SOILS

Large areas of soils have high pH values. Most of these areas are in arid or semiarid regions where there has been insufficient rainfall to wash the basic cations from the soil, but some exist in humid or semihumid regions. The latter exist because irrigation waters used are high in carbonates and/or bicarbonate or the soil was recently formed from marine deposits rich in shells. Some high-pH soils temporarily exist in humid or semi-humid regions because of excessive use of liming materials or alkaline materials such as wood ashes.

TABLE 6.3. Acidity (A) or Basicity (B) Equivalent to Calcium Carbonate per 100 Pounds of Several Soil Amendments

Material	Analysis	Acidity (A) Basicity (B)
Ammonia, anhydrous	82-0-0	148A
Ammonium chloride	26-0-0	120A
Ammonium nitrate	33.5-0-0	59A
Ammonium nitrate, limestone Cal-Nitro, A.N.L.	20.5-0-0	neutral
Ammonium phosphate, mono	11-48-0	65A
Ammonium phosphate, di	18-46-0	69A
Ammonium phosphate, di	11-34-0	74A
Ammonium phosphate, nitrate	27-15-0	59A
Ammonium phosphate soln.	11-34-0	44A
Ammonium phosphate sulfate	16-20-0	91A
Ammonium sulfate	21-0-0	110A
Calcium nitrate	15-0-0	20B
Cal-urea	43-0-0	36A
Nitrogen solutions		
Aqua ammonia	21-0-0	38A
Ammonium nitrate-ammonia	37-0-0	67A
Ammonium nitrate-ammonia	41-0-0	74A
Ammonium nitrate-urea	28-0-0	50A
Ammonium nitrate-urea	32-0-0	58A
Urea ammonia	49-0-0	88A
Urea ammonia	51-0-0	92A
Nitric acid (100%)	17-0-0	40A
Nitric acid (65%)	11-0-0	26A
Phosphoric acid	0-83-0	59A
Phosphoric acid	0-72-0	51A
Phosphoric acid	0-68-0	48A
Phosphoric acid	0-54-0	38A
Potassium chloride (Muriate of potash)	0-0-60	neutral
Potassium nitrate	13-0-44	26B
Potassium hydroxide	0-0-84	92B
Potassium phosphate, mono	0-51-33	neutral
Potassium phosphate, di	0-40-44	20B
Potassium phosphate, tri	0-32-61	48B
Potassium phosphate, tri, poly	0-47-52	25B
Potassium sulfate	0-0-50	neutral
Potassium sulfate-magnesia (Sul-Po-Mag, K-Mag)	0-0-0 (18 Mg)	neutral
Sodium nitrate	16-0-0	29B
Sodium nitrate-potash	15-0-14	27B
Superphosphates	0-20-0	neutral
	0-46-0	neutral
Sulfur (99.6%)		316A
Urea	46-0-0	84A
Ureaform	38-0-0	68A

Source: Pierre, W. H. 1933. Determination of equivalent acidity and basicity of fertilizers. *Ind. and Eng. Chem., Anal. Ed.* 5(4).

The high pH soils in arid or semiarid regions are of three different types. (1) Saline soils are those that have excessive salts but have less than 15 percent exchangeable Na. The pH of these soils is less than 8.5. Water infiltration and exchange of gases make it relatively easy to correct the

TABLE 6.4. Crop Removal per Acre of Basic Cations

| Crop | Harvest | | Basic cations removed (lb) | | |
	Weight (lb)	Yield	K	Ca	Mg
Alfalfa hay	12000		225	168	32
Barley, grain	3840	6 T	20	2	4
straw	4000	80 bu	60	16	6
Coastal Bermuda grass	12000		150	33	22
Coffee	1800		45	58	64
Corn, grain	10500	150 bu	39	15	22
stover	9000		120	27	18
silage	64000		222	65	33
Cotton, lint & seed	4500	3 b	45	6	12
burrs, leaves, stalks	6000		105	84	24
Oats, grain	3200	100 bu	18	3	4
straw	5000		82	10	10
silage	16000		196	65	33
Pangola grass	23500		358	67	46
Peanuts, nuts	3000		30	6	5
vines	5000		66	88	23
Peas, vines and pods	5000		100	175	15
Pineapple		15000 plants	445	102	53
Potatoes, white, tubers	30000	600 bu	188	5	9
Rice, grain	4500	100 bu	10	4	5
straw	6000		68	11	6
Sorghum, grain	5000	100 bu	21	5	6
straw	7500		130	36	23
Soybeans, seed	3000	50 bu	53	9	9
hay	5000		63	56	25
Sugarcane stalks	200000*	100 T	280	110	40
Sugar beets		30 T	208	27	10
Tobacco, leaves	2000		100	75	18
Tomatoes, fruit	40000	20 T	133	7	11
Wheat, grain	3600	60 bu	20	2	9
straw	4500		44	9	5

* Two-year crop.

high pH and eliminate the excess salts. (2) Saline-sodic soils also have excessive salts and a pH less than 8.5 but have an exchangeable Na content greater than 15 percent. Their physical condition, however, is quite normal because of their relatively large content of total salts other than Na, thus making it still possible to reclaim them. (3) Sodic soils have a pH greater than 8.5 with more than 15 percent exchangeable Na and high salt content. Unfortunately, the high proportion of Na in the total salts makes these soils highly dispersed, resulting in very poor water infiltration and gaseous exchange. It is very difficult to mix potential helpful amendments in the soil profile.

As indicated above, some of the poor growth of plants on these high-pH soils can be attributed to lack of P and certain micronutrients, but on the saline, saline-sodic, and sodic soils, high salts also contribute to the problem. The presence of large amounts of exchangeable Na compound the problem in saline-sodic and sodic soils because of its effect on the air and water legs of the triangle.

In overlimed soils or those derived from marine deposits rich in shells, there often is another factor to consider. If the liming materials are calcitic in nature, these soils may have too little Mg for favorable growth. In addition to making these soils more acid by the use of acidifying materials or fertilizers, plants usually respond to large applications of Mg fertilizers. We have gotten good responses with a number of plants by using acid forms of N fertilizers (ammonium sulfate), including acid-forming materials (aluminum and iron sulfates), and increasing the Mg content of fertilizers to about 5 percent.

CORRECTION OF pH

Changing to a crop that tolerates either low or high pH is one way of raising good crops under less than ideal conditions. But often it is not a viable solution. Raising low soil pH values or lowering high ones have been worthwhile agricultural practices for centuries, and will be considered in some detail in Chapter 11.

SUMMARY

The pH of soil or other media affects crop production primarily by its effect on the nutrient leg of the triangle, but aeration and water also can be affected to lesser degrees. The effect of pH on solubility of various minerals accounts for much of its influence on nutrients. The nutrient leg is also affected by the effect of pH on microorganisms, but it is the effect on microorganisms that is largely responsible for the influence on the water and air legs.

The term pH, which is the negative logarithm of the hydrogen ion concentration, represents the relationship between hydrogen and hydroxyl ions on a scale of 0 to 14, with a pH of 0 (extreme acidity) representing nothing but hydrogen ions, a pH of 7 (neutrality) equal amounts of hydrogen and hydroxyl ions, and a pH of 14 (extreme alkalinity) all hydroxyl ions.

Soils become acid as cations are removed by crops or through leaching, with climate being the dominant factor. Acidification is most rapid in soils derived from minerals low in cations and in coarse soils that can easily be leached. The use of most nitrogenous fertilizers and additions of sulfur hasten the process while cation additions slow it.

In mineral soils, a pH range of 6.0 to 7.0 is suitable for many agronomic plants, but a few do better at 5.5 or less and some at values at 7.0 or higher. Many more plants are tolerant of either the low or high values. A list of preferred pH values is given but suggested values are not absolutes as the ideal pH varies somewhat with the soil composition. Slightly lower pH values are satisfactory on soils with high OM or those that are highly weathered.

Much of the poor growth with low pH is associated with an overabundance of Al, Mn, and in some cases Cu or other heavy metals and an insufficiency of P, Ca, and Mg. Lack of Mo also can be a factor, especially for leguminous plants. The poor growth with high pH values often is associated with excess Na and H_2CO_3, but insufficient amounts of P, B, Cu, Fe, and Zn.

The amounts of N are also influenced by pH largely through its influence on microorganisms as they fix N from the air, release it from organic materials, transform it to nitrates which are easily leached, or convert it to gaseous forms that can be lost through evaporation.

Chapter 7

Cation Exchange Capacity and Anion Exchange

INFLUENCE OF CATION EXCHANGE CAPACITY ON THE FERTILE TRIANGLE

The exchange of cations (ions with a positive charge), designated as cation exchange, has a direct bearing on the nutrient side of the triangle. The extent of a soil's capacity to hold cations and exchange them for others is an indication of its fertility. The ability to hold several essential nutrients (NH_4^+, Ca^{2+}, Mg^{2+}, and K^+) in readily exchangeable forms makes CEC a ready source of these elements, acting as a reservoir to replenish the soil solution as needed.

Cation exchange data provide an index to the amounts of cations that are held strongly enough to slow losses from leaching or volatilization but yet are readily available. Soils with high CEC have the ability to hold large quantities of cations, which can act as a reservoir of nutrients. Soils with high CEC naturally have the capacity for good nutrition, automatically increasing the potential for a long nutritional side that is maintained for relatively long periods.

The realization of this potential, however, will be decided by the saturation percentage of the CEC with desirable cations. If provided with proper amounts of Ca^{2+}, Mg^{2+}, and K^+, the soil pH will be in a suitable range for most crops, soil structure should be suitable for good root growth and extension, and there should be enough Ca, Mg, and K for the crop's needs.

As noted in Chapter 6, the basic cations (Ca^{2+}, Mg^{2+}, K^+) can be replaced by acidic ions (H^+, Al^{3+}), especially in humid climates. The replacement will make soils more acid.

Soils with adequate CEC can be limed and fertilized with relatively large amounts of bases, restoring exhausted soils to much of their fertility, which will last for relatively long periods. Much larger amounts of cations (NH_4^+, K^+, Ca^{2+}, and Mg^{2+}) can be applied to soils with elevated CEC compared to

177

soils with low CEC, with little danger of leaching or volatilization losses. Unfortunately, NH_4^+ can be converted to the more leachable anion, nitrate (NO_3^-) and be lost unless nitrifying inhibitors are added.

Soils with low CEC also can be made fertile by additions, but it will take much smaller quantities to do the job and, because of the small holding capacity, fertilization and liming have to be repeated at frequent intervals. Small amounts of cations need to be added to low-CEC soils to minimize leaching losses, making it necessary to use one or more fertilizer side-dressings on these soils if leaching is a problem.

The beneficial effect of high CEC in retaining cations is offset to some extent by the need of adding much larger amounts to obtain sufficient availability if levels are low. For example, it took two to four times as much K to provide a satisfactory level of K in the soil in a clay loam with high CEC than for a sandy loam with low CEC (see Figure 3.2). The CEC also affects the efficiency of several herbicides, requiring larger amounts on soils with elevated CEC to obtain satisfactory weed kill.

In fertile soils, the largest amounts of cations held are Ca^{2+}, Mg^{2+}, and K^+ in decreasing order. The NH_4^+ ion, which can be held in relatively large amounts after fertilization with ammonium fertilizers, is usually present for short periods. It may be taken up by plant roots, fixed within the exchange complex in nonexchangeable forms, or quickly converted in most soils with good aeration to nitrate-nitrogen (NO_3^-). The NO_3^- is weakly held by anion exchange and is either removed by plants or readily lost by leaching.

Problem soils tend to have large amounts of H^+ and Al^{3+} (acid soils) or Na^+ (sodic, solonetz soils) in the exchange complex. In some acid soils, Mn^{2+} also may be present in toxic amounts (See Appendix Table 1 for data on important soil ions).

The presence of large amounts of the basic cations (Ca^{2+}, Mg^{2+}, K^+, and NH_4^+) in the exchange complex tend to denote highly fertile soils, but the presence of large amounts of H^+ and/or AL^{3+} with lesser values of the basic cations is indicative of acid infertile soils.

The CEC also affects the air and water sides of the triangle, but more indirectly than the nutrient side. The kind of cations in the CEC and their relative percentages in the complex have a bearing on soil structure, thereby affecting air and water being held by the soil or moved through it. Soils largely saturated with Ca^{2+} or containing large percentages of Al^{3+} will tend to have good structures, while those with large percentages of Na^+ will be highly dispersed with poor water infiltration and gas exchange.

FUNDAMENTALS OF CEC

Cation exchange capacity refers to the capacity of a soil or medium to hold in a readily available form several cations that can be readily exchanged for others. The property arises from the attraction of electrical charges on colloid surfaces (primarily clay and organic matter) for mobile ions of opposite charge. The primary charge on soil colloids is negative, attracting positively charged ions (cations) of the soil solution, principally calcium (Ca^{2+}), magnesium (Mg^{2+}), potassium (K^+), aluminum (Al^{3+}), sodium (Na^+), and hydrogen (H^+). The ammonium (NH_4^+) ion may be held for short periods, or small amounts of copper (Cu^{2+}), ferrous iron (Fe^{2+}), ferric iron (Fe^{3+}), manganese (Mn^{2+}), and zinc (Zn^{2+}) may also be held. (Negatively charged anions are also attracted to positively charged colloidal materials but the amounts held in most soils are small and of little importance.)

Cation exchange capacity is expressed in terms of milliequivalents (meq) per 100 g of dry soil. A meq = one one-thousandth of an equivalent or combining weight (atomic weight or formula weight divided by its valence). Cation exchange denotes the exchange of equivalent amounts of one positively charged ion for another.

Not all cations are held by equal forces but follow the lyotropic series of $Al^{3+} > Ca^{2+} > Mg^{2+} > K^+ = NH_4^+ > Na^+$, which reflects both the cation charge and hydrated size. If present in equal amounts, Na^+ is exchanged most rapidly, with the Al^{3+} ion being the last to be exchanged. The cations are held until they are exchanged into the soil solution by other cations, which in turn can be exchanged for cations on the root surfaces. But the preferential exchange or absorption by plants from the soil solution does not follow the lyotropic series completely as the absorption is modified by mass action, which favors the cation that is more concentrated.

The interchange between soil solution and exchange sites of the soil or of the root takes place on an equivalent basis, i.e., one equivalent weight H^+ for an equivalent weight of K^+, Ca^{2+}, Al^{3+}, etc. (1.008 milligrams [equivalent weight of H] will exchange 39 mg of K or 20 mg of Ca or 8.994 mg of Al. Because of the single charge, the equivalent weights of H and K are the same as their atomic weights, but the atomic weight of Ca has to be divided by 2 [its charge] and that of Al by 3 [its charge] in order to obtain their equivalent weights.) The milliequivalent weights of the more common cations held by the soil and their corresponding weights in lb/ac are given in Table 7.1.

The ions attracted to and held on the exchange complex of virgin soils are largely a result of the parent material from which the soil was formed and the prevailing climate. With many parent materials and under dry

TABLE 7.1. Milliequivalent Weights of Some Common Cations and Their Corresponding Weights in Pounds per Acre*

Cation	Milliequivalent weight (mg)	Corresponding weight (lb/ac)
Aluminum (Al^{3+})	9	182
Ammonium (NH_4^+)	18	360
Calcium (Ca^{2+})	20	400
Hydrogen (H^+)	1	20
Magnesium (Mg^{2+})	12	240
Potassium (K^+)	39	780
Sodium (Na^+)	23	460

* For mineral soils, assuming 2,000,000 lb per plowed acre.

conditions, Na^+, Ca^{2+}, Mg^{2+} will predominate and there will be little or no H^+ and Al^{3+}. The Ca^{2+} and Na^+ are readily displaced even in partially humid climates. Under normal humid conditions, there is the tendency of H^+ ions in rainwater or formed from decomposition of OM or respiration of plant roots and Al^{3+} formed from hydration of minerals to replace more Ca^{2+}, much of the Mg^{2+}, and K^+ from the exchange complex. The replacement of Ca^{2+}, Mg^{2+} and K^+ with H^+ and Al^{3+} leaves these soils acid and infertile. If parent materials are rich in Ca, Mg, and K^+, sufficient amounts of these cations replace those that are lost, maintaining suitable fertility.

Cultivation of soils and the removal of crops tends to remove large quantities of cations (see Table 6.4), which also can leave soils too infertile to support adequate crop growth. Sufficient liming and fertilization are needed to replace those cations lost by leaching and crop removal if normal crop production is to prevail. In many cases, it is necessary to more than replace the lost cations but actually to increase the amounts held on the exchange complex if satisfactory crops are to be grown.

FACTORS AFFECTING CEC

Textural Class and CEC

The CEC varies with textural class due to the dependence of CEC upon the amounts of colloidal materials (clay and humus). Soils with more clay and/or humus will have higher CEC, as shown in Table 7.2.

TABLE 7.2. Typical CEC for Several Soil Textural Classes

Textural class	CEC (meq/100 g)
Sand	3-5
Loamy sand	5-8
Sandy loam	8-12
Silt	10-15
Loam	13-18
Silt loam	15-20
Sandy clay loam	14-29
Clay loam	16-28
Silty clay loam	18-30
Sandy clay	15-30
Silty clay	22-32
Clay	30-40
Organic soils	55-200

Source: All data, except that for organic soils, is derived from Information Capsule #137, Midwest Agricultural Laboratories, Inc., Omaha, NE. The organic soil data is supplied by the author.

Effect of Clay Type

The CEC is not only a reflection of the amounts of clay and OM making up the exchange complex but also the type of clay, which in turn has been affected by soil mineral composition and weathering. Hydrous oxides, which result from extreme weathering, have a CEC of 2 to 6 meq/100 g of soil; kaolinite clay 3 to 15 meq; illite 10 to 40; montmorollinite 80 to 150; and vermiculite 100 to 150 meq/100 g. Highly weathered soils with predominantly kaolinitic type clays can be expected to have much lower exchange capacities than soils with about the same clay content but with montmorollinite or vermiculite clays.

Generally, the minerals can be considered as consisting of sheets of silica tetrahedra joined with sheets of hydrated aluminum oxide octahedra. In some cases, iron and magnesium substitute for aluminum and also other cations may be in the crystalline lattice structure. There are two major mineral structures with markedly different CEC:

1. If oxygen atoms hold together a sheet of aluminum oxide octahedra with a sheet of silica tetrahedra and the structure is repeated with the

pair of sheets held together by hydrogen bonds, the mineral is considered to have a 1:1 structure. Minerals with a 1:1 structure, as is common with kaolinite, have very low CEC.

2. In the other group of minerals, there are two sheets of silica tetrahedras with one sheet of aluminum oxide between them. Instead of being held together by hydrogen bonds as with the 1:1 minerals, the sheets in 2:1 minerals are held together by other cations.

Clay type is influenced by the degree of weathering or its exposure to percolating rainwater and elevated temperatures. Percolating rainwater is especially effective because it removes cations—the extent depending on amount of rainfall, temperatures, and degree of percolation. The cation first leached away will be Ca^{2+}, but it will soon be followed by Na^+, Mg^{2+}, and K^+ in this order. With continued weathering, Si also will be leached away, leaving primarily Fe^{3+} and Al^{3+} or hydrous oxides behind as is found in lateritic soils. Hydrous oxides have little exchange capacity. It is interesting to note that the very low CEC of clay and clay loams reported in the above table were of laterite soils that had undergone extreme weathering, eliminating much of the bases in the minerals and reducing some of them to hydrous oxides.

Effect of Organic Matter

Cation exchange capacity is greatly influenced by soil OM. On a weight basis, the capacity of most organic materials is greater than that of common soil minerals, with common values for OM in the range of 100 to 300 meq/100 g. One percent of well-humified organic matter will supply about 2 meq CEC, whereas the average CEC of silicate clay is about 0.5 meq. In sandy soils or those dominated by the clay minerals kaolinite or illite, most of the CEC may be due to its OM. Unfortunately, the OM of sandy soils tends to be rather low, due to the greater aeration prevailing in these soils, resulting in low CEC values.

Effect of pH

The CEC is not only dependent on the type of clay and its base saturation but also varies with the pH of the clay and somewhat with the pH of the solution used for measuring it. It rises as the pH is increased from 4.5 to 7.0, but then falls again as pH values are increased over 8.0. The change is about 50 percent, but is greater if the soil contains primarily 1:1-type rather than 2:1-type minerals.

PERCENT BASE SATURATION

The ease with which cations are displaced, which affects their availability, depends in part on their percentage saturation in the exchange complex, usually being more readily available as their percentage saturation increases. Base saturation denotes the percentage of the total CEC that is occupied by base cations (K^+, Na^+, Mg^{2+}, Ca^{2+}, or cations capable of reducing acidity) rather than acid cations (H^+, Al^{3+}, i.e., cations contributing to acidity). The percent base saturation affects pH, nutrient availability, and leaching.

Plants tend to remove bases or base cations, replacing them with H^+ and increasing the acidity of the soil—so that soils with low base saturation are also very acid soils. There is a close relationship between percentage base saturation and a soil's acidity. A soil completely saturated with bases will have a pH of at least 7.0, its pH falling below 7.0 as some of the bases are replaced with H^+ or Al^{3+}. Mineral soils in humid regions as they become more acid in the pH range between 5.0 and 6.0 will generally lose about 5 percent base saturation change for each 0.1 drop in pH.

Availability of the cations tends to increase as their percent base saturation increases, providing that certain ratios of the cations are maintained. Ideal percentage base saturations of the major cations appear to be about 70 to 85 percent for Ca, 10 to 15 percent for Mg, and 4 to 7 percent for K. Attempts to fertilize or lime soils to these ideal percentages have not necessarily produced better crops. Apparently, soils that have these ideal percentages produce good crops, but good crops can be grown without ideal percentages providing that sufficient nutrients are added frequently so as to favorably affect the soil solution adjacent to the roots.

Calcium Saturation

The percent Ca saturation has an important bearing on crop production not only because of its effect on soil pH, but also because it has an important influence on soil structure and calcium availability. Yields and quality of muskmelons grown on soils of volcanic origin in Guatemala, Costa Rica, and Honduras have been related to percent Ca saturation of the soil. Because of time schedules, soils containing considerable clay may be worked when wet, leading to serious compaction. Best results with a minimum of problems were obtained with Ca percent saturation above 70 percent, slight problems with values of 60 to 70 percent, and severe problems as the Ca saturation approached 50 percent.

The compaction associated with low Ca saturation and with declining OM contents related to intensive cultivation has led to poor infiltration of irrigation water and restriction of root systems. Although soil structure improved as Ca^{2+} saturation increased from gypsum additions, these additions alone or combined with subsoiling by Paratill did not completely overcome problems associated with working wet soils. Further improvement has been made by additions of fulvic acids and the inclusion of cover crops between melon crops.

Sodium Saturation

Unlike the good physical condition of soils saturated with large amounts of calcium, soils having more than about 10 percent of the CEC saturated with Na^+ will usually have poor physical conditions with very poor permeability. It becomes very difficult to correct such soils because distributing soil amendments is not very satisfactory due to the poor workability of these soils. Also, their poor permeability makes the introduction of water needed for leaching a very slow process.

Such soils are more common in arid regions where Na has not been leached out of the profile, or if water tables are about 3 feet from the surface. The high table allows salts to rise with capillary water and remain in the upper profile due to lack of leaching. In some cases, excess Na is present in the CEC due to the use of irrigation water with high levels of Na^+ but inadequate Ca^{2+} and Mg^{2+} (high SAR—see SAR under water, Chapter 2). In a few cases, such soils may arise from the introduction of large amounts of Na^+ in fertilizer salts, such as sodium nitrate.

Soils with high saturation of Na^+ in the exchange complex are known as solonetz soils, in contrast to the solonchak soils (saline soils). The solonetz soils may exist as two different types:

1. The common type, known as saline-sodic soils, will have 15 percent or more of the CEC saturated with Na^+, but will have high levels of salts (>4 mmhos/cm in the saturation extract). Such soils can be reclaimed but a soil amendment such as gypsum or sulfur must be added before leaching. Failure to do so will convert the saline-sodic soil to a sodic soil, which is much more difficult to reclaim.
2. The second type, known as sodic soils, also has a high level of sodium saturation (>15 percent) but does not contain high levels of salts. The very poor infiltration of these soils and difficulties in mixing amendments into the affected soil make them very difficult to reclaim.

Both soils have high pH values (7.5 to 10.0), poor infiltration, and poor air and water exchange. The low salt content of sodic soils plus high pH values tend to deflocculate clay and organic matter, which leads to a water-unstable structure. The dispersed humic particles give a black appearance to some of them, which at times have been known as black alkali soils. The dispersed clay moves down into the profile, forming impermeable pans that further limit root development and efforts to reclaim them.

Values of less than 15 percent saturation with Na^+, while limiting their classification as solonetz soils, may still provide problems in growing some crops. Although some crops benefit from Na applications, many of them are sensitive to elevated Na concentrations. As little as 5 percent Na^+ saturation at times has given us some problems in germination of crops, while 10+ percentage saturation has seriously interfered with germination of several small-seeded crops. The problem, which may be related to sensitivity to Na, appears to be at least partly due to physical condition of the soil covering the seed. Seeds had more difficulty emerging from soils with the higher saturation of Na, especially as the saturation percentage was greater than 10 percent.

The extent of problems resulting from large percentages of exchangeable Na^+, usually associated with soil dispersion, varies with different soils. The sandier soils can remain stable with much higher percentages of exchangeable Na^+ (even up to about 20 percent) than the clayey soils, which will disperse at much lower levels of 8 to 10 percent.

The greater tolerance for Na by the lighter soils evidently is due to their good permeability and porosity, allowing for greater filling of the pores by dispersed particles and still having sufficient pore space for movement of air and water. The small pores of the clayey soils are quickly filled with the dispersed particles, making them largely unavailable for good transfer of air and water.

The type of clay also affects critical saturation percentages. Montmorillonite clays are dispersed with relatively low levels of Na^+ saturation; kaolinitic soils are much more stable, tolerating 25 to 35 percent saturation; and kaolinite-metal oxide clay mixtures can remain fairly stable up to about 50 percent saturation.

CATION FIXATION

The importance of base saturation on availability of cations is readily apparent from studies of cation fixation. Cation fixation, which decreases CEC, takes place as the cations react with the solid phase of the soil—probably reacting with octrahedral sheet between two tetrahedral Si-O sheets in 2:1 clay minerals.

Fixation of K^+ has perhaps been studied more than that of other cations, but the fixation of NH_4^+ can be as extensive as that of K^+. The extent of fixation appears to be related to the ionic radius. Fixation is of minor importance for the cations with small ionic radii (Ca^{2+}, Mg^{2+}, Na^+ and Li^+), but can be extensive for cations with large radii (NH_4^+, K^+, Ba^{2+}, and Rb^+).

Serious fixation of added K^+ takes place in soils with less than 4 percent potassium saturation—the degree of fixation being inversely proportional to the percentage saturation. The addition of K will yield large amounts of available K after the base saturation is raised above about 4 percent. At values less than 4 percent saturation, a good part of the added K is fixed—with greater amounts made unavailable as the percentages fall below this figure.

The fixed K can become available with time but this may not be of benefit to the crop receiving the K fertilizer, since it may be several seasons before large amounts of the fixed K become available to plants.

Ammonium also can be fixed in certain soils so that it is not replaced by other cations. The amounts fixed are increased with increasing concentrations of NH_4^+, increasing temperatures, and reaction time. As with K^+, fixation of NH_4^+ appears to be a problem with soils rich in 2:1 clay fractions low in exchangeable K^+ or other monovalent cations.

Of the 2:1 clay minerals, greatest fixation of NH_4^+ and K^+ takes place with vermiculite. Some fixation takes place with illite, but varies with degree of weathering and K saturation of the lattice. No appreciable fixation takes place with montmorillonites under moist conditions.

ANION EXCHANGE

The anion exchange capacity of most soils is relatively small compared to their CEC, but several soil minerals and amorphous colloids can adsorb anions. Anions are more strongly held as pH is low. The important ions NO_3^- and Cl^- are weakly held but HPO_4^{2-} and H_2PO^{4-} can be held rather strongly. A probable order of strength of adsorption is as follows: phosphate > arsenate > selenite = molybdate > sulfate $\geq$ fluoride > chloride > nitrate (Parfitt and Smart, 1978).

The weakly held NO_3^- and Cl^- are usually removed rather quickly from the soil by leaching, but phosphate ions held in the exchange complex can be an important reserve for plants. But maintaining an ideal level of about 0.3 to 3 ppm P in the soil solution, even with appreciable P held by anion exchange, is often difficult in some soils. Although the concentration of adsorbed P may be greater than ideal in the soil solution, the replenishment of the soil solution may be inadequate due to the rapid rate of removal by plant roots. Meeting the requirements is more likely if the concentration of adsorbed P is high and pH is not too acid, as the anions are more strongly adsorbed at low pH. Raising the pH above 7.0, however, is counterproductive as phosphate ions may be precipitated.

SUMMARY

Cation exchange capacity defines the ability of a soil to hold available cations that can be readily exchanged for others. The exchange complex, which is derived from the humus and clay, is primarily negatively charged. In fertile soil, it will be filled primarily with calcium (Ca^{2+}, 65 to 75 percent), magnesium (Mg^{2+} 10 to 15 percent), and potassium (K^+, 4 to 7 percent). Under humid conditions, the exchange is usually for hydrogen (H^+) and aluminum (Al^{3+}), which tends to leave the soil acid and infertile. The Ca^{2+}, Mg^{2+}, and K^+ displaced from the exchange complex into the soil solution are replaced with H^+ as they are taken up by plants. These cations can be replaced from soil minerals or soil additives such as liming materials, fertilizers, manures, etc. Replacement from soil minerals generally is inadequate or too slow, requiring additions from soil amendments if the soil is to remain fertile.

Soils with large CEC capacities (clays, clay loams, silt loams, organic soils) will have large reserves of essential cations, unless they have been weathered and the cations displaced with H^+ and/or Al^{3+}. But for availability to remain in adequate ranges, relative base saturation needs to be retained in approximate percentage ranges as outlined above.

The light soils with their low CEC values tend to have small reserves of cations, which are quickly displaced and need to be replaced often if their fertility is to be maintained.

In dry climates, sodium (Na^+) may be present in the exchange complex in large percentages. Large quantities of Na^+ may also be present in various climates due to its introduction from irrigation waters or certain fertilizers (sodium nitrate, sodium-potassium nitrate). Some problems can arise with as little as 5 percent Na^+ saturation, but various serious physical problems are usually present as the saturation of the exchange complex exceeds 15 percent.

Soils largely saturated with Ca^{2+} (about 70 percent) or Mg^{2+} (15 percent) tend to have better structures, allowing for much more effective use of air and water.

Anion exchange is much less than CEC in most soils. The NO_3^- and Cl^- ions are very weakly held and are soon removed from the exchange complex by leaching. The sulfate ion is more strongly held and that of phosphate very strongly held on some soils. Adsorbed phosphates can be an important part of P nutrition, particularly if held in large amounts and if soils are slightly acid.

Chapter 8

Conductivity or Salts

INFLUENCE OF SALTS ON THE FERTILE TRIANGLE

For agriculturists, the term "salts" refers to the total cations and anions rather than sodium chloride (table salt) present in soil or water. Salts are important because many plant activities are regulated by the amount and kind of salts in soils or other media.

The quantity and composition of total soluble salts affect all three sides of the fertile triangle. Since salts comprise all cations and anions, low levels indicate serious nutrient shortages of N and/or K (the predominant fertilizer ions in most soils). Usually the low levels of salts are associated with poor water infiltration, affecting the water side of the fertile triangle. High levels of salts also can affect the nutrient and water sides of the triangle, since they interfere with both water and nutrient uptake. High salts can also affect the aeration side of the triangle if they contain large quantities of Na^+ and/or HCO_3^-.

Low Salts

Low salts are often caused by inadequate fertilization. They occur most often on sandy soils in humid regions. The low natural fertility of these soils is difficult to raise to adequate levels for any length of time because heavy rains can leach out N, K, and Mg. The result is a low-salt condition, since these elements contribute much of the salts usually present in these fertilized soils. The low salts, as such, may not be harmful, but the shortages of N, K, and possibly Mg, usually indicated by the low salts, adversely affect plant nutrition.

Extensive leaching from irrigation also contribute to low salt conditions, which can exist fairly frequently in irrigated sandy soils or pot cultures.

Rarely, low salt conditions exist because of fixation of N and/or K. The former can readily occur when large amounts of organic materials with wide C/N ratios are incorporated into soils or potting media. The condition tends to correct itself with time or if extra N is added with the organic materials. Additional nitrogen in the form of NH_4^+ can be fixed within the exchange complex. Fixation of K^+, similar to the fixation of NH_4^+, also may be considerable on certain soils. Availability of the fixed NH_4^+ and K^+ is very slow. Although large amounts of NH_4 and K may be present, much of it is so tightly held that there is little to enter the soil solution and increase the salt content.

Besides being indicative of low levels of several important nutrient elements, low salt levels can denote serious difficulties with water infiltration. Irrigation waters low in salts tend to infiltrate poorly into soils because such waters maximize swelling and dispersion of both soil minerals and OM and increase the ability to dissolve and leach out calcium. The swelling and dispersion as well as the lowered calcium decrease soil porosity with a net result of decreasing soil permeability and reduction of both water and air movement.

Excess Salts

The harmful effect of high salts is primarily due to the large osmotic effects of the salts in reducing the movement of water into the plant and the effect of salts on plant metabolism. Sometimes the osmotic effects are strong enough to reverse the process of water uptake to the point of causing some water loss from the cells.

Soluble salts increase the osmotic pressure of the water and the effect is similar to raising the tension of water held by the soil. For example, plant roots removing pure water from a soil at a suction of 3 atmospheres will have to exert a force of 7 atmospheres to take up the water if the osmotic concentration of it has been increased to 4 atmospheres. In some experiments conducted a number of years ago by C. H. Wadleigh and co-workers, the addition of 0, 0.1, 0.15, 0.25, and >0.25 percent sodium chloride to a soil watered at increasingly dry conditions increased the free energy needed by alfalfa, beans, and corn to extract water from the soil, from a low of 2 atmospheres for the soil without added salt to 8 to 9, 10.5 to 11.5, 12 to 13, and 16 to 17 atmospheres for the treated soils (cited by Russell, 1966).

Some adaptation to high salt level is made with time by osmotic adjustment, whereby the higher salt content increases the rate by which the salts are taken up. The higher salt intake lowers the water potential in the roots and improves water uptake, with a resulting increase in plant turgidity.

This adjustment and the movement of excess salts to vacuoles where their effects are minimized take considerable energy. The reduced energy appears to affect several energy-requiring processes of N and CO_2 assimilation and protein and cytokinin synthesis. Plants grown under conditions of high salts show poor CO_2 assimilation so that carbohydrate formation is impaired. Respiration rates are usually higher and carbohydrate storage is depleted. The combination of reduced carbohydrate formation and its accelerated loss by an induced higher respiration rate tend to produce plants with reduced growth and small, dull bluish-green leaves. Wilting is possible but not very common.

High salts also indirectly affect the aeration legs of the triangle. The presence of high salts makes it difficult for plants to take up water, and much higher levels of moisture have to be maintained for plants to get sufficient quantities. The maintenance of sufficient water may make it difficult to keep good levels of aeration at all times. Excess salts also may be indicative of high levels of Na^+ and HCO_3^-, both of which can have harmful effects on soil structure and so can affect the aeration and water sides of the triangle.

In addition to the harmful osmotic effects and the adverse effects on soil structure, porosity, and water infiltration, high salts may contain excesses of certain elements that can seriously limit growth by injuring plants. High levels of salts may have one or more ions at concentrations harmful to plant growth. Common ions are Na^+, Cl^-, HCO_3^-, SO_4^{2-}, HBO_3^-, with Na^+ and HBO_3^- being most troublesome. The Na^+ ion, when it exceeds about 10 percent of the exchange capacity, tends to disperse soil, making it much more difficult for water to move in and through it. High levels of both Na^+ and Cl^- in the salts can lead to undue accumulations of these elements in certain plants (stone fruits, grape, rose, strawberry) in quantities sufficient to cause toxicity. The boron ion at very low concentrations (1 ppm) can induce serious physiological disorders in some plants.

Uptake of Nutrients

There is considerable evidence that high salts have an adverse effect on the uptake of several nutrients, depending on the concentration or makeup of the salts. Calcium is commonly affected and often reduced to deficiency levels for crops such as tomato (blossom end rot). The poor uptake of Ca^{2+} appears to be related to the relatively high content of Na^+ normally present with many high salt conditions. High concentrations of SO_4^{2-} also reduces Ca^{2+} and K^+ uptake while increasing that of Na^+. High salt levels commonly lessen the uptake of several micronutrients, especially Fe^{2+}. Deficiencies of Fe are relatively common, especially if the salts are relatively high in HCO_3^-.

SOURCES OF SALTS

Weathering Process and Lack of Rain

A good portion of the soluble salts in many soils is due to weathering of soil minerals in a climate with limited rainfall that fails to leach out of the profile most of the salts formed in the weathering process. If sufficient Ca remains in the soil, structure remains open and allows a number of plants to grow despite the relatively high salt content. The presence of large amounts of Na and its transformation to sodium carbonate (Na_2CO_3) as it combines with CO_2 released by roots and microorganisms provides for a highly dispersed soil. The dispersed soil restricts root development and complicates the normal movement of water. (See Chapter 5 for additional information about these high-pH soils.)

Other Sources of Salts

In humid regions, there may be little left of the salts released from weathering, but crop production often is restricted because of excess salts. The excess of soluble salts in many of these soils owes its presence to ions released from soil OM, or added by manures, liming materials, fertilizers, waters, certain spray residues, and organic materials. Anything that may release cations or anions is a potential contributor to the total soluble salts. Of the various materials added to soils, irrigation water and fertilizers are the primary sources of high salts.

Some of the salts introduced by various materials are absorbed by plants and removed from the soil. In fact, plant growth is not possible unless the plant takes up sufficient salts to meet its nutrient requirements. However, a great many ions released from the soil or introduced by additions of manures, fertilizer, and waters are left behind and can accumulate. If salts are removed or leached from the soil, they are of no immediate concern to the growing plant but they can accumulate to toxic concentrations if: (1) there is insufficient water for leaching; (2) the water has excess salts; and/or (3) the soil has poor permeability, poor drainage, or a high water table. Salts can be concentrated further by capillary action when they are moved to upper layers as the water evaporates. This accumulation, in addition to what may be currently introduced by water, fertilizer, or other soil amendments can exceed tolerable limits, reducing the ability of the roots to function in the normal uptake of nutrients and water.

Fertilizers and Other Soil Amendments

Salts introduced by fertilizers and other amendments, while less of a problem in humid regions, can cause problems, at least temporarily, because of excess applied or improperly placed. Larger amounts of fertilizers can be safely used in both humid and arid regions if fertilizers supply minimal amounts of salts per added nutrient elements. (See Table 8.1 for a list of common fertilizers and their contributions to salts.)

TABLE 8.1. Relative Salt Indices of Some Agricultural Chemicals Based on Sodium Nitrate As 100

Materials	Analyses	On major nutrient basis	On weight basis
Ammonia, anhydrous	82-0-0	9.4	47.1
Ammonium nitrate	33.5-0-0	49.3	104.0
Ammonium nitrate, limestone	20.5-0-0	49.2	61.1
Ammonium phosphate, mono	12-62-0	6.6	29.9
Ammonium phosphate, mono	11-48-0	7.3	26.9
Ammonium phosphate, di	21-53-0	6.7	34.2
Ammonium phosphate, di	18-45-0	7.6	29.9
Ammonium polyphosphate soln.	10-34-0	—	20.0
Ammonium polyphosphate soln.	11-37-0	—	22.0
Ammonium sulfate	21-0-0	53.0	69.0
Calcium nitrate	12-0-0	72.8	52.5
Calcium phosphate, mono	0-55-0	10.0	15.4
Gypsum	23% Ca, 18% S	3.2	8.1
Limestone, high calcium	40% Ca	1.9	4.7
Limestone, dolomitic	20% Ca, 10% Mg	0.4	0.8
Nitrogen solution	28-0-0	—	64.0
Nitrogen solution	30-0-0	—	69.0
Nitrogen solution	32-0-0	—	73.0
Nitrogen solution	37-0-0	33.5	77.8
Nitrogen solution	40-0-0	28.2	70.4
Nitrate of soda-potash	15-0-14	51.2	92.2
Potassium chloride (muriate)	0-0-50	35.0	109.4
Potassium chloride (muriate)	0-0-60	31.0	110.3
Potassium nitrate	13-0-46	20.1	73.6
Potassium sulfate	0-0-54	14.1	46.1
Sodium nitrate	16.5-0-0	100.0	100.0
Sulfate of potash-magnesia	0-0-21	32.5	43.2
Superphosphate, ordinary	0-20-0	6.4	7.8
Superphosphate, treble	0-45-0	3.5	10.1
Urea	46-0-0	26.7	75.4

Source: Raeder, L. F. Jr., L. M. White, and C. W. Whittaker. 1943. The salt index—a measure of the effect of fertilizers on the concentration of the soil solution. *Soil Sci.* 55:210-218. Copyright © Williams and Wilkins. Reprinted by permission of Waverly, Baltimore, MD.

Slow-Release Fertilizers

In addition to using low-salt soluble fertilizers the grower can limit damage from high salts due to fertilizers by using slow-release sources. Organic sources of nutrients and coated or encapsulated fertilizers allow for the incorporation of much larger amounts of nutrients without problems of excess salts. The release of available N and/or K is slowed so that at any one time, the salts resulting from those which dissolve are relatively low.

Various organic, slowly soluble, or coated materials can serve the purpose of supplying fertilizer nutrients with relatively low salts. Several listed in Table 5.8 (castor pomace, cottonseed meal, dried blood, fish meal, and hoof and horn meal) were longtime favorites, but are difficult to obtain today because of high values for these materials in the pet food market. Manures, listed in Table 5.6, can still serve the purpose but they may contribute considerable salts and need to be applied with care. Manures do have the advantage, though, of increasing the MHC of the soil, and higher concentrations of salts can be tolerated if larger amounts of water are present.

Ureaform, also known as UF (sold as Uramite or Nitroform), or isobutylidene-diurea, known as MU and sold as IBDU, can slowly supply available N. The fertilizers MagAmp or K-Mag can supply slowly available Mg and P as well as N. All dissolve at slow rates so that the salt levels are seldom excessive. Growers generally have found their costs rather high and so limit their uses to specialty crops of high value. Results with the ureaforms have not always been satisfactory because of a very slow release in many cases.

Irrigation

Irrigation may be responsible for high salt conditions, particularly in arid regions where there is insufficient rainfall to leach out salts. The amounts of salts introduced by irrigation will vary with the quality and amount of water supplied, which in turn depends upon the crop, climate, and soil. It is not unusual to apply a foot of water in a crop year in many areas. Even relatively high-quality water (1 mmhos/cm) contains about 640 ppm of salts and a foot of this high-quality water will contribute about 1,500 lb of salts per acre.

This amount combined with salts derived from fertilizers, manures, and other amendments will soon create intolerable conditions for crop growth unless there is sufficient rainfall or irrigation to wash out of the soil sufficient quantities of added salts. Damage is more severe on soils having low OM or CEC because of their much lower moisture content.

Excess Salts Due to Evaporation

The concentration of salts in the soil solution tends to rise as evaporation leaves the salts behind. Water loss by the plant, as transpiration, also tends to increase the salt content of the soil solution since the roots tend to take up water from the soil, leaving much of the salts behind.

Salts move with water fronts and are either leached from the soil if sufficient water is added or accumulate in the soil under conditions of limited moisture. That which accumulates in the soil will tend to move with capillary water, rising under dry conditions and accumulating at the high points in the row or bed. If plantings are made at the top of the bed, salt levels may become high enough to severely restrict growth. Satisfactory crops can often still be grown under such conditions by planting on the sides of the beds. The problem may also be resolved by using overhead irrigation to wash salts downward or by applying furrow irrigation in every other furrow to move much of the salts to the unused furrow.

Not to be overlooked is the temporary change in concentration as a soil dries. The toxicity of salts depends on the amount of water in the soil. A safe amount of salt when moisture is plentiful often may become intolerable for many plants as the soil dries. Even with low salt concentrations, it becomes more difficult for the plant to obtain sufficient water as the moisture approaches the wilting point. As was seen above, increasing the salt concentration greatly increases the difficulty of uptake. The point at which irrigation water is needed should consider the salinity of the soil as well as the moisture remaining in the soil.

A safe level of salts originally in the root zone may be increased to dangerous levels by the capillary rise of groundwater from a close water table and concomitant evaporation of some of the water. The situation is apt to be much worse if the concentration coming from ground waters is already high, as is often the case if the groundwater is fed by saline seeps or affected by drainage from high-salt areas.

The potential damage from evaporation is greater on the lighter soils and those with low OM because of the relatively small amounts of water remaining in these soils as the moisture approaches the wilting point. Whereas 15 to 20 percent of the moisture (volume) remains in the heavier soils at the wilting point, only 2 to 5 percent remains in the sandier soils. The extra water remaining in the heavier soils can keep the salts at a safer level.

The damage from salts induced by evaporating water is strikingly demonstrated with germinating seeds in the openings of plastic-covered beds (plastic culture), or on the tops of furrow-irrigated uncovered beds used in conventional culture.

The former occurs when slow-germinating seeds, such as peppers, are grown in plastic-covered beds during periods of little or no rain. The water from seepage irrigation, laden with nutrients from heavily fertilized bands, evaporates primarily from the small openings, about 2 inches in diameter, in which seeds have been planted. The evaporating water leaves the salts behind in the upper soil surface close to the developing plant. The salts can increase until they kill the seedling or more commonly do some damage to the stem so that the plant dies prematurely from disease or is toppled by strong winds.

High Water Table

High water tables are usually harmful as they can restrict the amount of oxygen available for the lower roots, limiting the root system. They tend to keep the soil moist, allowing extra compaction from machinery and foot traffic and in temperate climates tend to keep soils too cool for early planting and growth. Sometimes, a high water table can temporarily be beneficial, providing that it is not high enough to seriously restrict the root system so that the plant can still take advantage of limited water supplies. But in the long run, high water tables also tend to contribute to soil salinization. Evaporation of water allows an accumulation of salts on or near the surface. For rainfall or irrigation to leach out these salts, it is necessary that the water table be deep enough so that the leaching removes the salts to a level from which they can no longer rise to the surface by capillary action.

Saline Seeps

Saline seeps contribute to high salt concentrations in certain soils. The salts usually originate some distance from and often at higher elevations than the affected field. Water from irrigation ponds, melting snow accumulation, or rainfall may move laterally (because of poor internal drainage) or downward carrying salts with it. If the melting snow or rainfall came from areas with little or no plant cover or from fields in which salts have accumulated due to inadequate leaching, the drainage waters will be laden with salts. In case of seeps, the drainage waters coming from above are deflected by less permeable strata and exit in affected fields, which explains the large presence of seeps on sidehills of rolling topography. The large amounts of salts coming from these seeps overloads the capacity of the soil to nourish salt-sensitive or at times even tolerant crops.

SALTS AND SOIL TEXTURAL CLASS

Although it is usually the sands that salt out first due to the small amount of water they hold, the heavier soils are more likely to become more or less permanent salt-problem soils. The fine-textured soils tend to have poor permeability, making it more difficult to leach them properly, as the amount of water necessary for leaching often reduces aeration to dangerous levels. Improving their permeability by good cultivation practices, subsoiling, and the use of gypsum or polymers may help sufficiently to overcome this problem, but often it is necessary to install drains to facilitate leaching of excess salts.

MEASUREMENT OF SALTS

Conductivity

Soluble salts can be determined by measuring the conductivity of water extract of the soil. It is based on the fact that pure water is a poor conductor of electric current, the conductivity of natural waters depending on the presence of ions other than H^+ or OH^-.

Conductivity, a simple test that can be run by a grower, is an indicator of the total dissolved salts in either soil or water. Electrical conductivity, EC, can be expressed as mhos (reciprocal ohm) or as siemens. To express conductivity as whole numbers, mhos per centimeter (mhos/cm) have commonly been reported as millimhos per centimeter (mmhos/cm) or as micromhos per centimeter (μmhos/cm) and the siemens per meter (S/m) as decisiemens per meter (dS/m). Some of the relationships between these different designations are given below:

$$mhos/cm = 1000 \ mmhos/cm$$
$$mmhos/cm = 1000 \ \mu mhos/cm$$
$$mmhos/cm = 0.1 \ S/m$$
$$S/m = 10 \ mmhos/cm \ or \ 10{,}000 \ \mu mhos/cm$$
$$S/m = 10 \ decisiemens \ per \ meter$$
$$ds/m = mmhos/cm$$

Other useful relationships with conductivity are:

$$osmotic \ pressure \ (OP) \ in \ bars = -0.36 \times EC \ in \ mmhos$$
$$parts \ per \ million \ (ppm) \ ions = 640 \times EC \ in \ mmhos$$
$$milligrams \ per \ liter \ (mg/L) \ of \ ions = 640 \times EC \ in \ mmhos$$
$$milliequivalents \ per \ liter \ (meq/L) \ of \ ions = 10 \times EC \ in \ mmhos$$

Salinity may also be expressed as grains per gallon or tons of salt per acre-foot of water. The grains per gal can be converted to mmhos/cm by dividing grains by a factor of 37.45; tons per acre-foot can be converted to mmhos/cm by dividing tons by a factor of 0.87. Also, a ton of salts per acre-foot of water = ppm salts × 0.00136.

Conductivity measurements of a soil water extract or of a saturated paste can be used to determine soluble salts and are helpful in selecting suitable crops and fertilizer, amounts and timing of fertilizer and water applications, and the manner in which planting beds are to be constructed. The test of soil water extract is simple, rapid, and can easily be run in the field. The cost of meters capable of running the test start at about $60, with more advanced models in the $500 range. No grower or consultant should be without a meter and no crop started or fertilizer applied before soil conductivity is tested.

Various ratios of water to soil have been used to obtain the water extract, the most common being 5 water:1 soil or 2 water:1 soil, with recent tendencies to use the narrower ratio. To determine conductivity of a water extract, two parts of water are added to 1 part of well-mixed soil sample (by volume). Deionized or distilled water is preferred but tap water can be used, if its conductivity will later be subtracted from that of the soil test result.

The soil water extract is prepared by mixing soil and proportional volumes of water with a spatula or spoon, and allowed to stand with intermittent mixing for one hour. The extract can be filtered prior to measuring conductivity or conductivity can be measured in the supernatant liquid.

The conductivity of the saturated paste extract is more reliable but, unlike the soil water extract, its determination is confined to the laboratory. The saturated extract is prepared by adding water intermittently with mixing until the soil is a glistening saturated paste, and then filtering the soil under light vacuum. Conductivity is measured as indicated for the water extract.

The temperature setting of the conductivity meter is adjusted to the temperature of either liquid and conductivity determined. Meters need to be calibrated using standard solutions. A 0.1 N solution of potassium chloride (7.46 gm/liter) should have a a conductivity of 1.29 mmhos/cm at 25°C.

In-Soil Measurement

Measurement of salts can be made directly in the soil by porous matrix salinity sensors or four-electrode units. These types of measurements are useful for irrigation monitoring and determining drainage needs. The porous matrix salinity sensor measures salinity of soil water that has been imbibed into a 1 mm thick ceramic disc. The four-electrode unit measures the resistance to the flow of an electric current between two electrodes placed in the soil while current between a second pair of electrodes is passed through the soil. Soil salinity is determined from measurements made on the same type of soil but with different moisture contents.

Interpretation of Conductivity Readings

High conductivity readings reflect the presence of large amounts of dissolved solids and is indicative of potential problems, especially as fertilizers or other chemicals are added to irrigation water or to soils. But the readings designating satisfactory growth ranges depend on the method of determination as well as the sensitivity of the crop. Much higher values are considered satisfactory if the saturated extract or lower ratios of water to soil are used. Typical interpretations for 2 parts water to 1 part soil or 5 parts water to 1 part soil or saturated paste extracts are given in Table 8.2. Higher values than those given can be tolerated by plants tolerant to salts or if the soil contains large quantities of organic matter and moisture.

TABLE 8.2. Interpretation of Soluble Salt Readings[a]

		Water extract		Satu-rated paste extract	
		1 soil: 2 water	1 soil: 5 water		
		(by volume)			
Saline class	Rating	mmhos/cm			Effect on crop
none	too low	<0.15	0.08-0.15		lacks nutrients[b]
none	low	0.15-0.50	0.15-0.25	<2	negligible[c]
weak	OK	0.51-1.50	0.30-0.80	2-4	slight[d]
moderate	high	1.51-2.25	0.80-1.00	4-8	many crops restricted[e]
strong	excess	>2.25	1.0-1.5	>8	use only tolerant crops

[a] Interpretation is based on normal moisture content. High organic matter in soil increases tolerance by increasing amount of moisture held.
[b] OK for seed germination but low for seedlings. Amounts of available N and K probably too low for growth of most plants.
[c] Salts are satisfactory for most plants. Lower end of scale may indicate too little N and K for rapid growth.
[d] Upper levels are too high for sensitive plants and for most plants during germination.
[e] Sensitive and even moderately sensitive plants can be severely restricted.

Source: Based on the author's observations and data presented by Waters, W. E., J. NeSmith, C. M. Geraldson, and S. S. Woltz. 1972. The interpretation of soluble salt tests and soil analysis by different procedures. *Florida Flower Grower,* 9(4):5.

Using Conductivity Measurements to Guide Fertility Programs

Conductivity measurements of some soils are useful in indicating whether there is sufficient N and K for good crop growth. In noncalcareous, low-sodium soils, conductivity readings of an extract made from 2 parts water to 1 part soil can be used for a quick estimate of N and K in soils fertilized with both these elements in the past. Readings of 0.5 to 1.0 mmhos/cm are usually indicative of sufficient amounts for rapid growth; readings of 0.2 to 0.5 mmhos show sufficient for growth maintenance, and conductivities <0.2 mmhos indicate the need for supplements. If previous soil tests have shown the presence of ample K, then only additions of N need be made. In most sandy soils or media low in K, both N and K need to be added. Such indications are preliminary in nature and need to be confirmed by tests for the elements, but we have used them in some fields to get a quick reading on the availability of fertilizer for the crop.

Frequency of Determining Conductivity

Conductivity readings of field soil need to be made at least once a year but ideally before each application of fertilizer, and certainly if conditions warrant troubleshooting. Measurements of artificial soils or hydroponic media need to be made much more frequently. Although hard and fast rules are difficult to formulate, monthly evaluation of artificial soils probably is adequate whereas weekly determinations of hydroponic media would be more suitable.

PLANT TOLERANCE TO SALTS

Plants differ in their tolerance to salts, which may vary with different stages of growth. A number of plants are more sensitive to salinity at germination, but some plants, such as barley and rice, are more tolerant at germination than at the seedling stage. Corn and wheat are more tolerant of salts at the seedling stage than at germination or during later growth. In a few plants, sensitivity to salts increases or decreases as the plant advances from germination to maturity, but this change in sensitivity to salts varies with different varieties. The most sensitive type of plant probably is an unrooted cutting since most intact roots create barriers to the influx of ions.

Cultivars can differ in tolerance to salts and special selections of cultivars tolerant to salt probably will increase with time. The relative tolerance of different plants to salt levels is given in Table 8.3.

Root stocks also influence the tolerance of the plant to salts by affecting the ability of trees and vines to absorb and translocate sodium, chloride, and boron. The maximum amount of Cl concentration in soil saturation extract that can be used with several rootstocks for some fruits and vine crops are given in Table 8.4.

TABLE 8.3. Relative Tolerance of Various Crops to Salinity

Crop	Tolerance	Crop	Tolerance
Field Crops			
Barley, grain	T	Rape	T
Broad bean	MT	Rice	MS
Rice	MS	Rye	MT
Castor bean	MS	Safflower	MT
Corn	MT	Sesbania	MT
Cotton	T	Sorghum	MT
Field bean	S	Sorgo, sugar	MT
Flax	MT	Soybean	MT
Oat	MT	Sugarbeet	T
Peanut	MS	Wheat	MT
Rape	T		
Flowers			
African violet	S	Lisianthus	MT
Alyssum	S	Marigold	S
Azalea	S	Pansy	S
Camelia	S	Petunia	MT
Celosia	S	Pink	S
China aster	MS	Poinsettia	MS
Chrysanthemum	MS	Portulaca	S
Coleus	S	Rose, common	MS
Dianthus	MT	Rose, multiflora	MS
Dusty miller	MT	Salvia	S
Gaillardia	MT	Torenia	S
Gazania	MT	Verbena	MS
Geranium	S	Vinca	MT
Gladiolus	S	Zinnia	MS
Impatiens	S		
Forages and Hays			
Alkali sacatan	T	Canarygrass	MT
Alfalfa	MS	Clover	
Barley	MT	Alsike	S
Bermuda grass	T	Beerseem	MS
Bluegrama	MS	Hubam	MT
Bromegrass		Red	S
Mountain	MT	Sour	MS
Smooth	MS	Strawberry	MS

TABLE 8.3 *(continued)*

Crop	Tolerance	Crop	Tolerance
Forages and Hays (continued)			
Clover (continued)		Ryegrass, per.	MT
Sweet, yellow	MT	Saltgrass	T
Sweet, white	MT	Sorghum	MS
White	MS	Sudan grass	MT
Corn	MS	Timothy	MS
Dallis grass	MT	Trefoil	
Fescue		Big	MS
Meadow	MS	Bird's-foot	MT
Tall	MT	Vetch	MS
Foxtail		Wheat	MT
Millet	MS	Wheatgrass	
Meadow	MS	Crested	MT
Harding grass	MT	Slender	MT
Love grass	MS	Tall	T
Oat	MS	Western	T
Orchard grass	MS	Wild rye	
Reed canarygrass	MS	Altai	T
Rescue grass	T	Beardless	MT
Rhodes grass	T	Canadian	T
Rye	MT	Russian	T
Fruits and Nuts			
Almond	S	Cashew	MS
Apple		Cherry	
Common	S	Barbados	MS
Kei	S	Sour	MS
Rose	MT	Sweet	MS
Sugar	S	Coconut	T
Velvet	S	Date palm	T
Apricot		Fig	MT
Common	S	Filbert	S
Tropical	S	Grape	MS
Avocado	S	Grapefruit	S
Banana	MS	Guava	
Blackberry	S	Cattleya	MS
Black walnut	MS	Common	MS
Boysenberry	S	Pineapple	S
Carambola	S	Imbe	MT

Crop	Tolerance	Crop	Tolerance
Fruits and Nuts *(continued)*			
Jaboticaba	S	Pitomba	MT
Kumquat	S	Plum	
Lemon	S	Common	S
Lime	S	Governor's	MT
Longan	S	Jambolyan	MT
Lychee	S	Pomegranate	T
Macadamia	S	Quince	S
Mango	MT	Sapodilla	MT
Mulberry	T	Sapote	
Olive	MT	Black	MT
Orange	S	Mamey	MT
Papaya	S	White	MT
Peach	S	Strawberry	S
Persimmon, Jap.	S	Tamarind	MT
Ornamental Shrubs and Trees			
Acacia		Birch *(continued)*	
farnesiana	MT	Sweet	MT
gregii	MT	Yellow	MT
White	MT	Bougainvillea	T
Adam's needle	MT	Boxwood	
African milk bush	T	Compact	S
Alder, Splendid	S	Japanese	MT
Allamanda	MT	Cajeput	T
Apple, Siberian crab	MT	Camelia	S
Arborvitae	MT	Carissa	
Ash		arduina	MT
Green	MT	grandiflora	MT
White	MT	Cassia	T
Aspen, quaking	MT	Casuarina	
Bamboo, heavenly	S	equisetifolia	T
Barberry	S	stricta	T
Bauhinia	S	Catalpa	S
Basswood	S	Cedar	
Beech	MT	Bay	T
Birch		Red	MT
Canoe	MT	Century plant	MT
Gray	MT	Cereus, night blooming	MT
Silver	MT	Cherry, black	MT

TABLE 8.3 *(continued)*

Crop	Tolerance	Crop	Tolerance
Ornamental Shrubs and Trees *(continued)*			
Coontie	MT	Hemlock	
Cotoneaster	S	Canadian	S
Cottonwood	MS	Eastern	S
Crown of thorns	T	Hibiscus	MS
Currant		Holly	
Black, European	MT	Beaufort	S
Dalbergia	MT	Chinese	S
Dodonaea	MT	Oregon grape	S
Dogwood	S	Honeysuckle	
Dracena sp.	MT	Cape	T
Echiaum	MT	European fly	T
Elder, European red	S	Zabel's	T
Elm	MT	Hornbeam, European	S
Eucalyptus		Iceplant	
camaldulensis	T	Croeum	T
citrodora	T	Purple	T
cladocalyx	T	Rosea	T
ficifola	T	White	T
globulus compacta	T	Inkberry	T
lehmannii	T	Ivy, Algerian	S
polyanthemos	T	Jacaranda	MT
pulverulenta	MT	Jasmine sp.	T
rudis	T	Japanese pagoda	MT
sideroxylan rosea	T	Juniper	
tereticornis	MT	Chinese	T
Eugenia sp.	T	Connferta	MT
Euphorbia, lactea	T	Pfitzer	T
Euonymus sp.	T	Spreading	MT
Fatsia	S	Stricta	T
Fir		Kalanchoe	MT
Balsam	S	Kopsia	MT
Douglas	MS	Lantana sp.	MT
Prostrate	MS	Laurel, Indian	MT
Frangipani	MT	Larch	
Geiger	MT	Japanese	MT
Hawthorne	T	European	MT
Hedera	MT	Lilac	MS

Crop	Tolerance	Crop	Tolerance
Ornamental Shrubs and Trees *(continued)*			
Locust		Palm *(continued)*	
Black	T	Date	T
Honey	T	Key	T
Lysistrus		Latan	T
Japanese	MT	Malayan dwarf	T
Iu	MT	Natal	T
Magnolia		Saw palmetto	T
Deciduous	T	Silver	T
Grandiflora	S	Washingtonia	T
Mahogany	MT	Pandanus	MT
Mahonia		Philodendron sp.	MS
Aquifolium	S	Photinia	S
Nevinii	T	Pine	
Marlberry	MT	Australian	T
Maple		Eastern white	MT
Box elder	MT	Japanese black	MT
Hard	MS	Norfolk island	MT
Norway	MT	Red	MT
Red	S	Sand	MT
Sugar	S	Screw	MT
Sycamore	S	White	MT
Melaleuca		Pitch apple	T
Armillaris	T	Pittosporum	
Leucadendra	S	Crassifolium	T
Nesophila	T	Tobira	MT
Milkstripe		Plum	MT
euphorbia	MT	Natal	T
Oak		Pigeon plum	MT
Bur	MT	Podcarpus	S
English	MT	Poplar	
Live	MT	Gray	T
Red	MT	Italian	S
Oleander	T	Privet	
Olive		Glossy	MT
Black	MT	Texas	T
Russian	T	Rose, hybrid	S
Palm		Rosemary	MT
Brittle thatch	MT	Sanseveria sp.	MT
Cabbage	T	Sapodilla	MT
Coconut	T		

TABLE 8.3 *(continued)*

Crop	Tolerance	Crop	Tolerance
Ornamental Shrubs and Trees *(continued)*			
Satin leaf	MT	Tamarix	
Sea grape	T	africanus	MT
Sea hibiscus	T	aphylla	T
Seaside mahoe	T	parviflora	T
Silver buffalo berry	MT	Tecoma sp.	T
Silver button bush	T	Thumbergia	S
Silver thorn	T	Tropical almond	MT
Southern wax myrtle	MT	Viburnum	S
Spanish bayonet	T	Vinca	
Spirea	S	major	T
Spindle tree	T	minor	T
Spruce		Willow	
Blue	MT	Arctic blue	S
White	MT	Golden	MT
Squawbush	T	Xylosma	MT
Sumac	MT	Yucca sp.	T
Tabebua	MT	Zamia sp.	MT
Turf and Ground Covers			
Agave	T	Dichondra	MT
Aloe	MT	Fescue, meadow	MT
Bahia	MS	Fig marigold	T
Bent grass		Hottentot fig	MT
Colonial	S	Ivy	
Seaside	MS	Algerian	T
Bermuda grass		Boston	MT
Common	T	English	MT
Ormond	MT	Jasmine, Confederate	T
Sunturf	T	Kentucky bluegrass	S
Tifgreen	MS	Lily turf	T
Tifway	T	Liriope	MT
U-3	MS	Pampas grass	MT
Callots	T	Portulaca	MT
Carpet grass	S	Purslane	MT
Centipede grass	MS	Rubber vine	T
Coontie	MT	Running strawberry	
Creeping fig	MT	bush	MT
Creeping Euonymus	MT	Ryegrass, perennial	MT

Crop	Tolerance	Crop	Tolerance
Turf and Ground Covers *(continued)*			
Shore juniper	MT	Weeping lantana	T
St. Augustine grass	MT	Wheat grass, fairways	T
Virginia creeper	MT	Zoysia	MT
Wedelia	T		
Vegetables			
Asparagus	T	Pepper	S
Beet, garden	T	Potato	
Bean	S	Irish	S
Broccoli	MS	Sweet	S
Cabbage	S	Radish	S
Cantaloupe	MS	Spinach	MS
Carrot	S	Squash	
Celery	S	Scallop	MS
Corn, sweet	S	Zucchini	MT
Cucumber	MS	Tomato	MT
Lettuce	S	Turnip	S
Onion	S		

S = sensitive, injury <2 dS/m.
MS = moderately sensitive, injury at 2-4 dS/m.
MT = moderately tolerant, injury at 4-6 dS/m.
T = tolerant, injury at dS/m >6.

Sources: Barick, W. E. 1978. Salt tolerant plants for Florida landscapes, *Proc. Fla. State Hort. Soc.* 91: 82-84; Bernstein, L., L. E. Francois, and R. A. Clark. 1972. Salt tolerance of ornamental shrubs and ground covers. *J. Amer. Soc. Hort. Sci.* 97: 550-556; Haehle, R. 1981. *Salt Tolerant Plants,* Broward Co. Agric. Agents Office, Ft. Lauderdale, FL; Rauschkolb, R. and B. Foreman. 1968. USDA Bull. 217; L. Bernstein. 1964. *Salt Tolerance of Plants,* USDA Bull. 283; G. Joyner. 1972. *Salt Tolerant Ornamental Plants for South Florida,* Palm Beach Co. Agric. Agents Office; Lorenz, O. A. and D. N. Maynard. 1988. *Knott's Handbook for Vegetable Growers,* Third Edition. John Wiley and Sons, New York.

SUMMARY

Salts, which consist of all cations and anions, affect all sides of the fertile triangle. Very low salts are usually indicative of poor fertility because of the absence of N and/or K. They also can be associated with poor water infiltration. High salts also can adversely affect nutrition by limiting

TABLE 8.4. Maximum Chloride Concentrations in the Soil Saturation Extract That Can Be Tolerated by Different Rootstocks and Still Avoid Leaf Injury

Crop	Rootstock	Maximum Permissable Cl Conc.	
		mol/m^3	ppm
Citrus	Rangpur lime	25	886
	Mandarin	25	886
	Rough lemon	15	532
	Tangelo	15	532
	Sour orange	15	532
	Sweet orange	10	354
	Citrange	10	354
Stone fruit	Marianna	25	532
	Lowell	10	354
	Shalil	10	354
	Yunan	7	248
Avocado	West Indian	8	284
	Mexican	8	284
Grape	Salt creek	40	1417
	1613-3	40	1417
	Dog ridge	30	1098

Source: Hoffman, G. J. 1981. Alleviating salinity stress. In G. F. Arkin and H. M. Taylor (Eds.), *Modifying the Root Environment to Reduce Crop Stress.* ASAE Monograph No. 4, The American Society of Agricultural Engineers, St. Joseph, MI.

water and nutrient uptake, primarily due to osmotic effects, but often compounded by the effect of certain salts (bicarbonates and sodium) upon soil structure.

In arid regions, much of the salts may be derived from the weathering process, but problems are often aggravated by what is introduced in irrigation water and various soil amendments (fertilizers, manures, composts, liming materials, spray materials) or moved into the soil by seeps. There is considerable variation in the salt content of many soil amendments, making it possible to choose materials low in salts for conditions where excess salts may reduce yields.

Salts derived from the weathering process are usually of little concern in humid regions due to their removal by leaching. Salts derived from irrigation and farm amendments may also be washed out by rains, but often may temporarily cause problems because of the large amounts introduced or if they are poorly placed. Poor soil drainage can also cause serious problems as salts from various sources accumulate.

Plants differ in their tolerance to salts, making it possible to grow satisfactory selected crops despite excess salts in soil or irrigation water. The tolerance may vary with different stages of the crop or rootstock used for certain crops. Germination is usually the most sensitive period for most crops, allowing acceptable crops to be made with borderline salts, providing low-salt conditions are maintained long enough for the crop to germinate.

SECTION III:
THE EFFECTS OF FARM PRACTICES
ON THE FERTILE TRIANGLE

All three sides of the fertile triangle can be affected by various farm practices. Several, such as liming and manuring, usually have favorable effects on all three sides. Others, such as irrigating, can have beneficial effects on water and nutrients but, if overdone or if poor-quality water is used, can have harmful effects on the air side of the triangle.

In this section, we examine the effects of the more common farm practices on the three sides of the triangle and ultimately on crop yields. In Chapter 9, we examine the effects of tillage machinery in some detail and briefly cover the effects of some other machinery. The maintenance of organic matter is covered in Chapter 10; regulation of pH in Chapter 11; regulation of salts in Chapter 12; and the marked physical alteration of soil or media in Chapter 13.

Chapter 9

Effects of Machinery

GENERAL EFFECTS OF FARM EQUIPMENT

Various farm machinery can affect the three sides of the fertile triangle. Short-term use of tillage tools can benefit air, nutrients, and water, but long-term use may be harmful to one or more sides of the triangle. Other machinery also can have harmful effects—usually on air and water—but undue compaction caused by heavy machinery, especially when soil is wet, can also affect the nutrient side of the triangle.

Although some compaction occurs naturally (rainfall impact and sub-soil compaction due to load of topsoil), much of it is a result of modern farming operations. Crusting increases as the soil is deprived of cover or if it contains little OM to stabilize the soil aggregates. Many farm practices involving machinery, such as soil preparation, fertilizer and lime spreading, cultivation, spraying, and harvesting, can seriously compact soils. Vehicular movement, particularly on wet, heavy soils, compresses topsoil, although some of the force may persist to the subsoil. These forces plus slicks produced by plows and disks often produce compact layers or soles that can seriously affect water drainage, root penetration, and the health of the root system.

Some of the damage to soil structure results from tire slippage if pulling equipment is too large for the horsepower or if there is poor weight distribution. The resulting compaction can extend down into the subsoil. The combination of slow speeds and slippage can greatly increase compaction and should be avoided at all times. Some compaction can be reduced by greater speeds, although there are definite limits to this approach.

The weight of machinery at the compression points has a bearing on deterioration of structure by compaction. Here too, the problem of weight is aggravated as clay content and moisture are increased. Use of light equipment instead of heavier units can be a means of reducing compaction. Lighter units can often perform as well as heavier units if: (1) narrow, high-speed units requiring little drawpower pull are used; (2) power take-off is used to propel the implement rather than traction; and (3) traction is maximized by use of crawler tracks or all-wheel drive.

Several structural problems in soils can affect yields by limiting water infiltration, water drainage, the amounts of air and water for plants, and by greatly increasing the energy needed for root penetration. Crusting or capping of soil surfaces or the restricting zones of hardpans, slowly permeable layers in the subsoil, or even poorly permeable clays can seriously restrict crop production. These conditions often can be prevented and, if already present, be corrected—or at least ameliorated—by cultivation or tillage practices. Unfortunately, tillage in the long run can be detrimental to soil structure, and systems have been developed for growing many crops with little or no tillage. The common methods of tillage with their benefits and deficits are presented in the early part of this chapter, and minimum tillage practices in the later part.

TILLAGE EQUIPMENT

Cultivation by tillage equipment can be beneficial or harmful to one or more sides of the fertile triangle. In the long run, cultivation tends to oxidize OM with resulting damage to structure, adversely affecting aeration, infiltration, and available nutrients. In the short run, cultivation can either enhance or deteriorate structure. The net result depends much on (1) the type of soil and its moisture, (2) the type of machinery used, and (3) the weight of machinery and how it is distributed.

Cultivation machinery is used for several purposes including incorporation of organic residues, opening soils to permit easy water infiltration and the exchange of gases, destroying weeds, burying weed seeds and potential disease organisms, and perhaps most importantly for the improvement of tilth. The Soil Science Society of America defines tilth as "the physical condition of soil as related to its ease of tillage, fitness as a seedbed, and its impedance to seedling emergence and root penetration."

Tilth is enhanced as clods (large aggregates formed when wet soil is worked) are broken up by machinery. Large clods in soils with considerable sand can be readily broken. Those in heavier soils may require considerable power. Such clods can be broken down readily by machinery if they are slightly wetted before working them. Slow wetting, such as is accomplished by a drizzle or light irrigation with small droplets, is helpful, while rapid wetting can be detrimental.

Tillage is one of the more important ways by which soil aeration can be improved. Soil preparation to form a seedbed and subsequent cultivation can result in better soil structure, which usually improves aeration—at least in the short run. The extent of improvement, if any, varies with the different types of equipment, and very importantly with the soil's moisture content.

Tillage practices for seedbed formation, which requires close contact of soil particles with seed, may reduce water infiltration and increase erosivity by reducing soil roughness and increasing soil sealing. Tillage implements vary in their effects on surface soil roughness (Eltz and Norton, 1997). Some effects of several modes of tillage on soil roughness are addressed in the section on tillage implements.

The ultimate benefits of tillage appear to be based more on the degree of equipment use—with greater benefits for soil structure, reduction of erosion, better use of water, and considerable cost savings resulting from reduced tillage. Evidently, much of the improvement in soil air resulting from the use of tillage equipment is temporary. Extensive tillage appears to limit the usefulness of crop residues as soil protectors and hastens the decomposition of soil organic matter, an essential key to the preservation of a soil's ability to provide air and water for the crop.

The type of tillage has an important bearing on structure. No-till or reduced tillage systems also have both positive and negative effects on structure. The concentration of organic residues on the surface with reduced tillage systems reduces water loss by evaporation from the soil, and lessens the damage caused by the impact of large water droplets. Reducing the impact of water droplets diminishes the harmful effects of wind and water erosion. Capping is largely eliminated, and plow sole pans also are eliminated or reduced. However, compact layers within the topsoil may be worse than with conventional culture due to the lack of shattering by plowing and because of greater damage by tractor traffic on the usually wetter soils present with reduced tillage. Some of this compaction can be corrected by use of chisels or occasional turning by plow.

Some general effects of different implements are listed in the following sections both for conventional soil tillage and reduced or no-tillage.

CONVENTIONAL TILLAGE

Primary Land Preparation

Moldboard Plow

The moldboard plow opens soils, greatly increasing aeration and water penetration. It is probably the most practical tool for loosening compacted soil, reducing mechanical impedance, and covering crop residues. It is an effective means of providing adequate soil roughness for improving water infiltration and reducing erosivity of cultivated soil, especially if the direction of the plow follows the land contour.

The depth of penetration is usually about 7 inches but plows capable of turning soil to a depth of 36 inches have been used for special situations. At first, water penetration is greatly enhanced, sometimes creating puddled soils if heavy rains follow soon after plowing. If rain does not follow immediately, soil aeration is increased, usually to such an extent that upward movement of soil moisture through the capillaries may be seriously impaired. Capillaries are reestablished naturally by water or by other machinery used for secondary tillage. Organic matter incorporated by the moldboard plow quickly decays, releasing soil nutrients and substances capable of clumping fine soil particles together, thereby improving soil structure temporarily within the plow layer.

Organic residues are incorporated into the soil more than with most other tillage machinery, which can be a mixed blessing. Incorporation of residues reduces later problems of seeding, planting, and cultivation. It can reduce certain plant diseases, insects, and weeds while temporarily increasing soil aeration. The benefits of handling a soil whose OM is incorporated rather than lying on the surface are often outweighed by the lack of mulch, which can decrease water infiltration while increasing surface compaction and erosion.

Another negative effect of the moldboard plow at constant depth is its tendency to produce a plow sole, which in time can impede drainage. Much more serious pans can be formed by various tillage machinery or heavy equipment. These compacted layers are more serious in sandy and silty soils and those containing considerable kaolinite (oxisols and ultisols). There is limited swelling and shrinking in these soils and so less chance of weakening the pan. Much of the porosity gained by use of a moldboard plow or disk can be nullified by the excess use of subsequent tillage implements.

Smearing moist soil on the furrow sole and the bottom of the furrow by plowing has harmful effects on gaseous exchange. The smearing evidently reorients soil particles in a layer that is very thin but of sufficient thickness to interrupt the continuity of the transmission pores. The problem is worse on undrained clay soils where the natural tendency is for lateral movement of water. The smearing further separates the topsoil with better transmission pores from the poorly aerated subsoil.

By far the greatest flaw of the moldboard plow, and probably to a lesser extent in other land preparation machinery, is its tendency to hasten the depletion of OM within the prepared layer. Long-term use of such machinery, by greatly increasing the temperature and aeration deep within the soil, tends to deplete much of its organic content. Aeration and such properties as water infiltration and water-holding capacity are actually decreased in the long run, although they may be greatly improved for short periods.

Disk Plows

The standard disk plow, with disk blades about 26+ inches in diameter and spaced 10+ inches apart, is much less efficient than the moldboard plow in porosity improvement. It also tends to create a compact layer just below tillage depth. Because the disk plow is forced into the surface by its weight rather than by suction as is the moldboard plow, it tends to be heavy, thereby increasing compaction. Use of the disk plow tends to leave a rougher soil surface than a moldboard plow. Although burying of residues is less than with a moldboard plow, there is sufficient inversion so that planting, seeding, or cultivation are not interfered with. Usually, more tillage needs to be used after disk plows than moldboard plows to obtain desired seed beds. The disk plow is used where the moldboard plow may be inefficient (land with rocks, stumps, or tree roots and in extremely hard or sticky soils) or where fast tillage with shallower penetration is satisfactory.

Vertical Disk Plow

The vertical disk plow with smaller disks (20 to 24 inches) is primarily used for preparing shallow soils (3 to 4 inches) for grain growing. In the process, crusts are destroyed, weeds are killed, stubble is mixed with soil, and soil is left with sufficient surface roughness for good water infiltration and resistance to erosion. It is a very fast method of preparing certain soils for crops that do not require deep tillage. A compact layer develops below the depth of penetration. Although it is usually a much cheaper method of initial soil preparation, vertical disks provide considerably less improvement of deep soil aeration than moldboard or standard disk plows.

Disk Harrows

Tandem disk harrows, with narrower disk spacing of 7 to 8 inches, are used for primary soil preparation of seed beds but more often may be used for many secondary operations such as incorporating lime, fertilizers, and herbicides. The common final preparation with disk harrows can create a rather loose seedbed as well as incorporating organic materials into the soil. About 50 percent of surface residues are incorporated with each pass of the disk. Unfortunately, most disk harrows pulverize the soil excessively, reducing soil surface roughness and decreasing their ability to reduce rainfall erosion, but this can be lessened by braking the disks so that they no longer roll free.

Tillage depth of disk harrows is about 6 inches but much of their advantages of aeration improvement may be nullified because of a compact layer (1 to 2 inches deep) just below the tilled depth. Such compaction may be minimal on clay soils or those with good organic matter contents but can be serious in certain sandy soils. The compaction can seriously limit root growth because of its effect on air exchange and movement of capillary water from lower depths. Compaction throughout the disked layer can be greatly increased if soils are wet when disked—a practice often used to dry soils. Repeated disking, especially at high speeds, at such times can be highly counterproductive.

Rotary and Power Tillers

These implements leave a soil open and fluffy but such conditions may not last long enough to benefit crop production. Overly pulverized soil can develop a surface seal that crusts excessively as the soil dries and has little surface roughness, thereby limiting infiltration and increasing chances of rainfall erosion. Unlike the rotary tiller, the power harrow with its horizontally rotating, oscillating, or reciprocating tines does not mix the surface with the lower layers of soil.

Chisel Plows

A series of short chisels, sweeps, or twisted shovels, mounted 15 to 18 inches apart on a draw bar, are used effectively for soil preparation, particularly on sloping lands. These chisel plows leave a rough surface and considerable surface residues that are helpful in reducing erosion and favoring water infiltration.

An advantage of the chisel plow as compared to a moldboard plow is the wider tillage with the same tractor, allowing some time and energy savings. Some disadvantages are potential plugging in heavy stalk residues unless these are shredded and disked prior to plowing, the additional secondary cultivation that is usually required, and the tendency to leave soils wetter and cooler in the spring.

The plugging with heavy residues can also be avoided by combining coulters or narrow (3 to 4 inch) moldboards mounted ahead of the chisel shanks. Both tend to reduce plugging but the moldboards also bury more of the residues. This may not be an advantage, but the added moldboards help to incorporate previously applied fertilizers and liming materials.

Chiseler or Subsoilers

Chiselers or subsoilers are the ultimate tools for breaking up compacted layers and providing maximum soil roughness. Compacted layers of various types, including soles formed by moldboard plows and disks, which tend to impede soil aeration, water infiltration, and MHC can be broken up at least for short periods with chisels or subsoilers. The effectiveness of subsoilers is greatly increased as the length of the chisel is increased, and if they can be used when soils are relatively dry. The design and form of the tine also change the effectiveness of the operation. The effective depth of subsoiling can be increased by using shallow tines ahead of the main chisel to loosen the surface and by attaching wings to the base of the ripper.

Most chiselers or subsoilers can effectively open soils to a depth of 18 to 36 inches. Improvement to greater depths can be obtained with rippers, trenching machines, and slip plows. Rippers do little mixing, but trenching machines, which can open a soil to a depth of about 54 inches, are also capable of mixing the disturbed soil. Slip plows can shatter pans to a depth of about 6 feet and also do a fairly good job of mixing the soil. The mixing done by trenchers and slip plows is especially beneficial if poor production is a result of poor water movement due to the presence of different-textured layers.

Subsoiling, especially at great depths, requires large amounts of power and so may not be cost effective on many soils. Reasons for poor returns are: (1) problems of compaction or root restriction may not be great enough to cause serious crop losses, (2) improvement may not last long enough to recoup costs, and (3) the subsoil opened may still not be conducive to root development due to excess Al and shortages of P and/or Ca. Adding gypsum, limestone, organic matter, or superphosphate behind subsoilers may help lengthen the period of improved structure as well as improving nutrient conditions in the subsoil.

The relative cost of preparing a soil to a depth of about 35 inches is about 1.0 for subsoiling at 39-inch centers, 1.2 for slip plowing every 79 inches, and 3.4 for moldboard plowing.

In Central America, improving soil porosities by subsoiling and the use of gypsum have given us melons that have withstood the long shipment to market (3 to 5 days) much better than melons grown on soils with compaction problems. Melons from the treated soils had higher K and Ca contents, which we feel are responsible for the improved quality.

Often, subsoiling under the row after conventional soil preparation can open up plow or harrow soles sufficiently to allow rapid root development. Forming a ridge or bed over the subsoil area helps mark the row for planting. The soil over the cut needs to be gently firmed with a tool such as a Liliston or by allowing the field to overwinter to allow soil to fall into the voids left by the subsoiler. Such settling will prevent seeds or young seedlings from dropping into voids following heavy rains or irrigation.

Field Cultivators

Usually used with disks for land preparation, field cultivators provide a high-speed operation that is useful for undercutting weeds and leaving them to dry at the surface. There is less inversion of crop residues than with disk-prepared soils so that there is good retention of soil moisture. They have low draft requirements allowing for speedy soil preparation, although, similar to dual-disk preparation, it tends to be shallow.

Secondary Tillage Implements for Land Preparation

Final preparation with light implements that do not slide or scrape soils will provide open soils with fewer ill effects than those of disk harrows. It may be advantageous to replace the disk harrow, at least for some of the operations, by rollers, spring-tooth, chain, Meeker, or spike-tooth harrows. Rollers are used primarily for breaking up clods and packing the surface soil by lowering high spots and filling in low spots. The spring-tooth harrows are satisfactory for light tillage and shallow penetration. Chain harrows are useful for breaking clods but are limited in depth penetration. The Meeker harrow with its straight disks is useful in breaking up clods and for the final preparation of seedbed for small-seeded crops. Spike-tooth harrows can be used to pulverize cloddy soils or to break up a soil crust. One or more of these lightweight tools can be used in tandem with a disk harrow for one-pass soil preparation with much less damage to soil structure and tilth. Generally, these implements tend to firm the upper soil so that water movement by capillaries is greatly favored over aeration. But soil roughness is reduced, which tends to decrease water infiltration and increase rainfall erosivity.

The comparative porosities resulting from the use of several different primary and secondary tillage implements are given in Table 9.1.

Post-Planting Equipment

Often, the well-prepared soil for seeding, with its almost perfect aeration and water movement, deteriorates to form crusts or compacts enough

TABLE 9.1. Total Fractional Porosity of Tilled Layer As Affected by Several Methods of Preparing Several Loam and Clay Soils

Tillage method		Total fractional porosity measured after spring tillage	
Fall	Spring	Range	Mean
None	None	0.52-0.53	0.52
Plow*	—	0.54-0.96	0.72
Plow**	—	0.48-0.68	0.60
Plow	Disk and harrow	0.58-0.66	0.61
None	Plow	0.55-0.84	0.66
Tandem disk	—	0.54-0.69	0.61
Chisel	—	0.68-0.69	0.68
None	Plow, disk, and harrow	0.53-0.82	0.67
None	Plow and wheel tracked	0.52-0.56	0.54
None	Rotary tilled	0.64-0.67	0.66

* Measured after fall tillage, before overwintering.
** Measured in spring before secondary tillage operations.

Source: Voorhees, W. B., R. R. Allmaras, and C. E. Johnson. 1981. Alleviating temperature stresses. In G. T. Arkin and H. M. Taylor (Eds.), *Modifying the Root Environment to Reduce Crop Stress*, p. 251. ASAE Monograph #4. American Society of Agricultural Engineers, St. Joseph, MI.

to seriously inhibit air and water penetration. The causes for such deterioration are many and have been partially covered under "Compaction" in Chapter 4, but it behooves the grower to alleviate these conditions wherever possible.

The Liliston cultivator or lightly set spike-tooth harrows can be useful in breaking soil crusts or caps formed soon after seeding. Such crusts interfere with the emergence of many plants. Timely improvement of these crusts allows for good emergence and greatly improves upper soil aeration and also infiltration of rains or overhead irrigation.

Cultivators of various types are used after the emergence of row crops up until about the time that most crops form a canopy covering most of the soil. In the past, much consideration was given to the surface mulch produced by these implements as benefiting crop production, but later evaluation considers them of little benefit for reducing soil moisture losses. Their greatest benefit accrues from weed control, although some benefits probably are gained from crust reduction, better surface aeration, and ease of water penetration on certain soils.

CONSERVATION OR REDUCED TILLAGE

The long-term effects of most tillage methods, especially when combined with monoculture, on the depletion of OM, with its detrimental effects on aeration, water infiltration, and erosion, has prompted examination of various methods of growing crops with reduced tillage.

No-Till

It is now possible to grow crops with little or no tillage. The old crop can be destroyed by herbicides or mechanically shredded so that it causes little or no impedance in establishing the new crop. Special planters equipped with a variety of coulters have been developed to cut through the surface debris and enable effective establishment of the new crop. Coulter selection is vital to provide proper cutting action and allow the planter opener to move unimpeded through the soil opening and place the seed at the proper depth. The narrow furrow needs to be closed and firmed to provide proper contact with the seed, which will allow even and rapid seed emergence. The residues remaining on the surface reduce the amount of emerging weeds, which can still be further controlled by use of additional weed killers.

The system has great advantages. Costs of working the land are greatly reduced. Residues left at the surface help infiltrate and retain moisture and are helpful in reducing wind and water erosion. Such mulches tend to favor good soil moisture and structure and reduce soil temperatures, especially close to the surface. In the long run, reduced temperatures and the reduced aeration at lower depths reduces the total amounts of OM loss, providing for longer terms of good soil structure. In well-drained soils, no-till allows for greater development of channels left by earthworms as the surface residues help provide better moisture conditions. The increase in channels and the improved water infiltration resulting from worm activity is cumulative since channels are not periodically destroyed by moldboard plow or other soil preparation machinery.

Greatest responses to no-tillage are on low OM soils because of the great improvement in OM close to the surface. The improvement in surface OM has a marked effect on upper soil structure, allowing better water and air management. Responses to no-tillage appear to result more often on light-colored loam soils that tend to crust easily rather than dark-colored silty-clay loams, which do not tend to crust but rather shrink while drying. Such shrinkage is beneficial to the latter soils as it tends to form cracks that allow air and water penetration. No-tillage is also well suited for sloping lands where it can markedly reduce erosion or for rocky fields where cutting coulter and planter double disk openers can override the rocks until normal soil conditions prevail.

No-till is not well suited for poorly drained soils. Infiltration, which is usually greatly improved with no-till on well-drained soils, is usually still poor on soils with clay pans or other impediments to water movement or are too wet.

There are disadvantages to the no-till system. Soils may be slow to warm up in the spring because of large amounts of residues left at the surface. This may delay early planting, which can markedly reduce certain crops such as corn grown in relatively short-season areas as in the northern part of the corn belt. The additional moisture also appears to be a detriment to no-till wheat and barley, especially if grown on clay soils (Rasmussen, Rickman, and Klepper, 1997; Legere et al., 1997). Removing the residues from the center of the row (Kasper, Erbach, and Cruse, 1990), or using zone tillage by in-row loosening with fluted covers (Pierce, Fortin, and Staton, 1992) improves grain production with no-till for some of these areas.

Limestone and slowly soluble fertilizers such as superphosphates applied in no-till tend to remain close to the surface. In time, this can result in acid and infertile soil layers relatively close to the surface (4 to 8 inches), which may limit the development of a more effective root system. Some of the harmful effects of such a situation appear to be offset at least in some soils by more effective roots close to the surface resulting from better moisture and temperature situations.

Additional problems may result from certain pests, which can benefit from the accumulation of residues at the surface and the lack of soil turnover. Residues can provide overwintering sites for insects, slugs, and small invertebrates. Additional moisture present from residues may also increase the problems from some of these pests. Some of the insects that may be more of a problem are armyworm, black cutworm, southern corn billbug, corn root aphid, stalk borer, wireworm, sugarcane beetle, seedcorn maggots, grape colaspis, maize weevils, southwestern cornstalk borer, and the southern cornstalk borer. Noninsect pests that can be more serious in no-till systems are slugs, prairie voles, and mice. Mice can burrow in the slot opened by planters in no-till systems, eating the seeds for some distance in the row. In orchards, mice often can burrow near young trees and effectively destroy enough bark and roots to seriously interfere with tree vigor so that they will develop poorly or winter-kill.

A combination of increased insect carryover and several viruses have increased maize chlorotic dwarf and maize dwarf diseases of corn grown under conservation-tillage systems.

Compaction with time also can limit performance of no-till for certain soils. Much of the compaction results as soil particles settle due to gravity or from the use of various farm machines. Compaction resulting from farm equipment is usually worse in wheel traffic locations. Despite the benefits of soil residues, compaction from natural forces or various farm implements can compact soils enough to warrant periodic soil plowing or other tillage. Although not as useful as a moldboard plow in overcoming soil compaction, much of it can be corrected with chisel plows.

Reduced Tillage

The poor distribution of certain soil amendments, pest problems, and unacceptable compaction have prompted growers to periodically till soils to get better mixing of essential materials, lessen the effects of insects and other pests, and to overcome undesirable compaction. This can be accomplished by plowing the soil after a few years of no-till or by using a reduced tillage method.

The reduced tillage can be confined to strips or take in the full width. Both systems require equipment capable of leaving some residues on the surface.

Full-Width Tillage

Full-width tillage may be deep or relatively shallow. The deep tillage is possible using several different types of equipment: (1) chisel plows, usually used in the fall in drying soils that require a disking or stalk-shredding operation before use to prevent plugging between shanks; (2) combination chisel plows with coulters mounted ahead of the chisel plows; (3) heavy-duty plowing tandem disk harrows and heavy offset disks with much larger blades (about 26-inch diameter) and wider spacing (10+ inches) than conventional disk harrows, which allow preparation of moist soil with less chance of plugging by residues (unfortunately, the plowing disks leave relatively small amounts of residues on the surface); and (4) V-sweep plows with blades 48 to 72 inches apart operating at 3- to 5-inch depths, which retain practically all of the residues at the surface.

Several tools are also used for full-width shallow preparation. Tandem disks are used for such purposes but leave relatively small amounts of residues on the surface. Tandem disks often are combined with field cultivators. Other types of equipment used for shallow soil preparation are seedbed conditioners (which have several rows of cultivators followed by a rotary mixing blade and tine or peg teeth following the rotary blade), roller harrows (a front gang of rollers followed by center-mounted spring teeth), and small grain combination tillage seeders. Usually some shallow tillage is made before using the small grain combination tillage seeders.

As with no-till operations, special planters are needed to provide proper placement and coverage of the seeds with these methods, which can retain considerable residues at the surface. A coulter capable of cutting through clods and residues is needed so that seed can be placed in moist soil. Gauge wheels and firming wheels are needed to form a column of firm soil under the seed and press the seed into the soil for good seed-soil contact.

Strip Tillage

Reduced tillage, limiting tillage to strips, offers several advantages. Uncultivated strips between limited-tillage rows allow for confinement of wheel traffic, limiting compaction in the row. If planted on the contour, the uncultivated strips allow orderly water runoff with limited erosion.

Strip tillage may be left flat or build up a high ridge. In both cases, the new row is established over the old one. If heavy residues are present, a stalk reduction program can aid in soil preparation. Either a rotary tiller or sweep tillage with trash rods is used in flat soil preparation. Fertilizers are applied in late fall or early spring ahead of the planting. The rotary tiller is centered ahead of each row planter and can incorporate herbicides and fertilizers. The sweep tiller with trash rods or a disk row cleaner opens a small area, allowing the planter that follows to move relatively unimpeded. Cultivation using disks to move soil away from the plant and then back again plus a sweep in the row middle is usually enough to control weeds.

Ridge Tillage

Ridge planting has gained ready acceptance, especially in areas where soils may stay wet and cool late in the spring and commonly do not do well with no-till. It has the advantage of lower costs than moldboard plowing but on well-drained soils may provide lower yields of corn and soybeans than well-managed moldboard plow operation. Its best use is on medium- to heavy-textured soil that is level or slightly sloping. In no case should it be used up and down a slope greater than 3 percent.

Necessary equipment consists of a cultivator capable of making ridges and a planter designed to plant on ridges. The cultivator can use sweeps or heavy-duty disks to build a ridge about 8 inches high. A stalk chopper is also highly desirable for growing corn on ridges. Existing corn stalks need to be chopped in before planting in corn residue. If stalks are chopped in the fall, the residues help protect the soil from erosion while good soil structure tends to form beneath it.

Fertilizers are broadcast before ridges are formed, or anhydrous ammonia is knifed in between ridges. A sweep with trash bars is used at planting to remove 2 to 4 inches from the surface, allowing rapid planting. Cultivation using disks is used to move soil away from the plant and then back again. Sweeps added to heavy-duty disk cultivators rebuild the ridges at layby.

Cover crops of ryegrass and vetch have been used in the Northeast. Seeded by air in late summer or early fall, seeds tend to bounce or slide off ridges to germinate in the row. The cover crop is killed by herbicides in the spring. Sweeps with trashbars can be used to clean the surface and rebuild the bed for replanting.

COMPARATIVE EFFECTS OF DIFFERENT TILLAGE SYSTEMS

Although reduced tillage has long-term merits compared to the usual tillage methods, there are disadvantages as well as advantages. These aspects of reduced tillage, along with several conventional tillage methods are presented in Table 9.2.

EFFECTS OF NON-TILLAGE MACHINERY

Many farm practices involving machinery, such as fertilizer and lime spreading, spraying, and harvesting, can have an adverse effect on soil structure. Especially harmful in many areas are different types of machinery to harvest crops. Used when many soils are wet, harvesting equipment can seriously compact soils. While freezing and subsequent thawing during the winter may alleviate some of the ill effects, soil degradation may be so severe that full recovery may not be possible. Problems resulting from harvesting while soils are wet often are more severe in warmer areas where frosts do not occur.

The weight of some tillage equipment may increase compaction problems, especially as soil clay content and moisture are increased. The huge equipment and loaded trucks employed in "mule train" harvesting can seriously affect a soil's capacity for air exchange and water infiltration and drainage.

TABLE 9.2. The Major Advantages and Disadvantages of Different Tillage Systems

System	Advantages	Disadvantages
Moldboard plow	• Suited for most soils. • Excellent incorporation. • Well-tilled seedbeds.	• Poor erosion control. • High soil moisture loss. • High fuel and labor costs.
Chisel plow	• Less erosion than mold-board plow or fall disk. • Good incorporation. • Breaks up pans.	• Additional operation with increased fuel costs. • High moisture loss. • Residues can cause clogging.
Disk	• Less erosion than mold-board plow. • Good incorporation. • Little residue clogging.	• Additional operations. • High moisture and erosion losses. • Compaction in wet soils.
Rotary	• Excellent incorporation. • Excellent seedbed. • Excellent erosion control up to seeding.	• May lead to soil crusting and erosion after seeding. • Possible higher power requirements. • Can diminish soil structure.
Strip or zone tillage	• Good to excellent erosion control. • Good moisture control. • Fewer allelopathy problems than no-till. • Soil warms and dries quicker than no-till.	• Little incorporation. • Dependent on herbicide use. • More management needed.
Ridge till	• Low fuel and labor costs. • Adapted for heavy, wet soils. • Few allelopathy problems. • Excellent erosion control if on contour.	• No incorporation. • Creating and maintaining ridges. • Keeping planter on top of ridge.
No-till	• Maximum erosion control. • Conserves moisture. • Minimum fuel and labor costs.	• No incorporation. • Dependent on herbicides. • Not suited for poorly drained soils. • More management needed. • Possible allelopathy problems.

Source: Adapted from Burchett, L., 1984. Conservation tillage here to stay. *Solutions,* March/April, p. 53.

Making use of long booms on distribution machinery limits the area damaged, although the greater weight of the larger tanks or hoppers to support the long booms increases compaction. Where compaction problems can be expected (heavy soils having little montmorillonite clay, wet soils), the use of large floater tires or crawler tracks that spread the weight over larger areas needs to be considered. Of course, the load weight can be increased to a point where the beneficial effect of distributing the load is nullified. In such a case, the use of permanent roadbeds for machinery movement may be necessary.

SUMMARY

Machinery can have both beneficial and harmful effects upon the sides of the fertile triangle. Beneficial effects allow increased aeration, water infiltration, and improved nutrition resulting from improved conditions for microorganisms and better uptake of nutrients. But use of machinery can have harmful effects, primarily as they affect soil structure, limiting aeration and water infiltration. The harmful effects can result from compaction or long-term oxidation of OM as aeration increases deep in the soil. Soil compaction tends to increase as OM is decreased and as weight of machinery is increased, especially if machinery is used on wet soils.

Tillage equipment is a major source of potential improvement or harm. It can improve the triangle sides of air, water, and nutrients, but such improvement may be relatively short-lived as extra aeration induced by tillage hastens the depletion of OM.

Of the various types of machinery used for soil preparation, the moldboard plow has the greatest potential for temporarily increasing soil aeration while doing the best job of incorporating crop residues. But because of the aeration effects on OM, it may in the long run do the most harm to the soil's ability to move air and water in and out and to provide nutrients from microbiological sources. Other implements used for soil preparation (disk plows, disk harrows, rotary and power tillers) provide less aeration and residue incorporation.

Field cultivators and several implements used for post-planting operations tend to provide minimal aeration. These implements as well as disk harrows can compact soils that are too wet, reducing aeration and water infiltration.

The long-term harmful effects of tillage can be lessened by using no-tillage or reducing the amount. No-till, by keeping much crop residue on the surface, improves water relationships close to the soil surface. By limiting aeration deep in the soil, it reduces OM loss. It is also usually cheaper than conventional tillage. However, no-till can result in smaller crop yields if soils stay too wet and cold in the spring, become more compact from lack of tillage, or suffer from poor distribution of liming materials and fertilizer in the soil profile.

Reduced tillage may provide better yields than no-till by allowing some aeration from tillage equipment, better incorporation of lime and fertilizers, and reducing the problem of overly wet and cold soils in the spring. Ridge tilling, a form of reduced tillage, is especially suited for wet soils but is limited to fairly level land unless the ridges conform to the contour. Strip tillage, another form of reduced tillage, is better suited for slopes greater than 3 percent, but contoured ridge tillage provides excellent erosion control. Occasional tillage with no-till to incorporate fertilizer and liming materials may provide an answer to the problem of mixing fertilizer and lime through the profile while still providing much of the benefit of no-till.

Other equipment also can harm soil structure. Sprayers, harvesting equipment, or manure, fertilizer, and lime spreaders can seriously damage soils.

The damage from all equipment is potentially greater if soils are wet and as load weight increases. Damage is also greater if conditions are suitable for tire slippage.

Chapter 10

Adding Soil Organic Matter

The many benefits of soil OM in improving all three sides of the fertile triangle make it mandatory to increase the amounts added to the soil and maximize effectiveness of the additions in every possible way. Due to the continuous decomposition of organic matter, which is hastened as soils are cultivated, constant renewal is needed to maintain adequate levels.

The benefits of soil OM can be augmented by (1) increasing the amounts of organic materials returned to the soil, (2) placing them so that maximum benefits are obtained, and (3) optimizing the conditions favoring the conversion of organic materials to OM.

INCREASING THE AMOUNTS OF ORGANIC MATERIALS RETURNED TO THE SOIL

There are essentially two ways in which organic matter can be introduced into the system: (1) it can be added to the soil in the form of manures, composts, organic fertilizers, peats, various crop residues, sludges, or as organic wastes; or (2) it can be raised in place in the form of cover crops, sods, pastures, and hays or by utilizing residues of cash crops raised for other purposes. Growing it in place is by far the more economical method, but types of organic matter suitable for maintenance are more limited.

ADDING MATERIALS NOT PRODUCED IN PLACE

Manures

Close to a billion tons of manures are produced annually in the United States, but only a small fraction (probably less than 10 percent) of this amount is applied to cropland. Much of it is produced in feedlots in arid regions far from suitable croplands. That produced by dairy cattle, broilers, and laying hens is often closer to areas where relatively large amounts of manure can be used for a wide variety of crops.

Manures supply OM, reduce erosion, and improve soil structure, which results in an increase in MHC and aeration. Besides improving structure by reducing compaction, they are an excellent source of energy for micro-organisms, which further affect soil structure beneficially. They are also excellent sources of macro- and micronutrients, which were covered more fully in Chapter 5.

Use of manures by the farmers that produce them makes economic sense, but use by other growers may be questionable. It still can be viable if the manures are free and close by (within about five miles). Long hauls eliminate economic advantages of low-cost or even free manures because transporting the large amounts needed to maintain organic matter greatly increase the cost of growing a crop. Although the nutrient elements supplied are an advantage, manures are relatively low analysis, seldom supplying more than 30 lb N, 20 lb P_2O_5, and 20 lb K_2O per ton. When it is realized that the same amount of nutrients can be supplied by 200 lb of a 15-10-10 fertilizer, it is obvious that use of manures by most growers needs to be judged primarily on the physical and microbial effects of the organic matter.

More frequent use of manures is likely as environmental laws prohibiting dumping or runoff into streams, lakes, and rivers are fully enforced. It is no longer possible to dump or flush waste manures into such bodies of water. In addition to better farm use, more manure can be utilized by composting or drying. Both approaches reduce the volume and offensive odors, making the products more appealing to homeowners, plant nursery operators, and landscapers.

Sludges and Effluents

Tremendous amounts of sewage in the form of sludges (about 6 million tons) or effluents (about 2 billion tons) are available annually in the United States for potential agricultural use. Sludges as manures have substantial effects on soil structure, favorably influencing both water and air management. Effluents, because of large water content, have less influence on a weight basis than do the sludges, but could be a source of water for many crops.

The same economic argument made against manures by many growers is even more convincing for sludges. In addition to being low-analysis materials, sludges can contain appreciable concentrations of heavy metals. The Environmental Protection Agency has set concentration limits for ten heavy metals in sludge and sludge products (see Table 10.1) that may be used on soil. Long-term use may cause some problems for plant growth even with permissible amounts.

TABLE 10.1. Upper Limits of Heavy Metals for Sewage Sludge As Set by EPA

Metal sludge[a]	Upper limit in soil[b] (ppm)	Cumulative loading rate in sludge[c] (lb/ac)	Safe limits in soil[d] (ppm)	Annual loading rate (lb/ac)
Arsenic	75	36.6	41	1.79
Cadmium	85	34.8	39	1.78
Chromium	3000	2679.0	1200	133.95
Copper	4300	1339.5	1500	66.97
Lead	840	267.9	300	13.39
Mercury	57	15.1	17	0.75
Molybdenum	75	16.1	18	0.80
Nickel	420	375.0	420	18.75
Selenium	100	89.3	36	4.46
Zinc	7500	2500.4	2800	125.02

[a] Cannot be used on land if any one level is exceeded.
[b] Metal concentration in sludge must be known and when loading rate has been reached for any one metal, use of sludge must be discontinued.
[c] No limit to sludge use if metals are below these "clean limits."
[d] If any metal exceeds its "clean" concentration limit.

Source: Adapted from Wallace, A. and G. A. Wallace. 1994. A possible flaw in EPA's 1993 new sludge rule due to heavy metal interactions. *Comm. Soil. Sci. Plant Anal.* 25(1&2): 129-135. Permission granted by Marcel Dekker, Inc., Monticello, NY.

Sludge use, especially for nonfood production, should be satisfactory if low soil pH values that make the metals more soluble and leachable are avoided. Sludges could be produced with low metal content if industrial wastes and garbage were separated from human wastes. Although expensive, it may be a means of making tremendous quantities of sludge suitable for disposal on the land. This could be a satisfactory means of solving a serious disposal problem while ensuring that deposits of fertilizer minerals would not be exhausted in the foreseeable future. But to make this an economic possibility, much of the cost of producing suitable sludges would have to be underwritten by the communities that must dispose of them.

Effluents can be an important source of water or a base for nutrient solutions for many crops grown close to the source, but there are problems relative to their use. Applications at rates greater than the ability of the soil to absorb them can lead to ponding and/or to runoff to surface waters. Ponding will destroy most vegetation and runoff can eutrophy waters, leading to fish kills and algal growths. Effluents may contain sufficient salts or heavy metals to be toxic to plants. Heavy metals may also pose health problems to humans and animals. Also of grave concern is the potential of introducing pathogenic organisms in effluents. The organisms are reduced by drying and sunlight or can be controlled by treating the wastewater. Laws governing the use of wastewater to avoid transmission of diseases vary in different states, making it obligatory for the grower to consult with state public health departments before using effluents for growing crops.

Organic Mulches

Various organic waste materials such as bark, tree trimmings, sawdust, wood chips, wood fiber, spoilt hay, bagasse, rice hulls, cocoa shells, grain straw, salt hay, and grass clippings are often used as mulches. Generally, such use is justified for landscaping, nursery, and small fruit production, but is usually too expensive for less valuable crops. The practice of bringing in various organic materials to be used as mulches is also justified for stabilizing soils in construction sites, mine spoils, and highway construction.

Composts

The use of compost as a source of organic matter has been restricted largely to gardens, landscaping, and synthetic soils used for potting and interiorscapes. Organic wastes of almost every description have been used for composts, but certain items, such as weed seeds, should be avoided. The composting process, which reduces the volume of the organic material as it converts it to a safe soil amendment, is accomplished by microorganisms. For maximum activity of the organisms, moisture content is maintained at 50 to 70 percent; sufficient air is maintained by limiting the size of the pile to less than 6 feet and turning it every three to four days; and a suitable C/N ratio is maintained by adding 0.25 lb of N per cubic foot of dry material. Materials added to the pile are reduced to no more than 6-inch lengths, and the water and N are added in layers. The temperatures resulting from decomposition are allowed to build up for several weeks. The completion of the process, which takes three to four weeks, is marked by cooling, dark color, lack of odor, and a crumbly texture.

Peats

The peats, some of which are used for lowering pH, are also used as a source of OM for gardens, turf and playing fields, lawns, and ornamental plants. They may be used as mulches or incorporated in the soil at a rate of about 20 6-cu ft bags per 100 sq ft. It is used extensively in artificial soils, especially for acid-loving plants. The addition of peats improves water infiltration and retention, soil structure, increases buffer capacity, increases microbial activity, and reduces the weight of potted plants.

Generally, the high cost of peats limits their use to production of soil mixtures used for growing potted plants, and to landscaping and production of a few specialty horticultural items of high value. The different kinds of peats and their suitability for different uses are given in Table 10.2.

TABLE 10.2. Value of Different Peats for Agricultural Use at Common Rates of Application

Use of peat humus	Rate of soil mix	Sphagnum	Hypnum	Reed	Peat
Soil conditioning	2-in layer*	fair	good	good	good
Top-dressing lawns, golf courses	1/8-1/4-in layer*	fair	good	good	good
Surface mulch	2 in. layer	excellent	fair	fair	poor
Potting soil mix	50% peat 50% vermiculite or soil	excellent	good	good	fair
Golf green soil mix	80% sand 10% clay loam 10% peat	poor	good	good	excellent
Rooting cuttings	50% peat 50% vermiculite	excellent	good	good	poor
Seed flat germination	pure mulled peat	excellent	good	fair	poor
For acid-loving plants	25% mixture in soil	excellent	N.R.	good**	N.R.
For acid-intolerant plants	25% mixture in soil	OK if limed	good	good***	good
N source	Soil mixes Top-dressing	poor	good	fair	good

N.R. = Not recommended
* worked into soil
** good if pH < 4.8
*** good if pH > 4.8

Source: Adapted from Lucas, R. E. 1982. Organic Soils (Histosols). *Michigan State Research Report #435.*

Other Organic Materials

Various waste materials can be used to augment soil OM, but their profitable use depends on the composition of the material and the quantities used. Animal wastes such as dried blood, meat scraps, and fish meal are seldom used for soil purposes because of their higher value as sources of nutrients in pet foods. Even when affordable, these materials are used in relatively small quantities—a few hundred pounds per acre—because of their relatively high nutritional content, and so are of little value for maintenance of OM. Of much more practical value for providing very large amounts of organic matter are the various wastes from trees, such as pine, fir, and cypress bark and sawdust, which have been used extensively for artificial soils in the production of high-valued ornamental crops. Usually, these materials are composted prior to their addition to lessen problems stemming from shortages of N and Fe or excesses of certain harmful products in the wood. These materials will be examined in some greater detail in Chapter 17.

Recently there has been encouragement to use small quantities of humic acids derived from leonardite (10 gal/ac) as a partial substitute for OM. The material is said to be useful in improving soil structure and aiding the uptake of several micronutrients through its chelating effects.

GROWING ORGANIC MATTER IN PLACE

For most growers, adding extra organic matter will only be economically possible if they grow it in place, avoiding costs of hauling and handling.

Ideally, plants that provide considerable amounts of organic matter can be grown to benefit water infiltration and storage as well as promoting sufficient gaseous exchange in the soil.

Sods, Hayfields, and Pastures

One of the best ways of increasing soil organic matter is by growing long-term sod crops such as pastures and hayfields. Many of these plants have heavy root growth that extends deeply into the soil. Grasses can add a ton of dry matter per acre per year from roots alone, and many grassland soils will contain over 5 tons per acre of roots while having only 1 to 2 tons of aboveground portions. Channels opened by these roots tend to move moisture and air deep into the soil. The movement and storage of water deep in the soil accounts for much greater water retention under sod than under tilled crops, and is probably responsible for a good part of the increased yield of cultivated crops following sods compared to that of continuously cultivated crops.

Various plants have been used for such purposes. Alfalfa, coastal Bermuda grass, ryegrass, timothy, and meadow fescue are being used extensively in temperate zones, while grasses such as bahia, guinea, napier, para, and pangola are common in tropical or subtropical areas. The temperate zone grasses are often mixed with white clover.

Rotations

Ideally, the use of hay or pasture fields should be combined with cultivated crops. The advantages in organic matter gained from the long-term sod crops could be utilized by several years of row crops, when again the organic matter could be replenished by the sod crops. Such arrangements were rather common on family farms up until about 50 years ago, but the short-term economics of monoculture eliminated such programs on vast acreage. The shift to monoculture was hastened by the elimination of farm animals as power sources, freeing large amounts of land needed to grow forage for the animals. Today, it is impractical for most growers to use sod crops as a means of optimizing soil OM, but some of them can use these or other beneficial crops in a rotation with cash crops.

Rotations of crops even without sods can be an effective means of maintaining soil OM and increasing yields. Part of the yield increases obtained with rotations is due to suppression of pests as well as from better water and air conditions. The comparative effects on OM in some Ohio soils by two different rotations with continuous wheat or corn are presented in Table 10.3.

TABLE 10.3. The Effect of Rotations on the Organic Matter of Some Ohio Soils

	Organic matter %
Initial content	4.4
30 years continuous corn	1.6
30 years continuous wheat	2.8
Rotation of corn, oats, wheat, clover, timothy	3.4
Rotation of corn, wheat, clover	3.7

Source: Adapted from Salter, R. M. and Green, T. C. 1933. *J. Amer. Soc. Agron.* 25: 622. The American Society of Agronomy, Madison, WI.

The rotation of cash crops without hay, sod, or cover crops may have other benefits, but it may not increase soil OM. Continuous cropping results in depleted OM but the amounts remaining vary with different crops. Most of the benefits from rotating cash crops appear to be reduction of disease and insect outbreaks, but slowing the depletion of OM with cash crops that provide greater residues or require less cultivation appears to be helpful. Rotations that include cover crops and/or sods add more organic matter to the system by reducing the extent of cultivation as well as adding extra OM.

Various types of rotations are being used. A rotation of rye or tall fescue used as winter forage with corn during the summer effectively reduces erosion and allows good yields of corn, providing all the rye or fescue is killed prior to planting the corn. A rotation of three years alfalfa with two years of vegetables has been practical in several areas. A number of two-crop rotations are being used extensively: corn/sorghum in the Midwest; small grains/canola in the prairie provinces; wheat/soybeans in the East and South; wheat/sorghum in the Plains. A three-crop rotation of corn/wheat/sorghum is popular in North Carolina. Rotations in the Corn Belt may include corn/wheat/clover; corn/oats/clover, or two years corn/oats/clover. Two years of pangola grass in a rotation with two years of stake tomatoes has greatly increased tomato production in Florida. Long-time pastures of bahia grass or native grasses have been also been used in Florida for one or two years of tomatoes or watermelons.

The practicality of including pastures, sods, or hay in the rotation depends on the utilization of these crops by animals. This option may not be suitable for most growers, but there are very few growers in the temperate zones who would not profit from the introduction of off-season cover crops in the rotation. Off-season cover crops utilize the land at times when no cash crop is grown. They greatly reduce erosion, help catch any left-over fertilizer, and can add considerable organic matter. Greater amounts of organic matter are produced if the cover crop is sown before the cash crop is harvested, which can be done in the last cultivation or by air. Costs of such plantings are low. The benefits are accentuated because of an early start at a time when there is little competition to the cash crop. Yet the period of growth is extended, allowing the production of additional OM and fixation of N by legumes.

Hairy vetch and rye have wide appeal as winter cover crops for many different areas of the country. Hairy vetch fixes large quantities of N (75-100 lb/ac), it can be interseeded in the cash crop before it is completed, and works very well with ridge planting. Rye, being a nonlegume, does not fix any N, but provides longer-lasting OM. It has allelopathic properties against some weeds, making it desirable for no-till operations.

In some areas of the country, rye and hairy vetch are interplanted, providing high N fixation with longer-life OM. Other choices include crimson clover, ryegrass, bur clover, and Austrian winter peas. Although less adaptable for most growers, there are times when a summer cover crop such as Sudan grass, cowpeas, or Rhodes grass can be utilized in the rotation.

The direction of OM change depends on the type of crop. As indicated in Table 10.3, continuous cash crops can yield different soil OM contents with continuous wheat far superior to continuous corn. Rotations of corn and wheat with other crops were far superior to either crop alone.

Strip Cropping

Although most rotations utilize the entire area for either the cash or organic matter crop, large areas are devoted to strip cropping. The strips may vary in width from a few feet (primarily grown for wind control) to about 200 feet. Minimum practical width for most crops is about 50 feet. A cultivated crop is usually alternated with crops that need little or no cultivation (sods, hayfields, small grains). The small grains allow annual crop changes in the strip, while a rotation with sods and hayfields may require changes every two or three years.

Field strip cropping is laid out in approximately parallel strips usually crosswise of the general slope. Unless applied to regular slopes, they are much less effective in reducing erosion than strips that generally follow the contour of the land. The latter method is often combined with various mechanical methods of erosion control, such as terraces and diversion ditches, to minimize the destructive effects of water. Both methods of strip cropping offer advantages in conservation of water and maintenance of OM when cultivated crops are rotated with sod or forage crops, hay, or small grains, but contour strip cropping is usually far superior.

Plant Residues

Crop residues are the major sources of OM in most crop areas. Roots, stover, leaves, stems, orchard trimmings, and other plant parts contribute to the total. For many crops, the roots will be the major part of the residue but it and other parts are increased as inputs of lime, fertilizers, etc. are maximized for the previous crop. Burning such residues to ease preparation of the next crop or for insect and disease control needs to be discouraged. Burning destroys a major part of OM with a simultaneous loss of the volatile elements N and S, and should be avoided unless absolutely necessary for pest control. Often, shredding the plant material before incorporation, adding N to speed its decomposition, and allowing enough time between incorporation and planting will accomplish the same purpose as burning without the attending losses.

The amounts of crop residues produced can be large. They have increased with better fertilization practices and introduction of newer high-yielding varieties. Some high-yielding corn crops have produced as much as 3.5 tons of stover and roots per acre, and some double cropping systems have yielded almost 6 tons of residues per acre annually. The average amounts of residue produced by several crops are given in Table 10.4.

Just saving the residues fails in many cases to add enough organic matter to avoid the problems caused by intensive agriculture on many soils. Residues can be increased by good fertilization practices, especially as enough N is introduced into the system (see Chapter 5). The effectiveness of residues can also be increased by proper handling of them. The increased effectiveness of residues maintained as mulches is covered in the section, "Maximizing the Mulch Effect," p. 243.

TABLE 10.4. Amounts of Residues Left After Harvesting Good Yields of Several Crops

Crop residue	Residue per acre (lb)
Barley straw	4,000
Corn stover	9,000
Cotton burs, leaves, stalks	6,000
Oat straw	5,000
Peanut vines	5,000
Rice straw	6,000
Sorghum straw	7,500
Soybean vines	5,000
Wheat straw	5,000

Source: Adapted from the table "Amounts of Nutrients Removed by Good Yields of Various Crops," *Liquid Fertilizer Manual.* 1980. National Fertilizer Solutions Association, p. 15-2. Reprinted by permission of Agricultural Retailers Association, St. Louis, MO.

Cover Crops

Green manure or cover crops are OM sources suitable for short growing periods. They are used between cash crops to help reduce erosion, and to save nutrients that may have been leftover from the cash crop. Leguminous cover crops also add appreciable quantities of N for the succeeding crop.

The value of cover crops is influenced by the kind of crop, its age when plowed under or terminated, amounts of nutrients left over from the previous crop, soil type on which it is grown, climate, and in the case of the legumes, the soil pH, and its content of Ca, P, K, Mg, and B.

Their importance in increasing organic matter is also subject to appreciable variation. Generally, these crops are in for such a short duration that there is little or no change in soil OM content. They can have a positive effect by helping to hold existing levels or by reducing the rate of loss. Their effect on soil OM is improved if the crop is allowed to approach maturity before it is terminated, as older crops are slower to decompose. The nonlegumes have an advantage over the legumes in resisting decomposition but the lack of N may need to be compensated by additions of manures or fertilizers.

Catch or cover crops cannot be expected to benefit structure as do the longer pasture, hay, or sod crops, but temporary effects especially in cool climates can be helpful. Evidently, the organic matter provides enough cementing materials to produce some extra aggregates of silt and clay, providing extra air spaces at least for short periods. The benefits of short-duration crops may be partly due to their preservation of existing structure by lessening the harmful effects of heavy rains or strong winds.

Among their other attributes, cover crops serve a useful purpose in mopping up leftover nutrients from the previous crop. Without their use, such nutrients could be lost to leaching or erosion and in many situations increase pollution problems. Fortuitously, in some cases the nutrients scavenged by the cover crop are sufficient to provide for its good growth. But often, particularly on poor soils, nutrients left from the previous crop are insufficient to grow a satisfactory cover crop. The problem is aggravated with nonlegumes since lack of N is usually responsible for poor performance of some cover crops.

A partial solution for this problem is to plant legumes with nonlegumes, but this may not be possible in many cases because of the late planting date of the cover crop. In such cases, cool-weather cover crops, such as rye or ryegrass, can make a much greater contribution to OM if the cover crop is fertilized. On the sandy soils of southern New Jersey, greater response was obtained if potassium was included with nitrogen applications. In several studies, the author found that some of the fertilizer (10-10-10) intended for the succeeding cash crop could profitably be applied to the cover crops (rye or ryegrass). The production of organic materials in most cases was more than doubled with the yield of the cash crop that followed (English peas) slightly increased. Yields may have been increased by the additional organic materials produced, but the conversion of inorganic N in the 10-10-10 to organic forms that could withstand leaching spring rains on these light soils may also have played a role.

PLACEMENT OF ORGANIC MATTER

The manner in which organic matter is added to soil has an important bearing on its effectiveness. Added to the surface where most of it is retained as a mulch provides maximum moisture gain and retention while contributing greatly to reduction in surface compaction and lessened erosion. Mulching with straw allowed deeper penetration and greater storage of water than disking or plowing under the straw (Table 10.5). Mixed with most of the soil, as is usually done with moldboard plowing, provides most

TABLE 10.5. Water Storage As Affected by Placement of Straw and Type of Tillage

Treatment Straw	Tilth	Precipitation* stored		Depth of Water Penetration
		in	%	(in)
+	mulch	9.7	54.3	71
+	disked in	6.9	38.7	59
+	plowed in	6.1	34.1	59
—	disked	3.5	19.6	47
—	plowed	3.7	20.7	47
—	basin listed**	4.9	27.7	59

* A total of 17.9 in. of water fell during the trial.
** Diked or dammed furrows.

Source: Adapted from Unger, P.W., G.W. Langdale, and R. I. Papendick. 1988. Role of crop residues—Improving water conservation and use. In W. L. Hargrove (Ed.), *Cropping Strategies for Efficient Use of Water and Nitrogen,* p. 71. ASA Special Publication No. 51, Amer. Soc. of Agron., Crop Sci. Soc. of Amer., Soil Sci. Soc. of Amer., Madison, WI.

rapid release of nutrients while helping with internal structure. Placement in the bottom of furrows or behind chisel plows offers maximum improvement of structure at lower levels.

Maximizing the Mulch Effect

The benefits of adding organic materials as mulches were outlined in Chapter 5. Plant residues are usually most effectively used when they are left as mulches, but greater benefits are obtained as more of the surface is covered and the coverage lasts for long periods. The percent of surface covered as the crop is harvested is affected by the crop and its yield, but its durability is affected by the crop, the way it was harvested, and the manner in which the soil is prepared.

Extent of Coverage

Protection of the surface is largely dependent on the extent of coverage by materials on the surface, which can be quite different for different crops.

There often are marked differences in benefits of different residues. For example, that of wheat straw in preserving infiltration is greater than that obtained with German millet which in turn is greater than that obtained with sorghum residues. Some of the benefits may relate to differences in weight produced by various crops but often relate to other factors, such as different densities and diameters of the residue and their ability to last over longer periods.

The extent of surface coverage after harvest, which is an important factor in determining the effectiveness of mulch, is affected by yields as well as the type of crop. Amounts of residue increase substantially as yields are increased and if the crop is grown in continuous no-till rather than conventional till. In Illinois, the residue from prior crops tended to build up under continuous no-till systems, adding about 7 percent surface cover for nonfragile crops such as corn and about 3 percent for the fragile soybean crop (see Table 10.6).

Fragile versus Nonfragile Residues

The benefits of mulched organic residues increase as their length of persistence increases. Residues that decompose quickly (fragile) cannot be expected to give the same benefits as those that persist over longer periods (nonfragile). The fragility of the residue is affected by the manner in which the crop was harvested as well as by the nature of the crop. Some fragile and nonfragile crops are listed in Table 10.7.

TABLE 10.6. The Influence of Corn and Soybean Yields Upon Percent Surface Covered by Residue After Harvest

Crop	Yield (bu/ac)	Surface cover	
		Conventional till (%)	Continuous no-till* (%)
Corn	<100	80	87
	100-150	90	97
	>150	95	100
Soybeans	<30	65	68
	30-50	75	78
	>50	85	88

* Residue tends to accumulate in continuous no-till, adding about 7 percent for nonfragile crops such as corn, and about 3 percent for fragile crops such as soybeans.

Source: Farnsworth, R. L., E. Giles, R. W. Frazes, and D. Peterson. 1993. The residue dimension. *Land and Water #9*, September. Coop. Extension Service, University of Illinois at Urbana-Champaign, p. 6.

TABLE 10.7. Classification of Different Residues by Their Persistence (Nonfragile or Fragile)

Nonfragile residues		Fragile residues	
Barley*	Millet	Beans, dry	Mustard
Buckwheat	Oats*	Canola/Rapeseed	Peanuts
Corn	Popcorn	Corn silage	Peas, dry
Cotton	Rice	Cover crop	Potatoes
Flaxseed	Rye*	fall seeded	Safflower
Forage silage	Spelt*	Flower seed	Sorghum silage
Forage seed	Sorghum	Grapes	Soybeans
Hay	Sugarcane	Guar	Sugar beets
alfalfa	Tobacco	Lentils	Sunflowers
grass	Triticale*	Mint	Vegetables
legume	Wheat*		

* Consider fragile if a rotary combine is used for harvesting.

Source: Farnsworth, R. L., E. Giles, R. W. Frazes, and D. Peterson. 1993. The residue dimension. *Land and Water #9,* September. Coop. Extension Service, University of Illinois at Urbana-Champaign, p. 7.

Implement Effects on Mulch

The extent of coverage with different materials is also affected by tillage implements. The effects of different tillage tools on the amounts of surface residues of fragile and nonfragile crops with variable original coverage are presented in Tables 10.8 and 10.9.

TABLE 10.8. Residue Cover Left by Several Tillage Systems Incorporating Different Crops

Tillage system	Previous crop, % residue remaining			
	Corn	Soybeans	Small grain	Sod
Moldboard plow, disk, field cultivate	5	2	5	10
Chisel (4-inch twisted points)				
• disk twice	15	2	10	20
• disk once, field cultivate once	20	5	15	25
Chisel (2-inch straight points)				
• disk twice	20	5	15	30
• disk once, field cultivate once	30	10	25	30
Primary tillage disk (deeper than 6 inches)				
• disk twice (standard tandem)	10	5	15	20
• disk once, field cultivate once	20	10	20	—
Shallow disking (less than 6 inches)				
• once (standard tandem)	40	20	45	50
• twice (standard tandem)	20	10	25	30
Field cultivate once	—	30	—	—
Till plant in ridge	30	20	—	—
In row subsoil, plant	70	50	80	85
No-tillage, plant	80	60	90	95

Source: Griffith, D. H., J. V. Mannering, and J. E. Box. 1986. Soil and moisture management with reduced tillage. In M. R. Sprague and G. L. Triplet (Eds.), *No-Tillage and Surface-Tillage Agriculture.* John Wiley and Sons, New York, p. 24.

TABLE 10.9. The Effect of Tillage Tools on the Percentage Residue Remaining of Fragile (F) and Nonfragile(NF) Crops Having Variable Original Soil Cover

Primary tillage tools	Original cover													
	100%		90%		80%		70%		60%		50%		40%	
	F	NF	F	NF	F	NF	F	NF	F	NF	F	NF	F	NF
							% remaining							
Chisel plow w/12" space														
16" med. crown sweep	55	80	50	72	45	65	38	55	33	50	7	40	22	32
w/2 x 14" chisel point	50	70	45	65	40	56	35	50	30	42	25	35	20	28
w/4 x 14-1/2" shovel	45	60	40	55	35	50	30	42	27	35	22	30	18	25
w/3 x 24" concave twisted shovel	45	60	40	55	35	50	30	42	27	35	22	30	18	25
Mulch tiller w/coulters 15" spacing														
w/18" low crown sweep	50	75	45	70	40	60	35	55	30	50	25	40	20	32
w/2 x 14" chisel point	50	75	40	55	35	50	30	45	27	40	22	32	18	25
w/3 x 24" concave twisted shovel	40	55	35	50	30	45	28	38	25	33	20	27	15	22
w/4 x 24" concave twisted shovel	35	50	30	45	28	40	25	35	20	30	17	25	14	20
Mulch tiller w/disk gangs 15" spacing														
w/18" low crown sweep	45	62	40	55	35	50	30	43	27	37	22	30	18	25
w/2 x 14" chisel points	40	55	35	50	30	45	28	38	25	33	20	27	15	22
w/3 x 24" concave twisted shovel	35	50	30	45	28	40	25	35	20	30	17	25	14	20
w/4 x 24" concave twisted shovel	30	45	27	40	25	35	21	30	18	27	15	22	12	18
V-Ripper														
30" spacing	55	70	50	65	45	60	38	50	33	42	27	35	22	28
20" spacing	50	60	45	55	40	50	35	40	30	35	25	30	20	25
Disks w 11" spacing														
3" deep	35	70	30	65	28	60	25	53	20	45	17	37	14	30
6" deep	20	40	18	35	16	30	14	28	12	25	10	20	8	15
Disks w 9" spacing														
3" deep	30	65	27	60	25	55	21	50	18	42	15	35	12	28
6" deep	18	35	16	30	14	28	12	25	10	20	9	7	7	14

Secondary tillage tools	Original cover													
	75%		60%		50%		45%		40%		35%		25%	
	F	NF	F	NF	F	NF	F	NF	F	NF	F	NF	F	NF
	% remaining													
Field cultivator														
4.5″ spacing w/4″ S-Tine sweep	*	*	*	*	*	*	*	*	22	30	20	28	13	20
6″ spacing**	*	*	*	*	30	42	27	38	24	34	21	30	15	21
9″ spacing***	*	*	40	55	32	45	30	40	25	36	22	31	16	22
Roller harrow														
w/2 1/2″ sweep	*	*	*	*	22	35	20	30	18	28	15	24	11	17
w/3/8 × 7 1/2″ reversible shovel	*	*	*	*	17	30	15	27	14	24	12	21	8	15
Mulch finisher, 8″ spacing/9″ sweeps														
4″ deep	40	45	33	35	27	30	25	27	22	24	20	21	13	15
Disk 4″ deep														
w/7 1/4″ spacing	16	37	15	30	12	25	11	22	10	20	8	17	6	12
w/9″ spacing	22	40	18	33	15	27	13	25	12	22	10	20	7	13

* Residue levels too high for equipment to function properly.
** With 7″ or 9″ sweeps.
*** With 10″ or 12″ sweeps.

Source: John Deere Tillage Tool: Residue Management Guide © 1991. Datalizer Slide Charts, Inc., Addison, IL. Reprinted by permission of Deere and Co., Des Moines, IA.

Deleterious Effects of Organic Mulches

Although largely beneficial, mulches can be detrimental due to the high level of moisture maintained at the surface, reduced warming of soils, and the extension of the harmful effects of certain herbicides, allelopathic substances, and both insect and disease conditions. Some of these adverse effects, which may be worse with large amounts of organic materials no matter where they are placed, have been covered in Chapter 5, in the section, "Harmful Effects of Organic Matter Additions." But placement as a mulch increases some of the harmful effects.

The increased moisture at the surface may physically delay the planting of crops. The increase in moisture by lowering soil temperatures can also delay emergence of the crop. Such delays can substantially lower the yields of some crops such as corn, when grown in the northern United States with its limited growing season.

The effects of reduced temperatures can also lead to plant damage as cold fronts lower temperatures to the freezing point. Mulches are effective insulators, shielding the upper roots from excessively high temperatures. This is a great advantage during the summer months, but can be detrimental during other seasons when heating of the surface soil during the day can warm it sufficiently to reduce cold damage during the night as temperatures drop to dangerous levels. In such cases, removing the mulch early to allow for maximum warming during the day and replacing it at nightfall can save sensitive plants if temperatures should drop too low. Although it is not commercially practical, homeowners may want to use this approach to save valuable plants.

By retaining the added organic matter on the surface, decomposition is slowed, thereby allowing a longer period of some possible problems. The slower breakdown of mulch allows certain herbicides to persist for longer periods, affecting the germination and early growth of sensitive crops. Relatively little work has been done on the subject, but it appears that the problem may be more serious for mulch materials brought in shortly after herbicide is applied, when little time for normal degradation has elapsed. The recent emphasis on proper disposal of yard waste has prompted studies with mulches of grass clippings previously treated with herbicides. The results indicate that clippings previously treated with 2,4-D, Dicamba, and MCPP are harmful to growth and development of tomato, cucumber, salvia, and marigold (Bahe and Peacock, 1995).

Plants contain many different allelopathic substances that selectively inhibit growth of certain soil organisms, limit germination of many different seeds, and restrict growth of many successive plants. Mulching can extend the effective life of some of these allelopathic substances.

The reduction of decomposition also may cause carryover of some diseases and insects. Growers often resort to burning organic residues for insect and disease control. The practice, while effective, needs to be discouraged wherever possible because of its negative effect on soil OM. While shredding organic material and incorporating it into the soil is not the ideal way of benefiting from it, it may be a suitable compromise where disease and insect carryover may be a problem.

Composting organic materials or allowing them to break down before the new crop is established seems to negate the effects of many herbicides and allelopathic substances, while limiting the destructiveness of certain plant insects and diseases. Composting pine bark prior to its incorporation into a soil medium eliminates much of the restriction in crop growth sometimes encountered in mixtures of bark, sand, and peat.

Maximum Incorporation of Organic Materials

Incorporating organic material, usually by moldboard plow or disk, provides for rapid decomposition with quick release of the nutrients and improvement of soil structure throughout most of the plow layer. However, the advantages are short lived. The more rapid destruction of OM leads in time to increased erosion and poorer structure with all the resultant problems of compaction, poor water infiltration, poor air and water holding capacities, and poor air exchange.

Deep Placement of Organic Materials

Organic materials can be added in furrows opened by moldboard plows or subsoilers. Some benefits can be obtained in compact subsoils, as the organic materials can cause longer periods of open subsoil. But the practice usually slows the tillage process. Unless materials are shredded and amounts limited, it can also cause problems of water movement in the soil. The practice is seldom used because of these limitations.

Living Mulches

A number of plants have been grown that continuously provide materials which serve as a mulch while the cash crop is grown. Benefits of such mulches include all of those cited above for organic mulches as well as crowding out weeds and increasing predators for insect suppression.

Orchards and Vineyards

Various groundcovers are grown in orchards and vineyards to provide a living mulch. Crops of hairy vetch, common vetch, crimson clover, red clover, buckwheat, chewings fescue, tall fescue, timothy, Kentucky blue-grass, and ryegrass are commonly used in apple and peach orchards, while crops of Pensacola bahia, hairy indigo, common Bermudagrass, and centi-pedegrass are more suited for the fruit crops (pecans and citrus) grown in warmer climates. The annual crops are used only when the tree or vine crop is young, while the perennials can serve as continuous ground covers as the cash crop ages.

Such groundcovers, despite improvements in soil OM, water infiltra-tion, and utilization, may ultimately lower yields of the cash crop. Some of this reduction appears to be due to reduced root densities of the trees or vines under the living mulch, which probably are the result of competition for air, water, and nutrients as the cash crop ages. In young orchards or vineyards, competition by the groundcover can be reduced or eliminated by keeping a vegetation-free area around the individual trees or vines. The area needs to be increased as the tree or vine ages, but it may be difficult to leave significant living mulched areas that are not restrictive when some plants (pecans, apples, citrus) reach maturity. A common approach is to eliminate the groundcover and maintain a weed-free floor by the use of herbicides. This approach, while useful for many soils, has failed under many conditions as compaction often leads to poor water infiltration, and in some cases, erosion has caused serious crop loss. The answer may be better selection of soils and proper terracing before planting long-term orchards or vineyards although some combinations of chiseling, soil trenching, and limited sod cover with mulches from vines or trees have maintained satisfactory yields.

Living Mulches for Row Crops

Recently there have been efforts to evolve systems of living mulches for various row crops. The introduction of certain herbicides capable of markedly slowing the growth of the organic matter crop without killing it and the development of planters capable of planting through trash have spurred these efforts. Generally, these systems grow an organic matter crop and when it has sufficiently developed, kill narrow strips of it with weed killer, plant the cash crop in the cleared strip, and keep the remaining cover crop controlled with low doses of herbicide.

SUMMARY

It is in the best interests of most growers to maintain soil OM at suffi-cient levels to at least provide for optimum air and water. This can be accomplished by using at least one of the following options: (1) increasing the amounts of OM added by (a) growing it in place (cover crops, sods, hays, crop residues) or (b) bringing in additions (manures, composts, sludges, crop wastes); (2) slowing the rate of decomposition by reducing the amounts of air available by (a) restricting cultivation of the soil (reduced tillage, no-till, or using rotations with sods, hays, pastures, small grains), and (b) adding sufficient N so that microorganisms do not unduly attack existing humus for their N needs.

In many farm operations it will be impossible to utilize all of the above but adopting at least some of them can reverse the harmful effects of OM reduction. Often, the addition of some extra organic materials combined with the reduction in OM decomposition is enough to maintain healthy air and water legs of the triangle.

Chapter 11

Regulating pH

As was pointed out in Chapter 5, soil or media pH usually has a pronounced effect on the nutrient side of the fertile triangle. But pH, by reflecting the saturation with Na^+ or Ca^{2+} and its influence on certain microbial processes affecting soil aggregation, can alter the air and water sides as well. These multiple effects are responsible for the close correlation between pH and the productivity of many crops. While it is most desirable to select a crop that will do well at existing pH values, it is usually more profitable to change the pH to suit the crop.

INCREASING SOIL pH

In many humid and semihumid areas, soil acidity often increases to the point that it may seriously limit crop production. In such cases, it becomes necessary to reduce soil acidity (increase pH) in order to (1) reduce toxic concentrations of some ions (usually Al^{3+}, but at times HBO_3^-, Fe^{2+}, F^-, Mn^{2+}); (2) make the soil more suitable for beneficial microbial activity; (3) control disease; (4) reduce losses of N as NH_3; and/or (5) replace the bases Ca and Mg. The reduction of acidity is especially useful for acid-sensitive crops (see Table 6.1).

Determination of Acidity

Acidity can be determined by measuring the pH of a soil sample. The measurement is simple, can be run on the farm, and offers tremendous returns. Most growers will profit from its use, which should be repeated at least once a year in humid regions.

The pH measurement, commonly run in a water suspension, can be determined by indicator dyes or pH meters. The meters are more expensive but provide more accurate results. Costs will be in the $60 to $2,000 range. The $60 instrument lacks automatic temperature compensation and is not as sensitive as the more expensive instruments but is suitable for routine testing.

A suspension of 1 part soil to 2 parts distilled or deionized water (by volume) is prepared by stirring intermittently for one hour.

Appropriate test strips containing indicator can be dipped into the supernatant liquid of the suspension, or the suspension can be filtered and the filtate treated with a few drops of appropriate indicator. The most commonly used indicator dye is brome cresol purple, capable of measuring pH in the range of 5.2 to 6.5+. Other indicators can be chosen to test for pH below and above these values. Some indicator strips available are capable of measuring pH over common soil values (4.0 to 8.0).

Use of the meter requires calibration with known buffer solutions (pH 4.0 and 7.0) prior to measuring suspension pH. Calibration with soils of known pH can also be helpful. The electrodes are inserted into the suspension and adjusted for temperature prior to taking the reading. Stirring shortly before or during the measurement enhances accuracy.

Preparing a suspension can be eliminated for the color tests. A small amount of soil is placed in a depression of a white spot plate, the indicator solution is added dropwise until the soil is completely saturated, and the soil is stirred with a stirring rod. In a few moments, the color of the supernatant liquid at edges of the depression can be compared with a color comparator chart.

pH and Salts

Soluble salts affect soil pH readings. Salts can displace hydrogen (H^+) and aluminum (Al^{3+}) from the exchange complex, which lowers pH, but this effect can be masked by the use of neutral salt solutions of 0.01 M calcium chloride or 1 N potassium chloride instead of water.

The commonly used calcium chloride solution (0.01 M $CaCl_2.2H_2O$) is made by dissolving 1.47 g in 1000 ml of water. The soil suspension is prepared as above except that the calcium chloride solution is substituted for water. Measurement is made with a glass electrode meter, but it is necessary to use a different set of interpretative values than that used for the water pH.

Materials Used for Increasing pH

The H^+ ions produced by dissociation, hydrolysis, and ion exchange are neutralized by the addition of bases such as liming materials. A number of materials containing the cations Ca^{2+}, Mg^{2+}, K^+, and Na^+ and anions such as CO_3^{2-}, OH^-, or SiO_3^{2-} can be used for such purposes. The Na compounds are almost never used because of their negative effect on soil structure and the K compounds seldom because of costs. High-calcium limestone, primarily calcium carbonate ($CaCO_3$), or dolomitic limestone, which is mostly dolomite ($CaCO_3.MgCO_3$), in various purities are the more commonly used materials. The silicates, sold as slags, and the oxides and hydroxides of calcium, or calcium and magnesium, as well as some waste materials are also being used extensively in certain regions. A list of the more common liming materials is presented in Table 11.1.

Slags

Basic slag, also called Thomas slag, which contains silicates and some phosphates as well as calcium, can be substituted for liming materials derived from calcium and magnesium carbonates. The amount of P and some micronutrients contained in slags may give them some advantages over the carbonate materials on certain soils.

Relative Effectiveness of Liming Materials

Not all liming materials are equally effective in increasing pH, their efficiency depending on chemical composition and degree of fineness.

The neutralizing value of various materials is often given as its calcium carbonate equivalent (CCE) or its ability to raise the soil pH in comparison to pure calcium carbonate, which is given a CCE value of 100. The CCE values of a number of liming materials are given in Table 11.1.

The effectiveness of the more common liming materials depends on their contents of Ca and Mg. Pure calcium carbonate ($CaCO_3$) exists as calcite or aragonite, but most agricultural limestones are mixtures of $CaCO_3$ with varying amounts of dolomite ($CaCO_3.MgCO_3$) and impurities. The classification of various Mg-liming materials is given in Table 11.2. The effectiveness of pure dolomite is $1.19 \times$ that of pure calcium carbonate (relationship of the molecular weight of magnesium carbonate to that of calcium carbonate, i.e., $100 \div 84 = 1.19$).

Effective Calcium Carbonate Equivalent (ECCE)

Although materials with equivalent CCE should provide similar changes in the pH of similar soil, there are differences in the rate and amount of rapid pH change due to the size of the particles. Very fine

TABLE 11.1. The Calcium Carbonate Equivalent (CCE) of Various Liming Materials[a]

Liming material	Common name	Formula	CCE[a]
Calcium carbonate	Calcite, aragonite	$CaCO_3$	100
Calcitic limestone	High cal	$CaCO_3$[b]	80-100[b]
Calcium oxide		CaO	179
Burnt lime	Builder's lime	CaO[b]	150-179[b]
Calcium hydroxide		$Ca(OH)_2$	136
Hydrated lime	Slaked lime	$Ca(OH)_2$	120-136
Dolomite		$CaCO_3.MgCO_3$	109
Dolomitic limestone	Dolomite	$CaCO_3.MgCO_3$[b]	to 108[b]
Calcined dolomite	Burnt dolomite	$CaO.MgO$[b]	to 185[b]
Hydrated dolomite	Hydrated dolomite	$Ca(OH)_2.Mg(OH)_2$	to 166[b]
Calcium silicate		$CaSiO_3$	86
Basic slag[c]	Thomas slag	$CaSiO_3$[b]	to 86[b]
Blast furnace slag[c]		$CaSiO_3$[b]	to 86[b]
Open hearth slag[c]		$CaSiO_3$[b]	to 86[b]
Ashes, coal		variable	0-40
Ashes, wood[d]	Hardwood ash	variable	to 80
Marl		variable	70-90
Portland cement[e] kiln flue dust		variable	to 100
Sugar beet lime[f]		variable	80-90
Shells	Ground oyster, egg	$CaCO_3$[b]	75-90[b]

[a] Relative values with pure $CaCO_3$ = 100.
[b] These may have various impurities, modifying their composition and CCE.
[c] The slags contain various amounts of Ca, Mg, and P. Basic slag contains 27-42 percent Ca, 1-5 percent Mg and 5-10 percent P; blast furnace contains 26-32 percent Ca, 2-7 percent Mg and <1 percent P; open hearth contains about 16 percent Ca, 5 percent Mg and 1 percent P. They also may have large amounts of micronutrients.
[d] Unleached wood ashes can contain 1 percent P, 4-21 percent K; the leached ashes about 0.5 percent P, and 1 percent K_2O. Both contain various amounts of trace elements, primarily Fe.
[e] Can contain varying amounts of K.
[f] Contains 0.05-0.6 percent P, with an average content of 0.38%.

Sources: Follett, R. H., L. S. Murphy, and R. L. Donahue. 1981. *Fertilizers and Soil Amendments,* Prentice-Hall, Inc., Englewood Cliffs, NJ; Wolf, B., J. Fleming, and J. Batchelor. 1985. *Fluid Fertilizer Manual.* National Fertilizer Solutions Association, Peoria, IL. Reprinted by permission of Agricultural Retailers Association, St. Louis, MO.

TABLE 11.2. Classification of Limestones Based on Their Magnesium Contents

Classification	as Mg (%)	as MgO (%)	as MgCO$_3$ (%)
High-calcium limestone	<0.5	<1.0	<2.1
Magnesium limestone	0.6-1.3	1.0-2.2	2.1-4.5
Dolomitic limestone	1.3-6.5	2.2-10.8	4.5-22.7
Calcitic dolomite	6.5-11.7	10.8-19.4	22.7-40.9
Dolomite	11.7-13.1	19.5-21.7	40.9-45.8

Source: NCSA Aglime Fact Book. 1986. National Crushed Stone Association, Washington, DC.

particles of less than 200 mesh (passing through a screen with 200 holes per inch) react very quickly. The fine particles that pass through a 100-mesh screen but are retained on the 200-mesh screen are still very rapid but less than those smaller than 200 mesh. There is further reduction in reaction rate as the particle size increases (through 60-mesh but held by 100-mesh screens) but these particles are still satisfactory for most liming operations. Particles of 20 to 50 mesh are extremely slow in reaction but tend to approach that of the much finer materials in about 18 months. Materials coarser than 20 mesh have little value for changing pH. The ECCE of several screen sizes are given in Table 11.3.

TABLE 11.3. Effective Calcium Carbonate Equivalent (ECCE) of Several Screen Sizes

Particle size	Effectiveness (%)	Mesh factor	Amount (%)	Screen sizes Passing	Screen sizes Retained	Product (factor x, % size)
>20	0 very slow	0	3		20	0
20-50	75 effective	0.75	10	20	50	7.5
50-60	90 effective	0.90	15	50	60	13.5
60-100	100 effective	1.00	45	60	100	45.0
<100	125 effective	1.25	27	100		33.75
					Total =	99.75

99.75 x 92% (% calcium carbonate in the material) = 91.77 ECCE

Source: Wolf, B., J. Fleming, and J. Batchelor. 1985. *Fluid Fertilizer Manual*. National Fertilizer Solutions Association, Peoria, IL. Reprinted by permission of Agricultural Retailers Association, St. Louis, MO.

Speed of Reaction

The reaction speed of the carbonates and silicates is often too slow to effectively change low pH values to benefit the current crop. The reaction time can be effectively shortened by grinding particles so that at least 80 percent of them will pass through a 100-mesh screen. Reaction speed of some of these materials also can be increased if they are heated to drive off much of the carbon dioxide to leave the oxides known as burnt lime (CaO) or burnt dolomitic lime ($CaO.MgO$). Burnt lime is difficult to handle, but its handling can be improved by slaking (adding water), which still produces fast-acting liming materials known as slaked or hydrated lime ($Ca(OH)_2$) or hydrated dolomitic lime ($Ca(OH)^2.Mg(OH)_2$).

The speed of reaction is also affected by the type of material. Coarse dolomitic materials are slower than equivalent calcitic materials but there is very little difference in speed of the fine or very fine materials. The oxides are the most rapid, hydroxides are intermediate, and the carbonates are the slowest of the three types. Slags react at about the same rate as equivalent fine limestones.

Amounts of Liming Materials

The amounts of liming materials needed to change a low pH value to a more satisfactory one depends on soil pH, crop, type of soil, and its OM and CEC (buffering capacity), the depth of incorporation, and the kind of liming material and its fineness (ECCE). Obviously a soil pH test is needed to determine whether liming materials are needed. But before adding liming materials to change a low pH to a more desirable one as shown in Table 6.1, it should be remembered that listed optimum values are not absolutes. Ideal pH varies somewhat with soil composition. Lower values than those indicated in Table 6.2 can produce satisfactory crops if the soil has little available Al and Mn but ample Ca, Mg, and Mo.

If the pH needs to be increased, the pH value alone fails to reveal how much material is needed, since this will be affected by the buffering capacity of the soil. The buffering capacity is closely related to soil CEC, which is largely dependent on soil texture and OM. Estimates of liming requirements can be made if the soil textural class, OM, and pH are known (see Table 11.4).

More accurate estimates of the amounts of liming materials needed can be determined by treating a soil sample with a known buffer solution and then measuring the pH of the resulting suspension. The several common buffer methods are not equally suited for all soils. The AE (Adams and

TABLE 11.4. Approximate Amounts of Limestone Required per Plowed Acre of Different Soil Textural Classes to Change pH to 6.5 (lb/ac)

Soil class	Organic matter (%)	Soil pH				
		4.0-4.4	4.5-4.9	5.0-5.4	5.5-6.0	6.1-6.5
		(lb/ac)				
Loamy sand	< 0.9	2000	1250	750	500	250
	0.9-1.6	3000	2000	1250	750	500
	>1.6	4000	3000	2000	1000	750
Sandy loam	<1.2	4000	3000	2000	1000	500
	1.2-2.0	5000	4000	3000	1500	1000
	>2.0	6000	5000	4000	2500	1500
Loam	<1.5	6000	4500	3000	2000	1500
	1.5-2.8	7000	5500	4000	2500	2000
	>2.8	8000	6500	5000	3500	2500
Silt loam	<1.8	7000	5500	4000	3000	2000
	1.8-3.2	8000	6500	5000	4000	3000
	>3.2	9500	8000	6500	5000	4000
Clay loam	<2.0	8000	6500	5000	4000	3000
	2.0-3.8	9000	7500	6000	5000	4000
	>3.8	11000	9000	7500	6000	5000

Source: Wolf, B., J. Fleming, and J. Batchelor. 1985. *Fluid Fertilizer Manual.* National Fertilizer Solutions Association, Peoria, IL. Reprinted by permission of Agricultural Retailers Association, St. Louis, MO.

Evans) method is suited for low-CEC soils (low clay and OM); the SMP (Shoemaker, McLean, Pratt) is better suited for soils with considerable exchangeable Al^{3+}, less than 10 percent OM, and high lime requirement (>2 tons\acre). The Woodruff buffer is suited for very acid soils with good contents of clay and OM. Many soil testing laboratories are equipped to determine the lime requirement using a suitable buffer.

Rate and Depth of Incorporating Liming Materials

Most lime requirements are based on changing the pH of the plowed layer ($6^2/_3$ inches) of mineral soils. If the liming material is to be incorporated to other than standard depths, the amounts needed are calculated by multiplying the lime requirements by an appropriate factor as given in Table 11.5.

TABLE 11.5. Lime Required As Affected by Depth of Incorporation

Depth of incorporation (inches)	Multiply lime requirements by a factor of
3 or less	0.43
4	0.57
5	0.71
6	0.86
6.75	0.96
7	1.00
8	1.14
9	1.29
10	1.43
11	1.57
12 or more	1.71

Source: Wolf, B., J. Fleming, and J. Batchelor. 1985. *Fluid Fertilizer Manual.* National Fertilizer Solutions Association, Peoria, IL. Reprinted by permission of Agricultural Retailers Association, St. Louis, MO.

Choice of Materials

Usually the choice of materials is based on composition, location, and price. If the soil or media is low in Mg (<10 percent Mg saturation) as well as Ca, materials containing considerable Mg (dolomitic) need to be chosen. Dolomitic materials are excellent slow-release sources of Mg. But adding dolomitic materials to soils high in Mg (>18 percent base saturation) can be counterproductive as continued use of dolomitic materials to high magnesium soils can lead to interference with Ca uptake.

The need to raise the pH quickly will warrant the use of quick-acting materials, such as the burnt or hydrated liming materials, or at least to consider very finely ground materials.

In assessing the true price of the material, its ECCE and the distance to be hauled must be considered. The high cost of some materials may actually be economical if their high ECCE permits hauling of appreciably less material over long distances.

The content of P and trace elements in slags and the presence of K and elements other than Ca or Mg in wood ashes can give these materials values beyond their ECCE. On the other hand, the presence of toxic substances in some low-cost waste materials needs to be properly evaluated before using.

Delaying Planting After Liming

The application of dry burnt or hydrated lime can injure crops if they are planted before the materials have had sufficient time to react in the soil. There is no need for waiting after the application of carbonates or silicates before planting. Planting needs to be delayed at least several days after applying the caustic oxides and hydroxides of lime. Additional time is needed if liming materials were applied to dry soils.

Fluid lime or the application of liming materials in water or with certain fluid fertilizers speeds the reaction rate. The use of fluid lime permits the safe and effective handling and more uniform distribution of very fine, dusty materials. It also makes the use of burnt or hydrated materials safe not only for the applicators but also for plants. The reaction time needed for safe application of dry burnt or hydrated lime in moist soil is usually less than a week but may be extended in dry soil.

Degree of pH change. The rapid high pH values obtained with either the oxides or hydroxides drop rather quickly and in time may be lower than that obtained with equivalent carbonates or slags. From a practical standpoint, limestones or slags are desirable for the maintainence of pH while burnt or hydrated lime are desirable for situations requiring a rapid change in pH. A combination of both the quick-acting materials and limestone works well for very acid conditions that need to be changed rapidly.

Placement of Liming Materials

Liming materials tend to move slowly in the soil, usually requiring thorough mixing with the soil for effective use. The concern with placement varies somewhat with the degree of change in pH required, being much more acute with low-pH soils. For very acid soils, it is often desirable to broadcast the material and disk it into the surface soil prior to plowing it under. Part of the limestone can be applied to the surface after plowing and disked in, or for very fast reaction, burnt or hydrated lime can be disked in after plowing in a large application of limestone, as suggested above.

The placement of liming materials is not as serious for soils with better pH values, where application is made for maintenance of pH rather than for elevating it. In such cases, the placement is much less critical, and the material can be effectively plowed under or disked deeply into the surface after plowing.

The slow movement of liming materials helps explain the relatively poor pH values of many subsoils despite extensive liming of the surface soil. For deep-rooted, acid-sensitive crops, it often becomes profitable to apply liming materials to the subsoil, and devices have been devised for applying materials to the subsoil in trenches opened by subsoilers. The operation is rather slow and expensive, and perhaps an equal and less costly effect can be obtained by deeper plowing but adding extra liming material to the very acid soil brought up to the surface. Often it is also necessary to add extra phosphorus and other elements to the surface soil to compensate for the poor fertility of most subsoils.

HIGH-pH SOILS

In many areas of the world, soil pH is too high for optimum crop growth. The high pH may be due to overliming or, in many arid regions, it may naturally have high contents of Ca, Mg, or Na.

Crop reduction at high pH values can be due to several reasons but primarily to the adverse effects of excess Na often found in arid regions or to the effect of high pH on the availability of N, P, K, B, Cu, Mn, Fe, and Zn. Lack of Fe, Mn, and Zn at high pH values is very common and accounts for tremendous losses.

The adverse effect of elevated pH on the severity of several soil-borne diseases is also relatively common. Whereas most of the problems of the availability of micronutrients usually become severe as pH values rise above 7.0, diseases of potatoes (white and sweet), tobacco, and watermelon may be serious at pH values >5.5, often making it desirable to grow these crops in the pH range 5.0 to 5.5.

There are several approaches to handling the problem of high pH. The simplest way is to change the crop sensitive to high pH to one that does well or at least tolerates it. In some cases, satisfactory crops can be grown if minor adjustments in chemicals are made by adding nutrients in low supply to offset shortages induced by low pH, or substituting acid-forming materials for others that may be neutral or leave a basic reaction (see Table 6.3). But in many cases, the high pH will have to be lowered.

Crops That Do Well at High pH

Rather than attempting to acidify the soil or to overcome chemical imbalances, it is often much simpler to grow crops that do well at high pH values. A number of such crops are listed in Table 6.1. In some cases, there are considerable differences in tolerance to the high pH among different cultivars, making it possible to grow good crops providing the correct cultivar is chosen.

Minor Adjustments to High pH

Extra Micronutrients

One of the major causes of poor yields of some crops grown on high-pH soils is the low availability of B, Cu, Fe, Mn, and Zn in these soils. Extra amounts of applied micronutrients can be helpful if care is taken to keep the applied element available. Mixing the element with considerable soil probably will provide very little extra available amounts, since the same conditions originally inactivating the microelement will once again make the added nutrients unavailable.

Banding Micronutrients

Some increase of available micronutrients can be obtained by adding the elements in bands applied to the soil so that there is a minimum of contact of the elements and soil. Availability is enhanced if the micronutrients are introduced with acid fertilizers so that an acid band of soil is maintained for a while.

Use of Chelates

The application of the metallic micronutrients as chelates is usually more effective than nonchelate sources, although there appears to be no extra benefit of applying Mn as a chelate. The form of Cu, Fe, and Zn chelates—usually more effective than nonchelated sources of these elements—can have a bearing on the effectiveness of the chelate. The EDTA form is not very useful for application to high-pH soils; the DTPA form is more effective at elevated pH but only at values 6.5 to 7.5. EDDHA is suitable for the higher pH values of 7.5 to 8.5.

Foliar Sprays of Micronutrients

The repeated application of these elements as foliar sprays can augment the relatively small amounts supplied by the soil and help make suitable crops despite the undesirable pH.

Shortages of Potassium and Magnesium

Other problems associated with some high-pH soils are the lack of sufficient Mg and sometimes K needed to help balance the excessive Ca. Extra Mg and K in soil-applied fertilizers or foliar sprays can help offset this imbalance. Some soils can fix such large amounts of K that the correction of the problem is greatly aided by extensive applications of foliage-applied K.

Lowering Soil pH

Often, there is no suitable choice for a particular crop other than lowering the soil pH. There are several practical approaches for lowering pH, the choice often depending on the nature of the high pH, how much and how fast it needs to be lowered, and the cost of materials.

Acid Peats

Although primarily used for moisture retention and improvement of soil structure, acid peats can be used to lower pH values. Because of costs, applications are limited to gardens, lawns, landscapes, rooting of cuttings, and certain acid-loving high-priced crops. It has been a standard addition to artificial soils used for golf greens, and for a number of crops grown in containers.

Sphagnum peat moss is the most acid (pH values 3.2 to 4.5); the pH of hypnum peat moss varies from 4.4 to 6.7; and that of sedge peat from 4.5 to 7.0. Because of considerable variation in the pH values of the hypnum, reed, and sedge peats, it is very important to have the material tested prior to use.

Soluble Phosphates

The application of partially soluble phosphates has been fairly effective for lowering the high pH of potting media. The treatment is started before planting and consists of repeatedly flooding the substrate for several hours with a solution of 50 lb treble superphosphate per 100 gal of water. If a recirculating system is used as for hydroponics, the solution is reused until its pH remains at about 6.8. After the substrate is flooded several times with the superphosphate solution, it is flooded with water. Such treatment will usually provide enough P to sustain a crop for about six months. As the phosphate is depleted, pH will rise and the treatment will have to be repeated. To avoid damage to a crop, the reflooding should be done between crops.

Banding Fertilizers

Attempts to lower the overall pH of field soils containing large quantities of coarse liming particles—shells or limestones—by means of acidic fertilizers, S, or S compounds are usually ineffective. In such soils, the change in pH tends to be very temporary as more unreacted calcium or magnesium carbonates are exposed. The effectiveness of such additions is greatly increased if the materials are added with fertilizers in bands.

Some increase in crop yield can be obtained on high-pH soils by banding acid fertilizers to create low-pH microspheres. The low-pH microspheres help to keep such elements as P, Mn, and Zn soluble and the restriction to bands reduces the contact of the element with soil, lessening the amount of fixation. The banding of acidic N fertilizers with phosphates is used extensively to help improve P availability. The addition of S or sulfur compounds to the fertilizers where possible lengthens the effectiveness of the acid microspheres.

Nitrogen Sources

The pH can be lowered with acid-forming N sources (see Table 6.2). Lowering pH by this approach is limited by the amount of N needed for the crop, and is of minimal value for short-term pH changes in most soils. But for sands and the low-buffered media used for hydroponics, the choice of nitrogen source can be an effective means of regulating pH. In such media, the use of predominantly NH_4-N over NO_3-N sources is sufficient to lower the pH in days. Ratios of 90:10 (NO_3-N/NH_4-N) tend to maintain existing pH values; ratios narrower than 80:20 acidify most media.

Sulfur and Sulfur-Containing Compounds

Larger pH changes can be brought about in most soils by the use of S or various compounds containing reduced S, although the presence of shells, coarse particles of limestone, or large excesses of liming materials in most media usually makes it difficult to make appreciable long-lasting pH changes. Some of the more common materials used for acidifying soils and other media and their relative effectiveness compared to S are given in Table 11.6.

Acidification with S is a slow process since the S has to be oxidized by microorganisms to sulfuric acid. The process may take a couple of years to complete. To speed the reaction, many growers have turned to the application of sulfuric acid, directly injecting it into the soil or through the irrigation waters. Sulfuric acid is a highly corrosive product and its handling requires extensive precautions to avoid damage to personnel or equipment.

As with the application of micronutrients, the pH of the entire soil does not have to be lowered below 7.0 to obtain the benefits of treatment. There

TABLE 11.6. Acidifying Materials and Their Relative Capability of Lowering pH Compared to Sulfur

Material	S (%)	Formula	Relative amount change as S*
Sulfur	100	S	100
Sulfuric acid (98%)	31	H_2SO_4	318
Aluminum sulfate	20	$Al_2(SO_4)_3.18H_2O$	694
Ammonium bisulfate soln	17	$(NH_4)HSO_3$	456
Ammonium sulfate	24	$(NH_4)_2SO_4$	277
Ammonium thiosulfate soln	26	$(NH_4)_2S_2O_3$	304
Ammonium polysulfide soln	45	$(NH_4)S_x$	177
Ammonia sulfur soln	10	$(NH_3 + S)$	199
Iron sulfate (ferric)	17	$FeSO_4.9H_2O$	588
Iron sulfate (ferrous)	18	$FeSO_4.7H_2O$	870
Iron pyrites (87%)	46	FeS_2	215
Sulfur dioxide	50	SO_2	200

* Basis of calculation: 1.8 lb pure calcium carbonate are needed to neutralize 1 lb NH_4-N; 3.125 lb calcium carbonate are needed to neutralize 1 lb S.

Source: Wolf, B., J. Fleming, and J. Batchelor. 1985. Fluid Fertilizer Manual. National Fertilizer Solutions Association, Peoria, IL. Reprinted by permission of Agricultural Retailers Association, St. Louis, MO.

appears to be marked benefits of treating small zones of high-pH soils with sulfuric acid. The benefits are greater in pot cultures than in the field because of the restriction of the root system to treated areas. In pot cultures, the creation of an acid zone with 15 ml of sulfuric acid per 2½ gal container prevents iron chlorosis (common in high-pH soils) of many plants. In the field, as little as 1000 lb/ac of sulfuric acid banded in a calcareous soil has given increased yields of soybean cultivars susceptible to iron deficiency, but rates of up to about 4000 lb/ac have failed to completely eliminate Fe deficiency (Frank and Fehr, 1983).

The amounts of S per acre necessary for acidification can be calculated from the formula presented below:

S needed = % base saturation (at present)
− % base
saturation at desired pH
× CEC
× equivalent weight of the acidifying agent
× 17.8

For example, if the CEC = 20 meq per 100 g of soil, and the base saturation of the soil at present = 95 percent, but a base saturation of only 75 percent is desired, the amount of sulfur needed per acre = $(0.95 - 0.75) \times 20* \times 16 \times 17.8 = 1139$ lb. If sulfuric acid was used instead of sulfur, the amount would be $(0.95 - 0.75) \times 20 \times 49** \times 17.8 = 3489$ lb. Because of two H atoms in the molecule, the molecular weight needs to be divided by 2 to obtain the equivalent weight.

An approximation of sulfur rate applications to lower the pH of soil having very little or no free calcium carbonate can be made on the basis of soil texture (see Table 11.7).

Various materials used for acidification and their relative effectiveness are given in Table 11.8.

Sodic Soil

Lowering the pH of sodic soils (those that have >15 percent Na) can present special problems, because their poor physical structure makes the incorporation of many dry materials very difficult. Washing out salts is almost impossible because of poor infiltration. Irrigation water treated with sulfuric acid, sulfur dioxide, ammonium, or calcium polysulfide waters improves infiltration and reduce the Na accumulation. Both sulfuric acid and sulfur dioxide reduces the pH of the waters by destroying carbonates and bicarbonates, both of which can have adverse effects on soil pH and structure.

Sodic soils that contain lime are benefited by additions of sulfuric acid, sulfur dioxide, sulfur, or other sulfur compounds. The sulfuric acid added or formed from added sulfur or sulfur compounds reacts with lime (calcium carbonate) to produce a fine gypsum, which is a more soluble form of calcium. The calcium from formed gypsum replaces Na which now can be leached from the soil by irrigation waters.

If the soil does not naturally contain lime or gypsum, some gypsum will have to be added to reclaim these soils. Large amounts of granular gypsum may be needed (2-10 tons/ac) for soil application. To improve permeability, applications are confined to the surface or mixed a few inches in the soil. Water applications require less material but the gypsum needs to be very fine (<0.25 mm) and applied continuously through the irrigation period.

---

*Equivalent weight of S.
**Equivalent weight of sulfuric acid. The molecular weight of sulfuric acid $(H_2SO_4) = 2 \times 1(H) + 1 \times 32(S) + 4 \times 16 (O) = 98$.

TABLE 11.7. Approximate Amounts of Sulfur Needed to Reduce Soil pH One Unit

Soil texture	Sulfur (lb)
Loamy sand	300
Sandy loam	500
Loam	800
Silt or clay loam	1200

Source: Wolf, B., J. Fleming, and J. Batchelor. 1985. *Fluid Fertilizer Manual.* National Fertilizer Solutions Association, Peoria, IL. Reprinted by permission of Agricultural Retailers Association, St. Louis, MO.

TABLE 11.8. Acidifying Materials and Calcium Sources Used for Reclaiming Alkaline Soils

Amendment	Chemical comp.	%Ca	%S	Relative effectiveness*
Gypsum	$CaSO_4.2H_2O$ 100	23	18	100
Aluminum sulfate	$Al_2(SO_4)_3.18H_2O$	—	20	
Ammonium bisulfate soln	NH_4HSO_4	—	17	
Ammonium sulfate	$(NH_4)_2SO_4$	—	24	
Ammonium thiosulfate soln	$(NH_4)_2S_2O_2$	—	26	
Ammonium polysulfide soln	$(NH_4)S_x$	—	45	
Calcium carbonate	$CaCO_3$	40	—	58
Calcium chloride	$CaCl_2$	36	—	86
Calcium nitrate	$Ca(NO_3)_2.2H_2O$	20	—	106
Iron sulfate (ferric)	$Fe_2(SO_4)_3.9H_2O$	—	17	61
Iron sulfate (ferrous)	$FeSO_4.7H_2O$	—	18	162
Iron pyrites (87%)	FeS_2	—	46.5	
Lime sulfur		9	24	78
Sulfur	S	—	100	19
Sulfuric acid (98%)	H_2SO_4	—	31.4	61

* Calcium sources are compared to gypsum as 100.

Source: Wolf, B., J. Fleming, and J. Batchelor. 1985. *Fluid Fertilizer Manual.* National Fertilizer Solutions Association, Peoria, IL. Reprinted by permission of Agricultural Retailers Association, St. Louis, MO.

Gypsum is very effective in removing sodium that helps maintain the high pH of some alkaline soils and also accounts for impermeability of the sodic soils. The amounts of gypsum and other chemicals needed to remove a milliequivalent of Na from a foot of soil are given in Table 11.9.

TABLE 11.9. Amounts of Soil Amendments Needed to Remove One Milliequivalent of Sodium from an Acre-Foot of Soil

Material	Formula	Amount (tons/acre-ft)
Gypsum	$CaSO_4.2H_2O$	1.78
Calcium chloride	$CaCl_2$	1.1
Iron sulfate (ferrous)	$FeSO_4.7H_2O$	2.78*
Limestone	$CaCO_3$	1.0**
Lime sulfur	24% S	1.33
Sulfur	S	0.32*
Sulfuric acid	H_2SO_4	0.98*

* These materials are useful only if sufficient limestone or other sources of Ca is present in the soil. The S would be much slower in reacting than the other materials.

** Limestone is very slowly soluble and will usually require additions of acid materials to hasten the reaction.

SUMMARY

Although it may be possible to choose a desirable crop that will do well at existing pH values, it often is necessary to modify the soil or medium pH so that the intended crops can produce satisfactory yields.

In humid areas, where leaching may have removed most soil bases, the pH will have to be increased before satisfactory yields of most crops are possible. The pH can be increased by use of many different basic materials, most commonly by use of calcium carbonate ($CaCO_3$) or dolomite ($CaCO_3.MgCO_3$) or mixtures of the two (dolomitic limestones) or the various oxides or hydroxides of these materials. Substantial use of the silicates (basic slag) and waste materials, such as sugar beet lime, also is made in certain regions.

A pH test, which is easily run on the farm, is the best means for determining whether the pH needs to be changed, but a pH test will not reveal how much liming materials need to be added. Since soil organic matter and clay are largely responsible for the buffer effects that modify the action of liming materials, an estimate of the amount of materials needed can be made based on soil class (approximate clay content) and organic matter. A much more accurate evaluation can be made by adding a suitable buffer to the soil and measuring the pH.

The amount of liming materials needed also will be influenced by the effectiveness of the material, which is largely determined by its calcium carbonate equivalent (CCE) and the fineness of the particles.

The choice of material is usually made by price, but other factors also need to be considered. If the soil or medium is low in magnesium, dolomitic materials should be used. The need for quick reaction will require oxides, hydroxides, or very fine particles, although a few extra days after applying the oxides or hydroxides are needed before planting in order for these materials to react and lose their caustic effect. Applying the oxides or hydroxides in water or liquid fertilizer can hasten the reaction time.

Soil pH values in arid regions will usually be too high for most crops. Occasionally, some soils or potting mixtures in humid regions also will be high due to overliming or the presence of shells or limerock pieces.

Often the adverse effects of high pH are due to a lack of one or more micronutrients that are immobilized at high pH, and can be corrected by adding them, either as foliar or soil applications or both. The effectiveness of the soil application is increased if these materials are banded with acid-forming materials or, in the case of Fe, Zn, and Cu, if they are applied as chelates.

Crop yields also can be improved if the high pH of soil or medium is adjusted to more favorable ranges by acidic materials or those that will yield acids after soil incorporation. Acid fertilizers can be used to lower pH. Their value is enhanced in hydroponic solutions or soils of low buffering capacity or if they can be banded in soils.

Sulfuric acid can be applied directly to soils but it is highly corrosive and requires considerable care in application. Sulfur, or sulfur compounds that will oxidize to sulfuric acid, are much easier to apply but take more time for reaction. The amount of sulfur needed for acidification can be calculated if the extent of present base saturation as well as the desired base saturation is known.

Long-term acidification is difficult if the soil or medium contains shells or limestone fragments. In such cases, it is probably more desirable not to treat the entire soil but rather to establish acid zones by placement of acid fertilizers or acidic materials.

Lowering the pH of sodic soils also presents problems because of the difficulty in mixing the materials adequately with the soil. If the soil contains free lime, sulfuric acid or materials that form can be added to react with the lime to form gypsum. If insufficient lime is present, gypsum is needed to help drive out the excess sodium so that soil will open enough to permit leaching of the excess bases.

Chapter 12

Regulating Salts

As was seen in Chapter 8, either low or high salts can cause problems with the sides of the fertile triangle, requiring strict control if production is to be satisfactory. The problems related to low salts occur frequently in the lighter soils in humid regions. High salts are more frequently encountered in arid regions.

LOW SALTS

As was pointed out earlier, low salts are usually due to inadequate fertilization, occurring most often on sandy soils in humid regions or in pot cultures that have been overirrigated. Occasionally, excess irrigation in arid regions, fixation of nitrogen by wide C/N ratio materials, or use of low-conductivity irrigation water may be responsible for low salts. Very low-conductivity irrigation water also can cause problems by corroding irrigation equipment and having an adverse effect on water infiltration.

Low salts are usually indicative of low nutrient levels (primarily N and K), which can cause reductions in crop yields. The adverse nutritional effects increase in severity as the conductivity of a soil suspension (1 part soil:2 parts water) falls below 0.5 mmhos/cm.

Correction of low salt levels is a simple matter. Low soil nutrition can be corrected by addition of fertilizer directly to the soil or through the irrigation water. Deciding which fertilizer and amounts to use can best be evaluated by soil tests that include available N as nitrate and ammoniacal forms. Sometimes, the test results will indicate that only N and K but often only the addition of N is needed.

The severity of corrosion and water infiltration problems caused by irrigation water increases as the conductivity of the water falls below 0.2 mmhos/cm. The addition of fertilizer to the irrigation water with suitable flushing will eliminate equipment corrosion and reduction of water infiltration into the soil. The corrosion and infiltration effects can also be controlled by passing the water over columns of gypsum stones or adding fine particle gypsum (>0.25 mm) continuously during the irrigation period. The addition of fine gypsum is not practical for microirrigation systems.

HIGH SALTS

Coping with Excess Salts

Several methods for coping with high salts are: (1) growing salt-tolerant crops; (2) preparing soil beds so that salts do not accumulate near the plants (see Figure 12.1); and (3) limiting the amounts of salts introduced by fertilizers, water, and other amendments.

Growing tolerant crops and preparation of soil beds to limit damage from salts has been covered in Chapter 8. In this chapter, emphasis is placed on limiting the amounts of salts introduced or reducing their toxicity by various procedures.

FIGURE 12.1. Variations in Salt Accumulation Due to Seedbed Formation

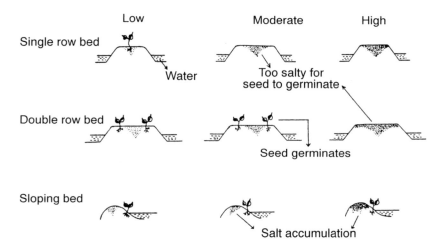

Source: Western Fertilizer Handbook, Seventh Edition. 1985. Interstate Printers & Publishers, Inc. Danville, IL, p. 36. Reprinted by permission of California Fertilizer Association, Sacramento, CA.

Fertilizers

Salt damage from fertilization can be limited by selecting fertilizer ingredients, avoiding applications of excess amounts, and by proper placement.

Choosing Fertilizer Ingredients

Damage can be reduced by selection of fertilizer salts to give the smallest amount of extraneous ions such as Cl or Na, which can accumulate in soils and add to the salt content. In most situations, choosing a fertilizer ingredient depends primarily on price, availability, and ease of handling. If, however, high salt is or may become a problem, the first consideration must be whether it provides the needed element(s) with the smallest increase in salts. Low-salt fertilizers should be the fertilizer of choice. For example, from Table 8.1 it is obvious that anhydrous ammonia contributes the least salts per unit of N of the various sources and would be the preferable source of N under saline conditions, if it were compatible with operations and suitable for the particular soil (sufficient CEC to hold the NH_3 and pH <7.0). With low CEC or elevated pH, NH_3 may be lost in large amounts. If anhydrous NH_3 is not practical to apply, it would be far better to use urea instead of sodium nitrate or even ammonium sulfate. Nitrogen can also be lost from urea when applied to high-pH soils but the loss is rather small if the urea can be worked into the soil soon after application. By the same token, potassium sulfate (sulfate of potash), although a more expensive source of K_2O than potassium chloride (muriate of potash), is the preferred source of K_2O if high salts may be a problem.

To reduce the potential problem of high salts from fertilizers, the first maxim probably should be to avoid any fertilizer nutrient unless it is needed. All soluble elements contribute to the salts, whether derived from fertilizers, manures, or other soil amendments. Adding an element already in good supply needlessly adds salts in addition to being a waste of money and a potential source of pollution. Close control of the soil nutrient status by soil testing becomes even more important under elevated salt conditions.

To avoid buildup of salts in borderline excess conditions, care must be taken to avoid adding large amounts of those elements that are needed only in small amounts, if at all. For example, in a situation where excess salts may become a problem, it may be far better to add potassium as potassium nitrate instead of potassium chloride or even the more benign salt (potassium sulfate). The potassium nitrate can be fully utilized by the plant (providing both N and K are needed), whereas with potassium chloride there is apt to be considerable chloride remaining, and in the case of potassium sulfate, considerable sulfate can be left in the soil. Plants are selective in their absorption of elements and much of the chloride or sulfate is apt to be left behind to contribute to the salt problem. Chloride and sulfate can be taken up in large amounts by some plants but because of the relatively small requirements for Cl and only moderate requirements for S, their presence in large amounts needlessly contributes to the salt problem.

Fertilizer Placement

There are many situations in both humid and arid climates where considerable salt damage occurs because there is an excess of salts in a zone, whereas most of the soil profile may be satisfactory. The high-salt zones may be due to poor placement of fertilizers or from salts left behind as water evaporates. It is not uncommon for emerging seedlings to be damaged by excess salts derived from fertilizer placed in contact with the seed or directly under the seed.

Broadcast applications of large amounts of fertilizer, particularly those high in salts, can easily cause damage to emerging seedlings, most of which are very sensitive to salts. Chances of injury are greater on soils with low CEC and OM, particularly if the soil already has a fair amount of salts prior to fertilization, and/or if incorporation of the added fertilizer was shallow.

Injury due to placement is more common as fertilizers are placed with seeds, too close to seeds, or directly under the seed row. Additions of fertilizers with seeds need to be limited to about 10 lb/ac of both N and K. Slightly larger amounts can be tolerated on heavier soils. Substituting side-placed bands for placement under the row appears to be a much safer procedure. Banding fertilizers at time of seeding to 2 inches from the seed line and at depths 2 inches lower than the seed line appears to be satisfactory for many crops, providing total salt is not excessive. Later, side dressings placed about 2 inches ahead of the furthest roots at a depth of 4 to 6 inches also is satisfactory. Up to a total of about 100 lb/ac of N plus K appears to be safe for many soils and crops.

Much larger amounts of fertilizers can be safely applied in bands if they are sufficiently ahead of developing roots. Up to about 200 lb of N and 300 lb K_2O per acre have been safely applied as near-surface bands placed about 6 inches from the seed row under plastic at or shortly before seeding. The distance of the fertilizer from the developing seedling is too great for early plant roots to reach the fertilizer, requiring a small amount of fertilizer distributed in the bed for early development.

Slow-Release Fertilizers

In addition to using low-salt-soluble fertilizers and placing them strategically, the grower can limit damage by using slow-release sources (see Chapter 8).

Water

Both rain and irrigation waters contribute salts. That from rainfall will introduce 12 to 35 lb per acre overall with amounts over 100 lb per acre per year in areas near the coastline. The amount of salts introduced by irrigation waters varies from about 300 to over 5000 lb per acre-foot of water. The low values represent waters very low in salts, which have conductivity readings less than 0.5 mmhos/cm or dS/m; the high values are irrigation waters high in salts with conductivity readings >4 mmhos/cm. Although all ions contribute to the total salts in irrigation waters, it is primarily the contents of chlorides, bicarbonates, and sodium that are major contributors.

The amounts contributed by rainfall and low-conductivity irrigation waters are usually not harmful if accompanied by sufficient rainfall to do some leaching. But high-conductivity irrigation waters can cause considerable harm. Waters with conductivities >1.5 mmhos/cm generally cannot be used for salt-sensitive plants. As indicated in Chapter 8, salt-tolerant plants can be selected that will give substantial yields even with relatively high salts (about 3 mmhos/cm). But such use must be accompanied by sufficient water to leach out excess salts, lest soil conditions deteriorate.

Chlorides

High levels of chlorides can be injurious to plants because of sensitivity and/or because of total salt effects. Plants vary in their sensitivity to chlorides. Azalea, bean, blackberry, camellia, citrus, lettuce, some legumes, rhododendron, potato, stone fruits, strawberry, and tobacco are sensitive; alfalfa, cotton, and milo are moderately tolerant; sugar beets, barley, Brussels sprouts, cabbage, cauliflower, corn, radish, spinach, and tomato are quite tolerant. Approximate soil concentrations of chloride tolerated by different crops are given in Table 12.1.

TABLE 12.1. Concentrations of Chloride in the Soil Solution That Provide About 75 Percent of Yield Compared to Yields of Several Crops Grown on Unaffected Soils

Field crops	Cl (ppm)	Vegetable crops	Cl (ppm)	Fruit crops[d]	Cl (ppm)
Beans, brown		Lettuce	595	Mulberry	350
and white	315	Strawberry	910	Grape	350
Pea	315	Beans,		Pear	700
Potato	910	dwarf, runner	910	Currant (blank)	1400
Broad bean	1190	Cabbage, red[a]	1190	Currant (red)	2450
Onion	1190	Potato	1785	Apple	2625
Flax	1785	Endive	1785	Blackberry	3150
Red clover	1785	Celeriac	2380	Plum	3150
Wheat, spring	2380	Cabbage, red	2380	Raspberry	3150
Spinach[b]	3010	Carrot	3010	Cherry (sour)	3325
Alfalfa	3605	Leek	3010	Peach	3325
Oat	4200	Brussels sprouts	3685	Cherry (sweet)	3325
Beetroot	4200	Cabbage[c]	3685	Gooseberry	3500
Barley, spring	5950	Cauliflower	3685		
		Spinach	3685		
		Chicory	3685		
		Kale	4795		
		Radish	4795		
		Purslane	4795		

[a] For keeping.
[b] For seed.
[c] Green savoy.
[d] 1951 data.

Source: Eaton, F. M. 1966. Chlorine. In H.D. Chapman (Ed.), *Diagnostic Criteria for Plants and Soil* (pp. 98-135). University of California, Division of Agricultural Sciences, Riverside, CA.

Leaf "burn" of many plants caused by the overhead application of water high in chlorides results from the buildup of salts as the water evaporates. The presence of Na or other salts increases this type of damage. Such injury is lessened if overhead irrigation is applied during periods of reduced evaporation (nighttime or the cool part of the day and periods of little wind), by increasing speed of sprinkler heads to make at least one revolution per minute, and by moving portable units in the direction away from the wind.

Sodium and Sodium Adsorption Ratio

Many plants are sensitive to Na while others respond favorably to its application. Sensitive plants were listed in Table 3.1; those that respond to Na, with and without sufficient K, are given in Table 12.2.

Almond, apricot, avocado, citrus, and plum leaves rapidly absorb Na from leaf applications and are injured by very low concentrations in overhead irrigation water. Bypassing the leaves allows safe use of concentrations 5 to 10 times that which would be harmful if applied directly to the leaves.

TABLE 12.2. Response to Sodium with and without Adequate Potassium

Response with ample K		Response with insufficient K	
Large	Slight to moderate	Large	Slight to moderate
Beet	Cabbage	African violet	Beans, white
chard	Celeriac	Alfalfa	Buckwheat
fodder	Coconut	Asparagus	Clover, red
mangel	Kale	Barley	Corn
sugar	Kohlrabi	Broccoli	Cucumber
table	Lupine	Brussels	Grass
Celery	Oat	sprouts	Bahia
Spinach	Pea, English	Carrot	Bermuda
Turnip	Radish	Chicory	carpet
	Rape	Clover, ladino	Pensacola Bahia
	Rutabaga	Cotton	Weeping love
		Flax	Lespedeza seicea
		Grass	Lettuce
		pangola	Onion
		Sudan	Parsley
		Horseradish	Parsnip
		Millet	Peppermint
		Rape	Potato, Irish
		Salsify	Rye
		Stock	Soybean
		Tomato	Spearmint
		Vetch	Squash
		Wheat	Strawberry
			Sunflower

Source: Lunt, O. W. 1966. Sodium. In H. D. Chapman (Ed.), *Diagnostic Criteria for Plants and Soils,* p. 411. University of California, Division of Agricultural Science, Riverside, CA.

Application of water containing high levels of Na can cause soil physical problems, but the extent of the problem is modified by the presence of OM, type of clay and its amount in the soil, and the amounts of Ca and Mg present in the water. The adverse effects are greater with increasing amounts of clay, particularly if the dominant clay mineral is montmorillonite rather than illite, kaolinite, or vermiculite.

The amount of Na that water can contain and still be suitable for irrigation is well expressed by the sodium adsorption ratio (SAR) ratio, presented in Chapter 4.

A ratio of <3.0 has been considered safe for most soils and crops, with problems increasing as the SAR rises above 3.0, becoming very serious if the SAR is >6.0. The amount of dissolved salts, however, has a bearing on the suitability of the different values, which is depicted in Figure 12.2. The four classes of Na hazard shown in this graph have the following characteristics:

1. *Low Na hazard.* Such waters can be used on nearly all soils with little danger of building up excessive Na saturation, but needs to be watched for an accumulation of Na that will injure sensitive crops.
2. *Medium Na hazard.* Can be used for coarse-textured or organic soils with good permeability. May present some problems on fine-textured soils if leaching is limited and there is no gypsum in the soil.
3. *High Na hazard.* Can expect Na problems on most soils with continued irrigation, unless soils contain considerable gypsum and OM and have good drainage.
4. *Very high Na hazard.* Usually unsatisfactory for irrigation except at low or medium conductivity, or if the soil contains large amounts of gypsum, or the water is treated with gypsum (Richards, 1934).

Irrigation and Excess Salts

As discussed in Chapter 8, irrigation is largely responsible for high salt conditions in many parts of the world—particularly in arid regions where there is insufficient rainfall to leach out salts. The quantities of water applied as well as the irrigation system can be used in keeping the salts at safe levels. The concentration of salts in irrigation water can be reduced by either allowing it to be diluted by rainwater in ponds prior to use or by the processes of ion exchange or reverse osmosis. Dilution by rainfall is possible only in certain regions. If used, efforts should be made to restrict evaporation by the use of deep rather than shallow ponds and/or flotation of plastic beads or films. Removal of salts by either ion exchange or reverse osmosis is rather expensive and can only be considered for high-value

FIGURE 12.2. Classification of Irrigation Waters Based on Their Conductivity and Sodium Contents

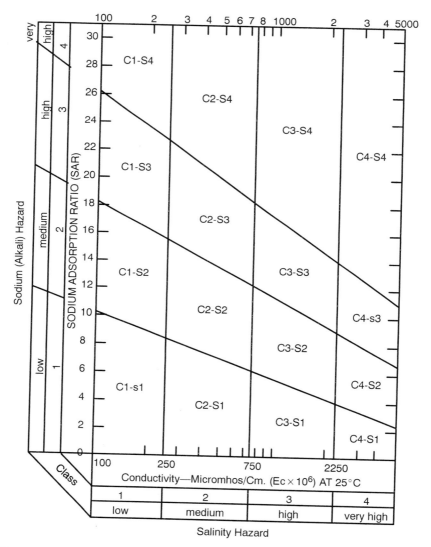

Source: Richards, L. A. (Ed.). 1934. Diagnosis and improvement of saline and alkaline soils. *USDA Handbook 60*, U.S. Salinity Laboratory, Riverside, CA.

crops. Even if used for such crops, water may need to be restricted to (1) propagation of cuttings, (2) spraying of sensitive crops, and (3) the final rinsing of leaves after overhead irrigation with high salt water.

It is desirable to use low-conductivity irrigation waters to avoid buildup of salts, but some water with relatively high conductivity can be used if the crop is salt tolerant and/or there is sufficient rain to prevent harmful accumulations of salts (see Table 12.3).

Salts in irrigation water can be compensated for (within limits) by altering the amounts of applied water (see Table 12.4).

TABLE 12.3. Permissible Number of Irrigations Between Leaching Rains Using Water Containing Varying Amounts of Salts and Applied to Crops with Different Salt Tolerances

Irrigation water		Number of irrigations for crops having different salt tolerances		
Elec. cond. (mmhos/cm)	Total salts (ppm)	Good	Moderate	Poor
1	640	—	15	7
2	1280	11	7	4
3	1920	7	5	2
4	2560	5	3	2
5	3200	4	2-3	1
6	3840	3	2	1
7	4480	2-3	1-2	0
8	5120	2	1	0

Source: Lunin, J., M. H. Gallatin, C. A. Bauer, and L. V. Wilcox. 1960. *Brackish Water for Irrigation in Humid Regions.* USDA and Virginia Truck Exp't Sta. Agric. Inf. Bull. 213.

TABLE 12.4. Percentage Increase of Applied Water with Different Salt Contents Necessary to Leach Enough Salts to Prevent Yield Reduction of Crops Having Different Tolerances to Salts*

Irrigation water		Tolerance to salts (%)		
Elec. cond. (mmhos/cm)	Total salts (ppm)	Good	Moderate	Poor
1.15	736	10	14	29
2.3	1472	19	29	57
5.75	3200	48	71	Not possible

*Percentages beyond the amounts of water normally applied to fill the soil MHC.

Source: Lyerly, P. J. and D. E. Longnecker. 1959. *Salinity Control in Irrigation Agriculture.* Texas Agric. Expt. Sta. Bull. 876.

Overhead Irrigation

As mentioned above, the overhead application of water containing excess B, Cl, F, or Na often allows too much of these elements to remain on the leaf surfaces. If irrigation waters contain excesses of these elements, it would be better applied by one of the systems that bypasses the leaves.

Overhead irrigation is useful for lowering the salt concentrations that accumulate near the soil surface from evaporation. By diluting the salts and moving them downward, overhead irrigation permits good growth of many plants that might suffer from the accumulation. If salts become excessive even with dilution, overhead irrigation can be used effectively to leach salts from most soils.

Furrow Irrigation

Some control of salts can be effected with furrow irrigation by placement of the plants in relationship to the irrigation furrow. As mentioned earlier, use of a double plant row per bed or placing the plants on the bed slope rather than the top permits use of irrigation water with elevated salts. The salts tend to rise to the highest point of the bed, limiting the salts around the plant.

Irrigating every other furrow also allows better use of saline water or high-salt conditions even if plants are in the center of the row. The alternate furrow watering permits addition of enough water so that the front and its accumulation of salts move to the unirrigated furrow between the two irrigated furrows (see Figure 12.3).

Systems that supply small quantities of water frequently are apt to cause less difficulty with salts than those that apply large amounts intermittently. Using short runs with furrow irrigation permits frequent irrigation. While not as suitable for leaching out salts as the systems that supply water to the entire surface, furrow irrigation, properly managed, can be used to prevent salt buildup and to leach out excess.

Drip Irrigation

Drip irrigation allows maximum use of soils with elevated salts or use of saline water because the salts tend to move with the water front and so can be moved away from plants if sufficient water is used. Salts are least troublesome if drippers are placed close to the plant.

Drip irrigation permits high-frequency application and thus prevents the rapid increase of concentration of the soil solution. It also can be effectively used in washing out excess salts.

FIGURE 12.3. Variations in Salt Accumulation Due to Irrigation Management

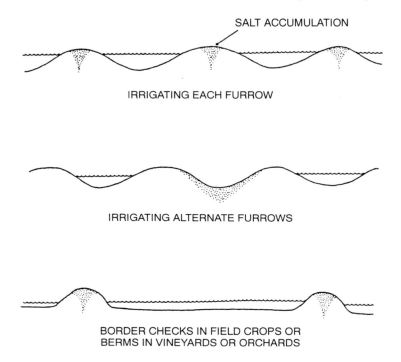

Source: *Western Fertilizer Handbook,* Seventh Edition. 1985. Interstate Printers & Publishers, Inc. Danville, IL, p. 36. Reprinted by permission of California Fertilizer Association, Sacramento, CA.

Subsurface Irrigation

This is probably the poorest way to deliver saline water as the salts tend to rise and accumulate close to the surface. Unless there are sufficient intermittent rains to wash the salts downward or dilute them, concentrations near the surface can soon build up to intolerable levels.

Flooding Irrigation

Flooding with irrigation water may present no more hazards than overhead irrigation, but in variable soils, salts can accumulate in certain areas, making for less uniform growth. Leveling the fields with lasers and introduction of organic materials to hasten water penetration can help overcome the poor distribution of water and the erratic accumulation of salts.

Flooding can be an effective way to leach out excess salts, particularly if the soil has been leveled. Effective use requires building barriers to hold water in place. A common practice of ponding a total of 2 to 3 feet of water requires extensive earthmoving. Intermittent flooding using only 2 to 4 inches of water at a time and keeping the soil wet 60 percent of the time will accomplish the same leaching with less total water use and will not require extra land preparation.

Water Management

Damage from high salts caused by evaporation of water can be alleviated by adding enough replacement moisture. With high salt levels, it may be necessary to keep the water level close to field capacity, particularly if the water table is contributing to the salt problem. Tensions, as measured by a tensiometer, need to be kept in the 10 to 20 range for sands and perhaps 25 to 40 for the heavier soils. Evaporation is apt to cause more harm on soils that retain the least water as the soil dries (sands or soils with low levels of OM). With salts in or approaching the dangerous range, it becomes doubly important to add water to these soils to prevent the rise of salts to damaging levels. In the long run, such soils profit from the incorporation of extra OM derived from residues, cover crops, and rotations including sods and hayfields, etc., because these additions increase the amount of water that can be held by the soil.

Reducing the amounts of salts introduced by saline seeps, fertilizers, and other amendments becomes especially important if evaporation can increase salt to dangerous levels, and if it cannot readily be reduced by adding suitable water.

High concentrations of salts due to evaporation that build up on openings in plastic culture can be corrected by adding water directly into the openings. Sufficient water must be added to dilute the salts and move them away from the collar of the plant. The additions may have to be repeated every several days until rain falls or the plant becomes large enough for the leaves to shade the opening.

Excess buildup of salts at the top of beds is a much more common occurrence. It, too, can be corrected by application of water as by overhead irrigation, but the problem can be avoided by planting two rows at the top of the bed or a single row on the side of the bed (see Figure 12.1).

The harmful effects of excess salts in irrigation waters as well as other sources can be reduced by the manner in which the water is applied. If there is insufficient rain, it is necessary to periodically add enough water to dilute the salts or move enough salts out of the root zone so that the growth of the particular crop is not reduced. Usually applying about 10 percent more water than is necessary for evapotranspiration can prevent an accumulation of salts, but greater amounts are necessary as salt content of the irrigation water increases (see Table 12.4).

Depending on the quality of water and the salts present, there comes a time when it is difficult or impossible to add enough water to provide a safe salt condition for the crop without leaching. The large quantities of water needed will make it difficult for plants to obtain sufficient oxygen. At or shortly before that point, sufficient water needs to be added to leach out a good part of the excess salts.

Removal of Excess Salts

Leaching

Excess salts in soils or other media can usually be removed by leaching. Obviously, leaching should be kept to a minimum to avoid water and nutrient losses and to prevent pollution of groundwaters or water downstream. The amount of water to add for leaching varies with different soils and conditions. The needs of media in benches or containers can readily be observed by the appearance of the drainage. Soils usually need slightly more than is necessary to fill the MHC. These amounts can be determined from MHC data, and then determining the amount of present soil moisture by one of the methods presented in Chapter 16.

The problem of leaching out salts may be more complicated than simply adding irrigation due to several conditions: (1) percolation may be impaired by soil dispersion, compact layers, and/or the presence of large amounts of clay; (2) uniform leaching is not possible because of uneven water application; and (3) salts in the water may limit infiltration while adding more salts.

Poor percolation should be ameliorated prior to adding the leaching water. That resulting from soil dispersion may be corrected by the addition of gypsum or sulfur or its various compounds. Compact layers and hardpans can be broken up by subsoiling or deep plowing. The uniformity of application can be improved by soil grading, changing nozzles, or cleansing irrigation delivery systems, or in changes of irrigation design.

Additions of water should not be considered unless water tables have been lowered enough by drainage so that effective leaching can take place. Sufficient water table depth will vary somewhat with the crop, soil, and salinity of the ground water, but the designations given in Table 12.5 can be used as a rough guide. The "good" category should be satisfactory without

TABLE 12.5. Classification of Water Tables

Rating	Range in water table depths
Good	Static water table below 82 inches but may rise to 46 inches for a period of about 30 days a year.
Fair	Water table at 82 inches but may rise to about 46 inches for a period of 30 days. No general rise.
Poor	Some alkali on surface; water table 46-82 inches but may rise to 36 inches for a period of 30 days.
Bad	Water table less than 46 inches and rising. Natural or artificial drains too far away to drain land.

Source: Hansen, U. E., O. W. Israelson, and G. E. Stringham. 1979. *Irrigation Practices and Principles.* John Wiley and Sons, Salt Lake City, UT. Reprinted by permission of John Wiley and Sons, New York.

lowering for drainage, and the "fair" category may be satisfactory for many situations. Generally, the "poor" and "fair" categories will need lowering of their water tables before leaching is attempted.

Amounts of water needed for leaching. The amounts of water necessary will depend on the crop, type of soil, quality of the irrigation water, the composition of excess salts, and the manner in which the water is applied. The total amount of salts in the soil profile has little effect on the amount of water necessary to leach out the excess.

If the levels of Na, bicarbonates, or carbonates are high enough to impede infiltration, some corrective materials will have to be added. Gypsum can be helpful in driving out Na, but if the water has high levels of bicarbonates or carbonates, sulfuric acid or sulfur dioxide might be better choices because of their reduction of bicarbonates.

The simplest calculation of leaching requirement, proposed by the USDA in 1954, is the minimum amount of applied water that has to pass through the root zone to avoid salt buildup. It is expressed as the fraction of water to be added to that being applied to keep salts in a satisfactory range. The leaching requirement (LR) is calculated from equation 1:

$$LR = \frac{ECw}{ECdw} \tag{1}$$

where ECw = conductivity of the applied water and $ECdw$ = conductivity of the drainage water in mmhos/cm.

A crop's salt tolerance needs to be considered in determining the amount of leaching. In equations 2 and 3 given below, the LR is calculated on the basis that remaining salts will give no more than a 10 percent reduction in yield. The electrical conductivity causing a 10 percent reduction of a tolerant crop such as barley can be almost ten times as great as that of a sensitive crop such as beans (8.0 versus 1.0).

It has been found that it may not be necessary to maintain a steady salt balance at all times and accumulation of some salts can take place in lower root zones for short periods, providing salt balance is reached in time and sufficient water is maintained in the root zone. By applying sufficient water in the upper root zone, the upper soil is leached with each irrigation and frequent reduction of the leaching requirement has minor benefits. More recently, two different methods for calculating the leaching requirement have been proposed. The first (see equation 2) is suitable for surface irrigation, including sprinkler irrigation that is not applied frequently. It is expressed by the following equation:

$$LR = \frac{ECw}{5ECe - ECw} \tag{2}$$

where LR and ECw have the same designations as in the above equation and ECe = electrical conductivity of a soil saturation extract expressed as mmhos/cm at $25°C$ that will cause no more than 10 percent reduction in yield (see "MS" in Table 8.2).

The second method of calculation is more suitable for high frequency sprinkler or drip irrigation and is expressed in equation 3:

$$LR = \frac{ECw}{2(max\ ECe)} \tag{3}$$

where the designations are the same except that ECe = electrical conductivity of a soil saturation extract that will cause 100 percent reduction in yield.

If the water used for irrigation has a conductivity of 2.5 mmhos/cm, the electrical conductivity of soil saturation extract is 5.0 mmhos/cm, and alfalfa is to be grown with low frequency irrigation, the

$$LR = \frac{2.5}{5(3.4) - 2.5} \quad \text{or } 0.17$$

If the same water is used to grow alfalfa except that water is frequently applied by overhead irrigation, the

$$LR = \frac{2.5}{2(15.5)} \quad \text{or } 0.08$$

Leaching Out Toxic Salts

At times, leaching is complicated because in addition to salinity, poor growth may be due to an accumulation of elements that are harmful to plant growth. The elements most commonly causing problems are B, Cl, and Na.

Boron accumulations toxic to plant growth can occur with relatively low concentrations, especially if the crop is sensitive to it. Unfortunately, it is more difficult to leach than several ions that may be causing problems because of excess salinity. Whereas the leaching of most ions follows the general rule that leaching with 12 inches of water through a 12-inch depth of soil removes 80 percent of the salts, it probably will take 36 inches of water to leach out 80 percent of the B in that soil.

Generally, B concentrations in the soil saturation extracts of the upper root zone are quite similar to those of the irrigation water. Problems are best avoided by choosing irrigation water low in B, but if impossible, selection of tolerant plants can permit its use for some time. The addition of sufficient water for leaching will aid in maintaining suitable levels for continued growth.

Chlorine as chlorides, on the other hand, is relatively easy to leach and the general rule readily applies. If accumulation of Cl is more of a problem than excess total soluble salts, calculation of the leaching requirement should be modified to one given in equation 4.

$$LR = \frac{Clw}{Cldw} \tag{4}$$

where Clw = Cl in the irrigation water and $Cldw$ = maximum permissible concentration of Cl in the drainage waters.

Unfortunately, the precise levels of Cl causing specific reductions of yield at various concentrations is not too well established for many crops. Some values provided in Table 12.1 can be used as a guide.

Manner in which water is applied. The manner in which water is applied for leaching has a pronounced effect on the efficiency of the process in removing the excess salts. The common practice of applying the same depth of water as the depth of the soil to be leached is a poor way of getting rid of excess, especially if the water is added at once as ponding.

Usually, leaching can be carried out by smaller additions of water applied more frequently, either by continuous sprinkling or intermittent flooding. The relative costs of the three methods, though, are generally in the order intermittent flooding (1.0) < continuous ponding (3.0) < sprinkling (3.6).

Studies indicate that more efficient use of applied water for leaching is made if the water is applied at a slower rate. Four inches of rain has been found to leach out more salts than 6 feet of water applied by ponding. A low water application rate by sprinklers of 0.5 inches per hour leached out substantially more salts than the same amount applied at 1.25 inches per hour. In these studies, by the time 0.5 inches of drainage water was collected, the 1.25-inch rate leached out only 65 percent of the salts present whereas the 0.5-inch rate had removed 78 percent (Keller, 1965).

The better leaching efficiency of rain or slower irrigation application rates appears to be due to the better diffusion of water into the small nonconducting pores of the soil and its movement of salts to the large conducting pores where it may be leached out. The diffusion takes considerable time and is not benefited by continued rapid application of water. The latter tends to move water largely through large pores and cracks, allowing little extra time needed for the diffusion process to work (Keller, 1965).

The usual practice of adding nutrients in every irrigation for potted cultures, supplying waters with elevated salts, usually makes it desirable to add enough water periodically (about every two weeks) to leach out the excess. Because of the open nature of most pot culture media, there is no physical problem with supplying this extra water since rapid drainage permits quick restoration of sufficient air around the roots. Leaching salts may cause problems with pollution of potable sources of water, requiring provisions to catch the leachates and reuse the water.

The situation is vastly different in many soils, where poor drainage is caused by compaction or results from considerable silt and/or clay in the profile. The slow drainage may keep enough water about lower roots so that lack of oxygen does become a problem. In such cases, removing the excess salts should be left for periods when the area is not cropped.

Proper handling of water on soils with drainage problems can limit the damage from excessive salts. Frequent small applications that can maintain sufficient moisture are preferred to heavy infrequent irrigations. An exception to this rule is the overhead application of water relatively high in B, Cl, F, or Na to crops sensitive to these ions. Infrequent heavier applications washes proportionally more of the harmful chemical from the leaves, which can be damaged by relatively small amounts of these elements. If plants not tolerant to these ions are grown, problems can be reduced or avoided by (1) applying irrigation to the soil rather than overhead, (2) using fewer but heavier irrigations as suggested above, and (3) applying irrigation during the evening hours to minimize stress from the environment.

Salt content of the water must be taken into account as foliar applications of nutrients or pesticides are made. In addition to the ingredients, it may exceed the tolerance of some plants.

Drainage

Except for humid conditions and well-drained soils, it is only a matter of time, regardless of the irrigation system used, for salt problems to become serious unless satisfactory drainage is present. Drainage is seldom a problem with a good water table, but becomes an increasing problem as the water tables are rated from fair to poor (see Table 12.5).

Use of water high in salts or containing Na, bicarbonates, or carbonates will hasten that time. Using good-quality water and fertilizers or other amendments low in salts slows the process. Good drainage and using sufficient water to periodically leach out the accumulation of salts allows long-term use with limited complications. If nature does not supply enough drainage, artificial drainage may be needed.

Often, hardpans or layers of clay can interfere with drainage, and salt problems can arise even with sufficient water for leaching. The drainage may be corrected by breaking up the layers with subsoilers or by deep plowing. Unlike the action of subsoilers or chisels, the effect of which is short lived, that of deep plowing or slip plowing can be more permanent. As indicated elsewhere, gypsum combined with subsoilers or chisels can improve soil infiltration and prolong its effect.

Drains of various types may be needed to move seepage water past impermeable layers, outcrops, decrease in grade, and various layers or hardpans that are too difficult to shatter. Ditches, drainage wells, "moles," perforated pipes, or tiles allow movement of water by hydraulic gradients existing in the soil. Systems for collecting, transporting, and disposing of the drainage waters need to be included with all artificial drainage systems. An outline of potential systems is presented in Chapter 13.

SUMMARY

Both low and high levels of salts can have adverse effects on the fertile triangle. Low salts, usually due to insufficient nutrients, tend to affect the nutrient side of the triangle. Low salts are often due to insufficient nitrogen and/or potassium and may be aggravated by use of low-salt irrigation water. Low-salt water can also seriously corrode irrigation pipe and can affect infiltration of water into the soil.

High salts are a much more common problem, especially in arid regions and on soils with low CEC and low OM or certain clay soils. The clay soils may take longer to become salty but are much more difficult to correct when they do.

Excess salts also adversely affect the nutrient side of the triangle, but if they are due to large amounts of bicarbonates and sodium, they can also affect the water and aeration sides of the triangle.

Excess salts can be tolerated within certain limits by using one or more of the following strategies: (1) crop selection; (2) planting on sides of beds; (3) using fertilizers low in salts and placing them properly; (4) using waters low in salts or enough water to leach out salts; and (5) drainage systems to aid in leaching out salts. Drip irrigation probably is most suitable for use with high-salt water or high salts in the soil because it tends to move salts out of the root zone.

All irrigation systems eventually lead to a buildup of excess salts unless nature or farm practices provide enough suitable water to drain away the excess. Leaching of salts is aided by irrigation, of which overhead and ponding are very effective, but an open soil with good drainage is necessary if the excess is to be washed out. If the soil is not naturally well drained, it needs to be opened by chemical and mechanical means including the installation of drains to provide good drainage to wash out excess salts. Rain or slow or intermittent irrigation will provide more leaching than the same amount of water applied quickly.

Chapter 13

Reducing Damage
from Excess Water

Excesses of water can be as harmful to crop production as shortages. Water can fill enough of the large soil pores so that there is insufficient gas exchange in the root zone. This results in insufficient oxygen so necessary for respiration. There is a concurrent accumulation of carbon dioxide and ethylene emanating from the roots, while a number of toxic substances can be produced from the soil in the low-oxygen environment. The effects seriously alter the aeration side of the triangle, with serious consequences for the nutrition and water sides as well, because the lack of oxygen interferes with nutrient uptake by the plant and the activities of microorganisms that affect plant nutrition.

Excess water can result from flooding, heavy rainfall, overirrigation, seepage from upper levels, or high water tables from the close presence of bodies of water. The possibilities of damage from rainfall or irrigation are related to the amount of water, the rate of its application, and soil infiltration.

Much of the damage from excess water is due to its poor movement through soils. In coarse soils or in media containing considerable perlite or other coarse materials there is usually enough movement of water through the large pores to maintain sufficient oxygen, but such movement in the heavier soils (medium and heavy loams), with their limited amount of large pores, is restricted to cracks and spaces between soil crumbs. The problem is compounded in heavy or coarse soils if the subsoil is compact, preventing rapid water movement. Often these can be opened by deep subsoiling.

REGULATING THE WATER TABLE

The height of the water table has a profound direct effect on both the air and water legs of the triangle and an indirect effect on nutrients (see Table 12.5 for description of water tables). The depth of the water table varies diurnally, falling during the day but rising during the night. The capillary fringe of the water table, or the portion of the soil that is kept wet by capillary rise, must be kept below certain levels to maintain satisfactory amounts of air and conditions for microbial and root development. If allowed to rise too high, air may be insufficient for root or microbial respiration and there is considerable danger that salts may accumulate at or near the surface, interfering with normal root development. If kept too low, the soil may become too droughty, requiring considerable irrigation.

The ideal level of the water table varies with different crops, soils, and salt levels of the ground water. Most agricultural crops will need a water table deeper than about 3 feet although a higher water table is satisfactory for pastures with their relatively short root systems and no need for cultivation. Higher levels also can be tolerated by most plants on sandier soils, and the higher levels may need to be maintained to slow the subsidence of peaty or organic soils by rapid oxidation of organic matter. Much lower water tables are needed for fruit trees, which have deeper root systems and need to be well anchored in the soil to prevent wind damage.

A range between 3 and 6 feet with good management can produce good yields of many crops. Generally, the maximum height of the water table plus the capillary fringe should not be allowed any higher than the normal one third of the root zone. Rooting depths of several crops given in Table 2.11 can be used to estimate desirable water table depths. But much lower water tables need to be maintained if ground waters contain enough salts to damage plants. In such cases, it may be desirable to keep the capillary fringe below the root system, using water tables below 10 feet and relying very heavily upon irrigation if rainfall quantity or frequency fails to maintain desirable moisture.

CORRECTING THE PROBLEM

Damage from excess water has been eliminated or sufficiently reduced to allow production of good yields by (1) increasing permeability of soils, (2) use of raised beds, (3) diverting or blocking out water that may flood the land, (4) slowing the application of irrigation so that water can infiltrate without flooding, and (5) installation of ditch or pipe drainage.

Increasing Permeability

In most sandy soils, there are sufficient large pores to make drainage quite rapid unless compact layers impede drainage, the water table is too high, or the soil consists primarily of fine sand. In the heavier soils, even light loams, there may be insufficient large pore space for effective drainage. Often the drainage of these soils can be improved by the additions of OM, growing of deep-rooted crops, deep tillage, addition of gypsum or polymers, and reduction of excess salts, which are covered more fully in various sections dealing with soil compaction (Chapter 4), OM additions (Chapter 10), and salt removal (Chapter 12).

Maintaining maximum water infiltration through optimum soil permeability often is still insufficient to provide optimum air and nutrients since poor drainage may be the result of (1) muddy water running over the surface and filling in pores, (2) rain or irrigation compacting the surface so that water does not infiltrate fast enough, or (3) compact subsoils that impede internal drainage.

Muddy water running over the surface often can be diverted by proper smoothing and grading. Pores filled by particles compacted by overhead irrigation can be aided by changing nozzles to provide smaller droplets or use of a mulch to break up droplets. Mulches are also effective against heavy rains. Compact subsoils can be opened up by deep plowing or subsoiling.

Frequently, these measures are inadequate to reduce damage from excess water. Raised beds may be the answer if the water table is too high or if there is an excess of rainfall only during certain periods.

Raised Beds

An effective means of reducing damage from occasional flooding is to plant the crop on raised beds. By providing a well-aerated medium for roots, raised beds essentially accomplish the same purpose as lowering the water table. An added advantage is that in the presence of excess salts, raised beds can be used to provide a relatively salt-free environment if plantings are made on the side of the bed where salts do not accumulate. (See Figure 12.1. Also see "Ridge tillage" in Chapter 9 for common methods of preparation and handling of raised beds.)

Height of beds used for such purposes varies from about 4 to 8 inches (10-20 cm). Higher beds have been used but present some problems in preparation and maintenance. The very high beds are useful for permanently elevated water tables but preparing beds in the presence of high water tables often leads to puddling and poor soil structure, making the installation of drains a more practical solution. The use of high beds to alleviate the harmful effects of occasionally heavy rains may make it difficult to maintain sufficient moisture between rains or if water tables drop appreciably.

Raised beds work well when excesses of water are due to occasional heavy rains. Water that falls too rapidly or over too long a period saturates the soil because runoff and soil infiltration cannot remove enough of it. Beds allow extra time to remove the water because enough roots in the beds have sufficient oxygen to function. If the soil is graded properly, the beds allow considerable water to be removed by the channels between beds.

Some disadvantages of raised beds are the costs of preparation and maintenance. Other problems are the difficulty of cultivating them and maintaining moisture between rains. The difficulty of supplying moisture is greater with the coarser soils and as the height of the bed is increased.

Raised beds are even more effective if crowned in the center with open ditches or drains between a series of raised beds. A series of raised beds are crowned by back furrows and the ditches can be made from the dead furrows. The crowns, 25 to 100 feet (8-30 m) wide, are raised 1 to 2 feet (30-60 cm) in the center and have ditches or drains placed between mounds.

Raised beds may not completely solve the problem, and it may be necessary to supplement other measures. If excess water is due to seepage movement from higher areas, it may be diverted from the target field by installation of ditches to divert the water. If this does not correct the situation, it may be necessary to remove some of the excess by internal moles, open ditches, or buried tile or buried perforated pipes. In many cases, measures to improve permeability or raised beds will probably still be needed even after water has been diverted and excess removed by ditches, moles, tile, or-perforated pipe.

Diversion

Potential flood water can be diverted by altering stream flow or by installing dikes. Supplemented with pumps to remove excess water and gypsum applications to facilitate salt leaching, dikes have been helpful in reclaiming much valuable farm land in the Netherlands.

Much of the excess surface water coming from elevations above target fields often can be moved over the field by proper smoothing to eliminate small depressions and grading to a uniform slope. Open ditches may be needed to help move excess water.

Surface or subsoil water coming from elevations above the target field can be diverted by a ditch perpendicular to the slope. Diversion of water is aided by raised beds that are on a contour and gently sloped.

Slowing Water Application

The slow application of water can be effective in reducing water excesses that otherwise may create oxygen shortages. Rapid addition of water can deteriorate soil structure, reducing drainage and limiting its ability to restore satisfactory aeration.

Drip irrigation is an ideal method of applying water at slow rates. The normally slow rate of 0.15 to 0.25 in/hr can be reduced to 0.1 in/hr if high pressure emitters delivering 0.5 gal/hr are used. Care must be taken to avoid uneven water distribution because the use of such low rates with low-pressure emitters may not provide enough volume to fully pressurize the lines.

Intermittent irrigation cycles are another means of avoiding water excesses that can lead to runoff and puddling in soils of low permeability. Intermittent cycles of 15 minutes on and 15 minutes off will cut the normally slow rate of drip irrigation in half and this usually will be satisfactory for light soils of poor permeability. Even longer cycles may be desirable for heavier soils with poor infiltration rates.

DRAINAGE

Soil drainage needs to be considered whenever water tables are too high or if, after increasing soil permeability, diverting water, and breaking up hardpans, the excess surface water still cannot be removed from the root zone in a reasonable period of time. In addition to the advantages of removing excess water, drainage can help materially in warming soils in the spring, reducing damage due to heaving of soils as they are frozen, increasing the depth of the root zone, reduced loss of N by microbial populations, and removing excess salts.

Water tables can be lowered by utilizing natural drainage systems, open ditches, moles, covered porous pipe or tiles, and pumps. Surface drainage, which utilizes open ditches and soil grading to promote rapid runoff, is the simplest approach and is suitable for all types of soil. Subsurface drainage, using moles, covered pipe, or tile, may not be suitable for some types of soil. It works well for soils that are porous enough to allow water to percolate down to the drainage device, but may not be suitable for the fine-textured clays, which have poor percent drainage capacity.

Percent drainage capacity can be used to estimate the suitability of a soil for subsoil drainage. Since pore spaces that are largely filled with water do not move water readily down to perforated pipes or tiles, percentage drainage capacity (Pdc) is equal to percent of soil volume that is pore space (Pvps) minus percent water-filled pore space (Pwfps), all on a volume basis, or

$$Pdc = Pvps - Pwfps$$

Adequate drainage is obtained with Pdc values of 10 percent or more, whereas very poor drainage through submerged pipes or tile can be expected if the moisture held is equal to or greater than the total dry soil pore space $(0 - Pdc)$.

Choosing a surface or subsoil drainage system will depend on the source of excess water, its direction of flow, amount of water, type of soil, and crop to be grown. A complete survey by knowledgeable personnel is usually worthwhile before choosing a system or attempting its design.

Surface Drainage

Natural Drainage

Freshets and streams need to be kept free of debris so that water flows freely at all times. Potential waterways can be graded to increase flow and kept in grass to avoid erosion. Water moved by these systems can be collected in ponds, but the surrounding soil needs to be high enough to avoid producing additional wetlands.

Open Ditches

Sufficient drainage is often made possible with open ditches, which are used mainly for moving water to remote outlets. They work well in intercepting water and diverting it from the cultivated areas. Water that has moved downward through the surface in irregular terrain flows laterally in the direction of the slope. By placing the ditch toward the bottom of the slope, perpendicular to the flow, much water can be diverted from land that otherwise might be flooded.

Open ditches can be used to divert water away from seeps that result from decrease in soil permeability or grade. If this water cannot be pumped out, drains can eliminate a considerable amount, providing there is a suitable outlet. Open drains are also used to remove excess water from broad valleys where there is little slope (relief-type drains). Depth and width of the ditches as well as the spacing between them are decided by the type of soil and the amount and intensity of rainfall that needs to be moved. Designs usually take into consideration the maximum amount of water that will fall in one hour in a five-year period. Clay or other impermeable soils will require much closer ditch spacing than the more open permeable soils. Often, a number of drains are needed for this type of drainage, making open ditches unduly wasteful of productive land and causing excessive interference with cultivation.

Side slopes of the ditches vary with type of soil formation from about $1/2$ horizontal to 1 vertical in stiff compact clays to to 3 horizontal to 1 vertical in loose open sandy soils. If machinery is to cross them, it is necessary to have side slopes of 8:1 (ratio of 8 inches linearly to every inch vertical). If machinery will not cross the ditches, side slopes of 2:1 can be built for loams, 1.5:1 for clays, and 1:1 for peat muck and sand. Bed slopes vary from about 0.05 to 0.15 percent, and depth of ditches need to be at least 1 foot, although depths of 5 to 10 feet are often used. Costs rise considerably as depth is increased beyond 6 feet. The depth of the drains, size, bed slope, and the slope of the sides is best decided by design engineers skilled in such construction.

Subsurface Drainage

Moles

Moles are small torpedo-shaped devices that are pulled through soils at about a 2-foot (61 cm) depth to create small ducts or channels that can move excess water out of the field into drainage ditches or other bodies of water. Considerable power is needed to pull the mole through the soil, but this can be reduced by using a shank with a knife edge. Usually the mole is moved through the soil every few feet. The mole drains created tend to be short lived and usually need to be reformed each year, but costs are considerably less than for buried tile or perforated pipe. Also, they can be used with perforated plastic liners that greatly extend the life of the mole drain. Mole drains are not suitable for all soils but do quite well for finer-textured soils and are very effective for organic soils.

Pipe Drainage

Drainage pipe, consisting of tile or perforated plastic, is also used for diverting water from upland areas as well as removing excess water from fairly level lands. It is often used to divert salts from upland areas or remove them from subsoils above hardpans, outcrops, or other impervious layers (see Drainage in Chapter 12).

As with open drains, spacing of perforated pipes or tiles and their depth is dictated by the type of soil and its permeability. Three main designs are used: (1) the herringbone system, which consists of long main drains with short collecting laterals; (2) the gridiron system, which has long parallel laterals that connect to a short main drain; and (3) the parallel system, consisting of a series of long parallel lines leading to a main drain. The system also needs manholes, sand traps, and observation wells for proper inspection and maintenance. Even more so than open drains, the design of covered pipe drainage should be left to people trained in this type of work.

It is necessary to envelop pipe drains in gravel or other porous materials. The envelope increases the effective diameter of the drain, but more importantly it (1) prevents fine particles from moving into the drain that eventually can clog it, and (2) it stabilizes the soil during drain construction and prevents later "wash ins." The material should completely envelop the drain, with enough under the drain, where the bulk of the drained water will come in.

Pumping Out Excess Water

Excess water may be due to artesian conditions of underground springs or movement of seepage water from above that is halted or slowed by impermeable layers. Drainage by open drains or pipe may not be practical. Pumping can be an alternative method of reducing the pressures of these aquifers. It should be a first consideration, if the wet area is rather localized, but the use of sumps and pumps to remove excess water should be reserved for areas that cannot be drained by gravity. Pumping also works well when combined with a diking system that reduces the amount of potential flood water.

Inverted Wells

Hardpans often impede water infiltration enough to seriously pond the surface water. Relatively shallow pans can often be broken up by deep plowing or subsoiling. At times, the pan is too thick or difficult to break up or may reform rather rapidly. In such cases, it may be amenable to a series of small-bore wells drilled through the hardpan into more permeable layers. These inverted wells are drilled with a screw auger about 9 inches (23 cm) in diameter and capable of drilling about 10 feet (349 cm) deep and then are filled with coarse gravel, except for the top several inches reserved for soil. The inverted wells can remove enough surface water to relieve some problems.

SUMMARY

Excess water can be harmful to crop production because it can reduce the amount of air available for plant roots and microbial functions, thereby affecting the aeration and nutrition sides of the triangle. Excesses may be due to rainfall or irrigation, but often are aggravated by poor soil permeability, high water tables, and seepage or runoff from higher areas.

Permeability problems occur more in the finer soils due to insufficient large pores or impermeable subsoils or, in the sandier soils, due to the presence of hardpans. Correction of the permeability problems of heavy soils can be made through the use of OM, liming materials or gypsum, or deep-rooted crops. Problems of impermeable subsoils in heavy soils or hardpans in the lighter soils at times can be alleviated by deep plowing or subsoiling.

The excesses from high water tables or occasionally heavy rains can at times be alleviated with raised beds or by diverting water from higher elevations by maximizing the natural drainage of freshets or streams or by means of grading and open ditches.

Excesses may still be a problem despite improvements in permeability, diversion of excess water, or use of raised beds. In such cases, it may be necessary to provide subsurface drainage by mole drains, covered tile or perforated pipe, or inverted wells. The suitability of these methods varies for different soils, and their use, as well as the designs for the buried tile or pipe, is best left to experts knowledgeable about subsurface drainage.

SECTION IV:
DIRECT INPUT TO TRIANGLE SIDES

It is possible to increase the three sides of the triangle by direct input of air, nutrients, or water, making marked changes that greatly increase the production potential.

The direct additions of air, presented in Chapter 14, have been limited primarily to liquid hydroponic cultures, although there have been some attempts to add substances that release oxygen in artificial media.

The additions of water, covered in Chapter 15, and nutrients in Chapter 16, have a much longer history and are much more used than the addition of air. The history of their use goes back to very early agricultural development. Direct inputs of nutrients and water are not only possible but are necessary in most soils and other media for good production of most crops. Without their additions, modern agriculture with its high yields would not be possible and in all probability much of today's population would be starving.

The literature dealing with the addition of water and nutrients is very extensive and increasing almost daily. Only the barest of outlines dealing with these subjects could be presented here. In presenting this material, I have attempted to summarize some of the salient features dealing with each so that the reader can get a basic understanding of their application.

Chapter 14

Adding Air

The addition of air to soil or solid medium occurs constantly as diffusion from the atmosphere. Air that is rich in oxygen is exchanged for soil air, which may be much lower in oxygen because of varying increased amounts of carbon dioxide, methane, and other gases. The exchange readily occurs in coarse-textured soils or media and in heavier soils with good structure, providing they have not been unduly compacted or overly saturated with water. Even those saturated with water soon return to good levels of aeration if the water can readily drain out of the soil or medium.

It is important that conditions be conducive to a ready exchange so that oxygen levels in soil or media are kept at satisfactory levels. As pointed out in preceding chapters, such exchange is facilitated by the additions of organic matter, liming materials, gypsum, and synthetic conditioners to soils, and organic matter and coarse materials to artificial media. The exchange is greatly facilitated by limiting compaction, promoting drainage, and avoiding excess or poor-quality water that can affect soil structure. Poor conditions can be corrected—at least temporarily—by tillage, additions of gypsum or liming materials, drainage, or elevating plant beds.

The addition of peroxides as seed coatings to rice in order to supply oxygen in water-saturated soils has been used commercially in Japan and the Philippines, and there has been some experimental work with hydrogen peroxide and several metal peroxides as seed coatings for barley, corn, grass, small grains, and soybeans planted in wet soils (Langan, Pendleton, and Oplinger, 1986).

Air is being injected directly in tile drains under some golf greens to increase air concentration at the lower roots. I have not been able to find scientific studies supporting this practice, but golf superintendents feel it is justified to compensate for possible compaction to which the greens are subject, especially during the rainy season when air may be naturally limited.

Aside from the use of seed coatings and tile drains under golf greens, there appears to be very little promise for direct inputs of air or oxygen to soils.

The possibilities for direct inputs in hydroponic cultures are much greater, and some measures for improving media oxygen levels currently being used are described here.

Most hydroponic cultures being used for various research projects are usually carefully given an abundance of air. The addition of air for aggregate culture is regulated by the frequency of the flooding (see the section, "Aggregate Cultures") and the size of the particles. Oxygen is added to liquid cultures by limiting the depth of the liquid, infusing the water with high levels of oxygen by various handling procedures, and the direct addition of compressed air.

AERATION IN HYDROPONIC SYSTEMS

Lack of air can be a problem in both liquid and aggregate cultures, but is usually more so in the liquid. The amount of oxygen in the liquid culture solution depends on the depth of the nutrient solution, its temperature, and the method of recirculation. Liquid cultures have been used without adding air, but generally satisfactory growth is limited to only a few weeks. By using large shallow tanks, placing a window screen about $1/2$ inch above the nutrient solution (roots grab the screen, grow rapidly, and branch out laterally to provide a large root system) and increasing air space between tank cover and nutrient solution (between 8 to 12 inches), satisfactory growth is increased for noncirculating hydroponic systems. Extra air can be introduced for liquid cultures by using compressors to bubble air through the solution and by handling the solution so that it falls some distance before returning to the header tank and by special recirculating procedures.

As a rule, more oxygen will be present in shallow nutrient solutions since larger amounts of air will diffuse into the solution with larger surface areas. But the amount of oxygen present is affected by the temperature of the solution, decreasing as the temperature rises—about 4 percent at $50°F$ ($10°C$) but dropping to only about 3.2 percent at $77°F$ ($25°C$) and about 2.7 percent at $100°F$ ($38°C$). Unfortunately, the need for oxygen increases as temperature rises due to increased respiration rates.

The volume, slope, flow rate, and frequency of application affect aeration of the solution. The volume needed for proper flow varies with different installations but about 2600 gal in the tank and 2400 gal circulating per acre are necessary to maintain aeration in the shallow flow solutions used for typical NFT (nutrient film technique) installations.

The NFT system—one of the more promising hydroponic systems—with its very thin film of circulating liquid is less prone to aeration problems than most liquid systems but it, too, will produce better plants if air is introduced positively into the system. This can be accomplished by (1) using a fish tank aerator for small systems; (2) providing a small air bleed in the suction side of the submersible pump; or (3) use of a nonsubmersible circulating pump that has 25 to 30 percent more capacity than is needed for circulating the solution so that a bypass can be fed. The bypass solution is then cascaded directly back to the catchment tank.

Adequate flow rates for NFT provide 2.1 qt per minute per gully. Flow rates greater than 4.2 qt per minute used for long gullies are satisfactory providing that the solution depth remains less than 10 mm (0.04 inch). Desired flow rates are obtained by carefully grading the slope or using raised systems with built-in slopes to provide a drop of 1:75 but never less than 1:100 for runs less than 67 feet. Longer gullies may increase aeration problems but this can be corrected by introducing the solution at several points along the run as well as the top. The frequency of application varies with the size of the plant and solar radiation. It is necessary to maintain a film of water around the roots at all times. At first, when plants are small and solar radiation is low, pumps may be switched on for one hour a day or as infrequently as once every seven days. The frequency of pumping must be increased as plants become larger. The system should be timed to full flow rates when weather conditions are favorable for plant growth and roots have become established out of the planting pots and into the gullies (Winsor, Hurd, and Price, 1979).

Some of the methods for aerating liquid NFT systems are used for other commercial systems. Fish tank aerators are used for small systems. Two other systems used for research purposes that are also suitable for small or intermediate-size hydroponic systems are (1) alternately filling and emptying a vessel so that it produces an intermittent supply of low-pressure air, and (2) a water suction pump is used to entrap air in a stream of water by arranging its discharge into a reservoir so that air can escape slowly through a capillary as the escaping water is slowed.

The efficiency of introduced air is increased by dispensing it as small bubbles. In this respect, introducing air into the cultures by sintered glass aerators, porcelain candles, or capillary glass jets bent into a P shape and placed horizontally with the jet directed horizontally tend to provide more efficient aeration (Hewitt, 1966).

Larger systems have used various methods of recirculating the solution to entrap as much air as possible. Generally the solution may be aerated prior to introduction, raised to a high level and then allowed to fall, or bypasses are used to recirculate the solution.

AGGREGATE CULTURES

To supply enough oxygen for aggregate cultures, it is necessary to replace most of the old solution with new liquid at each watering or feeding. This replacement is almost impossible with systems that supply the liquid from the bottom, allowing it to rise to the surface and then drain away (subirrigation). With subirrigation, the new liquid mixes with but does not displace the old. Even applying the liquid at the surface may not adequately replace the old due to the adherence of water films to the roots and the substrate after the solution or water is drained away. The problem of replacement varies inversely with the particle size of the aggregate. It may be worse with a mixture of particle sizes due to filling of the voids by different-size particles.

Generally, with suitable aggregate sizes that allow for complete drainage, sufficient air reenters the voids to allow normal growth. Sufficient air then depends on the frequency of the pumping and drainage periods. A 10- to 15-minute pumping period followed by a 15- to 30-minute drainage period has given very good growth of several plants in Israel (Schwarz, 1977).

Noncollapsible Aggregates

Aeration depends on maintenance of airspace. Vermiculite or other materials that collapse (bagasse, sawdust, composted wood products) in an irreversible manner are poor choices for media because of their relatively short life in which suitable oxygen levels can be maintained. To compound the problem, some of these materials lose water much more rapidly than gravel of sizes from 0.4 to 0.8 cm, requiring more frequent watering.

Particle Size

Larger particles require less aeration than the fine particles but need water more frequently. In one study, flooding was not necessary to supply enough oxygen for growing tomatoes on uniform particles and so it could be limited to every 28 hours to supply sufficient water (see Table 14.1). Mixing the particles required flooding every 2 to 2.5 hours to supply needed oxygen. With carnations, only the large gravel (1.5 cm) supplied sufficient oxygen so that frequent flooding was not necessary to maintain it, allowing flooding to be applied as infrequently as every 9 hours. However, with the 0.4 cm gravel, it was necessary to flood every 1.5 hours to supply enough oxygen, although flooding every 4 hours supplied enough water.

TABLE 14.1. Required Flooding Intervals to Supply Water or Oxygen for Carnation and Tomato Grown in Cultures of Different-Size Aggregates (hr)

Diameter of particles (cm)		Water		Oxygen	
		Carnation	Tomato	Carnation	Tomato
0.4	uniform	40	28	1.5	*
0.525	uniform	18	12	1.25	*
1.5	uniform	9	6	*	*
0.4-1.2	mixed	51	35	1.5	2
0.6-1.2	mixed	42	29	1.75	2.5

* Flooding not needed.

Source: Steiner, A. 1968. Soil-less culture. *Proc. 6th Colloquium of the International Potash Inst.*, Florence, Italy. International Potash Institute, Worblaufen-Bern, Switzerland.

Filippo or Netherlands Hydroponic System

This system, illustrated in Figure 14.1, ensures adequate aeration for aggregate culture. It is ideally suited for small operations with beds up to 20 feet in length (Figure 14.1, part A) containing gravel about 15 mm in diameter. The solution falls freely to provide optimum aeration. It is forced to stream through the bed because there is no pipe at the bottom. A leaking hole on the opposite side of the bed drains the solution at a rate of about $1/2$ that of the entry. The remaining half of the solution fills the culture bed until the level reaches the overflow. The solution is aerated again as it drains back to the tank. The pumping is regulated by a time switch so that the solution is pumped for a period slightly longer than it takes to fill the beds.

The method can be adapted to larger installations by use of special ducts made from glass, roofing tiles, etc. For larger installations (Figure 14.1, part B), the solution is directed across rather than through the length of the bed. The solution enters along the longitudinal wall of the bed via a duct wide enough so that the solution can stream freely and thus provide about the same rate of flow along the entire bed.

Hydrogen Peroxide

Theoretically, the addition of hydrogen peroxide (H_2O_2) could be used to supply oxygen at least to liquid cultures, as it simply decomposes to

FIGURE 14.1. Diagram of the Netherlands or Filippo System for Aggregate Culture

A. Diagram of a small installation

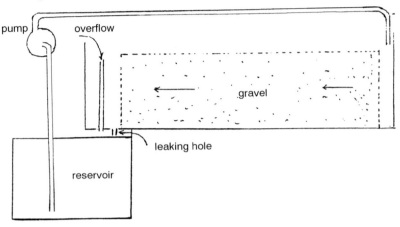

B. Diagram of larger installation

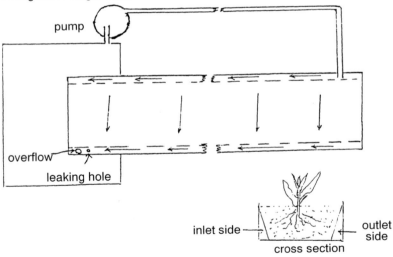

Source: Steiner, A. 1968. Soil-less culture. *Proc. 6th Colloquium of the International Potash Inst.,* Florence, Italy. International Potash Institute, Worblaufen-Bern, Switzerland.

water and oxygen. Small amounts of hydrogen peroxide (about 1 ppm in final solution) in water cultures has resulted in increased yields of cereals. Addition of hydrogen peroxide to supply 3 percent H_2O_2 in water cultures has reduced disease incidence in tea seedlings (cited in Hewitt, 1966). Evidently, the benefits have not outweighed the costs as the use of hydrogen peroxide to supply oxygen for the hydroponic production of various plants has not been adopted commercially.

SUMMARY

Although crop production is often curtailed on many field soils due to insufficient air, improvements generally have been brought about primarily by practices that allow better gas exchange rather than by direct inputs of air. Direct air inputs are being used to lengthen the aeration side for both liquid and aggregate hydroponic cultures. Aeration is increased in liquid cultures by limiting the depth of solution, increasing the flow rate, recirculating procedures, and direct aeration by various types of aerators. Sufficient coarse noncollapsible materials combined with frequent waterings from the surface are used to maximize aeration for aggregate cultures.

The Filippo system, which uses a solution that falls freely and is forced by special construction to stream through the bed before emerging, tends to provide very good aeration. The addition of hydrogen peroxide to solution cultures has been used in a very limited way, evidently due to rather high costs compared to mechanical means of adding oxygen.

Chapter 15

Adding Water

With increasing frequency, irrigation is being used to overcome temporary or long-lasting water shortages. Irrigation water affects the aeration and nutrition sides of the fertile triangle as well as the more obvious water side.

To obtain maximum benefits from the three soil influences on crop yield, the grower must pay close attention to amounts, quality, and timing of water applied. Insufficient water fails to provide full yields. Applying water when not needed or in amounts greater than the medium can hold leads to loss of nutrients by leaching, deterioration of soil structure, limits gaseous exchange within the soil, and contributes to pollution of water resources. Poor-quality water can cause failure of some irrigation systems, harm crops, and have long-lasting harmful effects on the soil. Unless the irrigation is timed to coincide with soil and plant needs, chances of obtaining the best crop yields with maximum quality are greatly diminished. The response to amounts and quality of water are closely connected to the irrigation system.

DETERMINING THE NEED TO ADD WATER

To provide suffcent water without excesses requires careful evaluation of the available supply. Water needs to be applied when available soil or medium moisture falls below a certain critical point. This point will vary with different crops and different stages of the same crop and has been outlined in Chapter 2. Having designated the critical point, it is necessary to determine the amount of available water and deliver additional amounts when these levels are suboptimal. The amounts delivered should not exceed the moisture holding capacity unless some leaching is desirable. At times, it will be necessary to limit the addition of water to levels appreciably below the MHC so that fruit quality may be enhanced.

The MHC can be estimated from soil class data as given in Tables 2.6 and 2.7 or more accurately by determining amounts of available water at the wilting point and subtracting from amount of water held at field capacity. This can be determined by some soil testing laboratories. The common method determines the amount remaining at field capacity by measuring water remaining in a saturated soil sample that has been subjected to $1/3$ atmosphere suction, and amount at the wilting point by measuring water remaining in the soil sample after it has been subjected to 15 atmospheres suction.

A less common method, requiring more time, determines field capacity by gravimetrically measuring the moisture remaining in a soil after it has been heavily irrigated and allowed to drain for a couple of days, while it is covered to reduce evaporation. The wilting point moisture is also determined gravimetrically from samples collected after growing dwarf sunflowers in moist but not rewetted soil, in a series of cans, until growth stops or permanent wilting takes place. Permanent wilting is indicated by rolling of the leaves that does not recover when plants are placed in a humid atmosphere.

The amount of available water can also be determined using bulk density measurements (bulk density can be determined by many soil testing laboratories) by the following procedures: (1) the percent water (weight basis) held at field capacity and at the wilting point is converted to volumes of water by multiplying the percent of water by the bulk density; and (2) the amounts of water held by a centimeter or an inch of soil are converted to the amount held by the effective root zone. For example, if the percent of water at field capacity = 25.6, that at the wilting point = 10.5, and the bulk density = 1.27 g/cm^3, then the volume percent of the available water = 0.256 − 0.105 × 1.27, or 0.192 cm^3 of water/cm^3 of soil. This is equivalent to 0.192 inches of water per inch of soil. The available storage capacity of an effective root zone can be obtained by multiplying a weighted average capacity by the root depth or adding together the amounts of water for all soil depth increments.

Evaluating the Adequacy of Soil Moisture

Once the storage capacity of the soil is known and the critical point has been established, it is only necessary to estimate the adequacy of existing moisture to determine whether irrigation is needed. Although simple in its concept, estimating the adequacy of existing water is neither simple nor very accurate, which probably accounts for the many approaches used.

Soil Moisture Measurements

Feel and Appearance

The simplest means of evaluating the need for irrigation employs plant appearance and feel of the soil. Wilting is an indicator of water need. Plant leaves are carefully examined during a sunny period in midday when transpirational losses are greatest. Leaves of plants needing water are flacid, lack turgidity, have a dark, dull appearance, and may be twisted or rolled.

The soil examination is made by removing a sample from the root zone by auger or soil tube. The number of samples examined varies depending upon depth of rooting and soil variation. In taking the soil sample, the presence of compaction or hard layers should be noted. The soil is evaluated by carefully rubbing it between the thumb and forefinger, squeezing a handful very firmly, and comparing the findings with values given in Table 15.1.

Although simple, the method has its limitations. It takes some time to train an operator to skillfully evaluate soil moisture status from the appearance and feel of the soil. Substantial yield may be lost by the time some plant symptoms appear. On the plus side, a great deal of useful knowledge about the soil regarding compaction and effective root zone can be gained by a skilled operator in making the soil examination.

Gravimetric

Soil samples are collected from the effective root zone, weighed, placed in an oven overnight at 212°F (100°C) and reweighed. The difference in weight divided by the dry weight × 100 = percent moisture in the soil. This value expressed as a decimal × bulk density = volume of water in cc per cm^3 or inches per cubic inch of soil.

The collection of sufficient samples for accuracy and the weighing process are time-consuming and tedious.

Tensiometers

A rapid method of determining water needs that has received a great deal of attention in recent years is the use of tensiometers. These devices consist of a porous cup and a rigid tube filled with water that is attached to a mercury manometer or vacuum gauge. Water can move freely through the porous cup in both directions. As the soil dries, water moves out of the cup, increasing the tension, which can be read from the manometer or the gauge. The system is relatively cheap, rapid, and adaptable for a wide

TABLE 15.1. Available Soil Moisture As Indicated by Appearance and Feel of Soil When Handled

Available water*	Feel or appearance of soil**			
	Sand	Sandy loam	Loam/silt loam	Clay loam/clay
Above field capacity	Free water appears when soil is bounced in hand.	Free water is released with kneading.	Free water can be squeezed out.	Puddles; free water forms on surface.
100% (field capacity)	Upon squeezing, no free water appears on soil, but wet outline of ball is left on hand. (1.0)	Appears very dark. Upon squeezing, no free water appears on soil, but wet outline of ball is left on hand. Makes short ribbon. (1.5)	Appears very dark. Upon squeezing, no free water appears on soil, but wet outline of ball is left on hand. Will ribbon about 1 inch. (2.0)	Appears very dark. Upon squeezing, no free water appears on soil, but wet outline of ball is left on hand. Will ribbon about 2 inches. (2.5)
75-100%	Tends to stick together slightly, sometimes forms a weak ball with pressure. (0.8 to 1.0)	Quite dark. Forms weak ball, breaks easily. Will not slick. (1.2 to 1.5)	Dark color. Forms a ball, is very pliable, slicks readily if high in clay. (1.5 to 2.0)	Dark color. Easily ribbons out between fingers, has slick feeling. (1.9 to 2.5)
50-75%	Appears to be dry, will not form a ball with pressure. (0.5 to 0.8)	Fairly dark. Tends to ball with pressure but seldom holds together. (0.8 to 1.2)	Fairly dark. Forms a ball, somewhat plastic, will sometimes slick slightly with pressure. (1.0 to 1.5)	Fairly dark. Forms a ball, ribbons out between thumb and forefinger. (1.2 to 2.5)
25-50%	Appears to be dry, will not form a ball with pressure. (0.2 to 0.5)	Light colored. Appears to be dry, will not form a ball. (0.4 to 0.8)	Light colored. Somewhat crumbly, but holds together with pressure. (0.5 to 1.0)	Slightly dark. Somewhat pliable, will ball under pressure. (0.6 to 1.2)
0-25% (0% is permanent wilting)	Dry, loose, single-grained, flows through fingers. (0 to 0.2)	Very slight color. Dry, loose, flows through fingers. (0 to 0.4)	Slight color. Powdery, dry, sometimes slightly crusted, but easily broken down into powdery condition. (0 to 0.5)	Slight color. Hard, baked, cracked, sometimes has loose crumbs on surface. (0 to 0.6)

* Available water is the difference between field capacity and permanent wilting point.
** Numbers in parentheses are available water contents expressed as inches of water per foot of soil depth.

Source: Goldhammer, D.A., and R.L. Snyder. 1989. *Irrigation Scheduling.* University of California, Division of Agriculture and Natural Resources Publication 21454.

variety of soils and crops, but it does have limitations, which are briefly covered below.

Standard tensiometers are not satisfactory for all soils. Because of an upper practical limit of 1 atmosphere, they are of limited value for measuring tensions on the heavier loams and clay soils. Tensions on these soils often can exceed this upper limit. Their use is also questionable for measuring tensions in very coarse soils or media containing considerable coarse particles (anthracite, letite, perlite, coarse sand, scoria) because of the uneven water distribution or poor contact between particles and cup. Successful production on coarse soils or media containing considerable coarse particles is possible only if frequent rain or irrigation can maintain low moisture tensions. But standard tensiometers are not suitable for detecting small differences in the 8 to 15 centibar range at which point irrigation may need to be started. A Low Tension Tensiometer (Irrometer model LT), effective in the 0 to 40 centibar range, overcomes this difficulty making it practical to monitor moisture in coarse soils or nonsoil media mixes.

Utilizing tensiometer data for starting irrigation works quite well (if special tensiometers are used for sands) but is less satisfactory for determining when to terminate it. Even for starting irrigation, it is necessary to know the critical point for the different crops and growth periods (see Tables 2.3 and 2.11) and it is helpful to remember that (1) saturation or free water is indicated by a zero reading, and (2) field or moisture holding capacity by readings of 5 to 10 centibars for sands to 30 centibars for clay soils. Unless there is a need to leach out salts, irrigating beyond field capacity is to be avoided to conserve water, eliminate leaching losses, and limit pollution of ground waters, lakes, and rivers.

Some idea of the time for terminating irrigation in very light soils can be gained by noting when lower soil tensiometer readings begin to fall. But for most soils this approach will lead to overwatering because of the long period required for the water to infiltrate the soil. It is better for these soils to add water in amounts that gave sufficient depth of penetration in the past.

Instrument readings are taken daily, preferably at the same time of day, keeping a record of the readings at various locations. Graphing the daily readings will enable rapid visualization of changes taking place and alert the grower to the need for irrigation. The approximate date of the next irrigation can be obtained by extrapolating the plotted readings ahead of the plotted lines. Irrigation should take place when the extended graph crosses the upper desirable tension.

Porous cup placement. The placement must ensure continuous contact of the soil and porous cup. The diameter of the hole punched or augered for the tensiometer should be equal to the diameter of the cup. The soil around the cup needs to be watered in to make good contact.

Difficulties in selecting locations for placing tensiometers arise if there is soil variability. Representative sites are selected avoiding unusually low wet areas or high dry areas. From a practical standpoint, selection of a site that represents a majority of the field probably is the best approach, although selecting several sites to compensate for considerable variability can be useful. In the latter situation, the additional tensiometers could show areas suffering from water shortages before the majority of the soils, often allowing for an earlier irrigation start without causing problems in the majority of the field.

Selection of several sites for tensiometer placement with variable soil conditions is also of the utmost importance when tensiometers are used to activate irrigation automatically. In such cases, the system needs to be engineered to irrigate the different sections separately.

Most accurate measurements are made within 2 inches of the cup, making placement in relationship to root distribution an important factor. The cup placement needs to vary with the depth of the root system (see Table 15.2). Instruments are available in standard lengths of 6, 12, 18, 24, 36, 48, and 60 inches. Usually, a single tensiometer is sufficient for shallow-rooted crops, if the porous tip is placed $3/4$ of the way down the root zone of roots less than 18 inches in depth. Deeper placements are needed for deeper-rooted crops such as orchards. Another placement at 36, 48, or even 60 inches may be desirable for very deep-rooted crops. In such cases, the shallowly placed instrument should have its porous tip at about $1/4$ of the depth of the root zone and the deeper instrument at about $3/4$ depth.

A correction is needed to compensate for the gravitational potential of deeply placed tensiometers. The correction can be made on some gauges by turning the calibration screw, but for others it will be necessary to subtract 3 centibars for each foot of depth.

Ideal placement of tensiometers also varies with type of irrigation. These placements are outlined in Table 15.3.

TABLE 15.2. Recommended Placement Depths for Tensiometers (in)

Depth of root system	Shallow instrument	Deep instrument
Up to 18	8-12	—
Up to 24	10-14	24
Up to 36	14-18	36
Up to 48	18-24	36-48
Over 48	24	48-60

Source: Data by Irrometer Co., Riverside, CA.

TABLE 15.3. Suitable Placement of Tensiometers for Different Irrigation Systems

Irrigation system	Number of stations*	Placement
Drip	2	One placed close to an emitter and in lower root zone. Second placed 10-18 inches** from emitter but in upper root zone.
Pivot	3	One in front of parked position between second and third towers. Second in same position 180°. Third behind parked position also between second and third towers.
Big gun tows	2	One between first and second sets; the other between next-to-last and last sets.
Wheel and hand-moved laterals	2	One between first and second sets; the other between next-to-last and last sets.
Furrow & flood	2	One near upper end of run; the other near lower end.
Solid set	1	Away from periphery of field and where full spray-pattern is received.
Subsurface	2	Lowest and highest parts of the field.

*A station is a tensiometer installation. It may consist of only one tensiometer for shallow-rooted plants or two or more for plants with deeper roots.
** Because of differential wetting patterns produced by drip irrigation in different soils, the tensiometer cup of the upper instrument is placed 10 inches away from a drip emitter on sandy soils but about 18 inches away for the heavier soils.

Tensiometers can be broken and must be placed so that they will not be disturbed by equipment or people. Placement within a row is usually satisfactory for annual crops that are irrigated by furrow or overhead sprinkler. Tensiometers should not be placed directly under a sprinkler head or permanent lines where the soil is usually wetter. Nor should they be placed where they may be shielded from the water by intervening plants.

In recently set orchards, the porous cup is placed in the root ball but is moved to other locations, and others added as the tree grows. New locations are chosen to reflect the changes in root distribution but placement must avoid areas that are heavily shaded or receive unusual amounts of water— either too little due to limb blockage or an excess from the drip line. For most situations in northern latitudes, ideal placement will be on the southwest side of the tree near but not at the drip line.

Tensiometers are sensitive instruments and need considerable preparation and maintenance. The manufacturer's suggestions for preparation and maintenance should be carefully followed to ensure accurate data.

Gypsum Blocks

Soil moisture can be evaluated by measuring the resistance between two electrodes planted in gypsum blocks. Because of rapid deterioration of the gypsum blocks in the soil, the all cast gypsum blocks have given way to nylon or fiberglass combined with the gypsum. More recently, gypsum blocks containing cast nylon units are being used. Water moves in and out of these blocks very much as it moves in the soil.

Gypsum blocks are more accurate than tensiometers in high-tension ranges (>0.33 bar) and, therefore, are better suited for medium and heavy soils. They are considered less reliable than tensiometers for the following: (1) sands or coarse-textured soils that release considerable water at low tensions; (2) drip irrigation, because of their poor sensitivity at high moisture levels; and (3) saline-alkaline soils, because salts affect resistance. Use of screen or cylindrical electrodes reduces the salinity effect and can make block use for the saline-alkaline soils more accurate.

Blocks are tested by soaking in water and hooking them up to a resistance meter before installation. A functioning, single block is placed in a hole only slightly larger than the block. Making the hole about an inch deeper than the intended placement allows for the addition of loose soil and a little gypsum plus lime under the block. The block is placed firmly in the hole, and the soil mixture with gypsum is added on the sides as necessary. Adding several fluid ounces of water around the block helps seal it in place. The block is covered with soil, gypsum, and lime mixture for several inches and the hole finally filled with soil, tamping carefully to avoid damage to the block or wires. Blocks are placed similarly to tensiometers at different locations in the field and at different depths depending upon the rooting system and type of soil. Wires from the blocks are brought to the soil surface.

Readings are taken with a portable resistance unit (soil moisture meter) that is specially calibrated and can be used to service many blocks. Meters are calibrated to read directly in percentages of available soil moisture. If not, the resistance in ohms can be converted to percent moisture by graphing the measurements of resistance as ohms of soil with varying amounts of readily available moisture from field capacity to the wilting point. A similar type curve is drawn of any new soil even if the instrument readings are given as percentages (see Figure 15.1).

Thermal Change Sensors

The conduction of heat by a soil is reduced as its moisture content decreases. At first, this principle was used to evaluate soil moisture by

FIGURE 15.1. Percentages of Soil Moisture Corresponding to Resistance in Ohms of Several Soils and Their Need for Irrigation

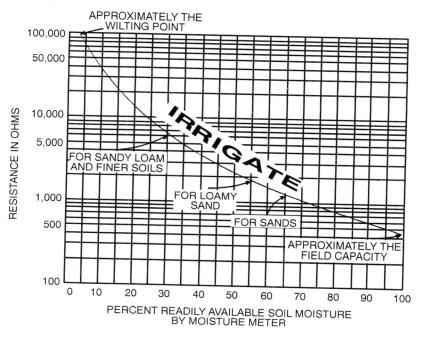

Source: Bouyoucos, G. J. 1956. Improved soil moisture meter. *Agricultural Engineering* 37(4): 261-262. Reprinted by permission of American Society of Agricultural Engineering, St. Joseph, MI.

winding enameled copper wire around pieces of glass tubing and burying the unit in the soil. A controlled current was passed through the wire for 1 to 2 minutes. The resulting rise in temperature was measured by changes in the wire's conductivity.

Newer units, using the same principles, have been enclosed in plaster of Paris blocks to reduce variability resulting from poor contact between soil and heating unit, but these blocks are less satisfactory than ordinary gypsum blocks in measuring high water tension. Further improvements utilize porous ceramic cups or disks to enclose the electrical unit. Water flows in and out of the cup or disk maintaining an equilibrium with soil water. The pore size of the disks can be varied for soils of different textures. Due to variation in pore size and electrical components, each unit needs to be standardized with a pressure plate apparatus or with other known standards.

The soil moisture is monitored with the following equipment: (1) a battery for power supply, (2) a precise 10 K resistor, (3) a $4^1/_2$ digit voltmeter capable of measuring microvolts in the 0 to 19 mv range, and (4) a timer. Several commercial units comprising these various parts are available at costs of about $2000.

Thermal units lend themselves to automatic irrigation control. For automatic use, additional electronic capability is needed to (1) automatically sample several sensors sequentially, (2) compare the sensors' output to threshold values for starting irrigation, and (3) to start the pump and open valves as needed.

A controlled current is passed through the wire for 1 to 2 minutes, and the resulting rise in temperature is measured by changes in the wire's conductivity. The readings are compared to those of soils with known moisture contents. A curve drawn of readings taken at field capacity and at various stages of drying until wilting occurs with moisture contents as measured by gravimetric means serves as an excellent standard.

Neutron Scattering

The loss of energy from fast neutrons, generated by radioactive cesium or americum, as they collide with slow neutrons from water forms a basis for measuring available water. A suitable instrument for measuring moisture will consist of a fast neutron source, a slow neutron detector, and a counting device. A scaler used with the soil probe is unaffected by the fast neutrons, reacting only to the slow ones. The fast-moving neutrons, colliding with the hydrogen in water, are gradually converted to slow-moving neutrons that can be counted by the detector. The number of slow-moving neutrons counted in a given time indicates the amount of water present because hydrogen in water molecules is the primary atom of low atomic weight in soils.

Readings need to be correlated with actual moisture contents of different soil types, collected at various depths and known volumes. The simplest way of doing this is to collect samples of soil at various depths as the neutron access tube is placed. The moisture content of these samples is then correlated with neutron probe readings taken at the same depths. Evaluation of available moisture is very rapid once the calibration curves are established from these readings. The primary disadvantages of this method are (1) the cost, which can be several thousand dollars; and (2) the expertise needed to handle radioactive and sensitive equipment.

Readings are taken by lowering the probe to the desired depth in previously placed access tubes. Tubes made from aluminum are preferred, although steel, thin iron, or plastics have been used. A series of tubes, 2-inch outside diameter, are installed vertically in holes bored by an auger of the same diameter to a depth 12 inches deeper than expected roots. The holes are located at strategic sites representative of the field to be irrigated, but should not be within 16 feet of any other neutron source. The tubes are tapped lightly and left in place extending 2 to 10 inches above the ground. Readings are quickly taken ($\approx$30 seconds) with a unit that can effectively service many locations. Readings are compared to standards set up at time of installation.

Capacitance

A nonnuclear soil probe using an access tube employs capacitance technology to measure the soil moisture by measuring changes in the dielectric constant of the soil. Most solid materials in the soil (sand, silt, clay, and OM) have dielectric constants ranging from 2 to 4, while that of water is 78. A 2-inch PVC access tube is used and the probe is lowered in the access tube.

Readings are taken at varying depths to determine the moisture profile. They can be compared with factory-preset calibrations or more accurately with values calibrated for the particular soil. These can be obtained by measuring soil moisture levels of samples collected at similar depths and determined by pressure plate or gravimetric means.

The method is very fast and works well over a wide range of temperatures (32 to 158°F) and moistures (0 to 60 percent volume). The Sentry 200-AP, manufactured by Troxler Electronic Labs, Inc. (Research Triangle Park, NC), uses a probe weighing a little over 2 lb and a control unit about 8 lb. The greatest disadvantage of the system is the relatively high costs, which for both units are about $3700.

Time Domain Reflectrometry

It is possible to measure soil water content (SWC) indirectly by noting the travel time of a short pulse of electromagnetic energy. Time domain reflectrometry (TDR) instruments transmit a high-frequency pulse and measure the travel time of the electromagnetic wave along the waveguide by measuring the voltage amplitude of the reflected pulse at known time intervals. The travel time through any given thickness of material is directly proportional to the square root of its dielectric constant, which varies greatly with the volume of soil water. The dielectric value for dry soils is very low (in the range of 3 to 6), while that of a saturated soil can be 50 to 60.

The dielectric value is quickly obtained by placing the instrument over protruding rods previously positioned to form waveguides. These rods are vertically placed 2 inches (5.1 cm) apart and parallel, with 0.4 inches (1 cm) protruding above the surface for attachment of the TDR instrument. A common arrangement for orchard trees uses four placements per tree. Instantaneous readings can be compared to graphs in the instrument or standards established by other means. The method is quick and reliable with no problems of legal regulation but costs for the instrument are high—about $5,000 to $10,000.

Evapotranspiration

Most soil water is lost by evaporation from the soil surface or by transpiration from the leaves. The two losses, which are difficult to measure separately, are considered as a unit designated as evapotranspiration (Et). Water lost as Et is controlled by heat, day length, air humidity, radiation, type of plant, and its coverage of the soil surface.

The net loss of water as Et can be used to determine when irrigation should start and the amount to apply. A rough estimate can be made using generalized Et data for particular crops in an area combined with specific rainfall and moisture-holding data for individual soils. A more precise method uses Et data gathered by measuring moisture loss from standard pans, located close to existing fields.

The second approach, although more accurate, is more expensive. It is not too reliable for estimating daily water needs but does provide a fairly good estimate of long-term (monthly) needs providing that water losses are carefully monitored. The measurement depends upon the use of standard pans and appropriate factors in converting water loss to water application, as considered in the next two sections.

Standard pans. Evaporation of water from a standard pan can be used to determine Et since the amount of water lost from the pan is directly proportional to evapotranspiration. In the United States, a standard pan (Class A) is circular, made of unpainted galvanized iron, with an inside diameter of 48 inches (120 cm) and 10 inches (25 cm) high. The unit is supported 6 inches (15 cm) above the ground by wooden planks to allow free air movement.

The water in the pan must be kept clean. A wire screen can be used to protect the pan from birds and animals but pan evaporation (Ep) values are lower by about 10.4 percent compared to an unprotected pan.

The depth of the water is maintained at 4 to 8 inches (10 to 20 cm) with fluctuations no greater than 2 inches (5 cm). The height of the water is accurately measured at the same time each day with a micrometer depth gauge accurate to 0.01 mm or 0.1 inch. This reading will indicate the 24-hour net change in water as affected by rainfall and evaporation.

The costs of the pan and its installation (about $1,500) can be depreciated over a number of years.

Factors affecting Et. The amount of water lost by evapotranspiration, although directly proportional, is seldom equal to that evaporated from a standard pan. Water lost from a standard pan equals evaporation only under the following conditions: (1) the soil is saturated, (2) plants cover the soil completely, and (3) the pan and plants are exposed to exactly the same conditions of wind and sunlight.

Since all of these conditions are seldom met, a correction factor is needed to convert Ep data to Et as presented in equation 1.

$$Et = Ep \times F \times C \tag{1}$$

Where F = crop factor; C = leaf coverage (solid cover = 1.00).

The correction factors for converting the Ep of pans placed in the open to Et varies from a factor of about 0.85 for conditions of high humidity and low wind to about 0.50 for conditions of low humidity and high winds. An average value of 0.7 can be used as a starting point if there is good penetration of water into the soil.

Correction factors have been calculated for different stages of growth of a number of crops in several localities. Often this data can be obtained from the horticultural or engineering departments of the state land-grant colleges or from state extension services.

The assumed factors (0.60 or 0.70 for good infiltration or 1.0 for ponded soils or those determined for particular crops by land-grant colleges) can be refined to represent specific conditions by combining pan data with that of tensiometers. The daily readings of tensiometers, placed per the suggestions above, are used to evaluate whether the assumed crop factor is maintaining satisfactory moisture. If moisture cannot be maintained in the 10 to 20 centibar range by irrigation based on the assumed factor, the factor is changed upward if the tensiometer readings increase and downward for lowered tensiometer readings. The assumed factor is correct if tensiometer readings are maintained in the desired range.

Water required and Et/Ep. If the 70 percent factor is correct, each inch (2.5 cm) of Ep will require 0.7 inches of applied water to keep the tensiometer readings in the 10 to 20 centibar range. The 0.7 inches of water needs to be applied overall only if there is solid cover. For irrigation systems that can supply water to limited areas, the percentage leaf cover will have to be calculated for less than solid cover. The percentage leaf coverage for orchard crops is the ratio of canopy area to total area of the tree spacings. For example, if the tree canopy is 10×14 ft and the trees are spaced 15×18 ft, the leaf coverage, given in equation 2, is as follows:

$$C = \frac{10 \times 14}{15 \times 18} = 0.518 \qquad (2)$$

For row crops, the coverage is computed from the ratio of the width of the leafy area to row spacing. For plants having a width of 12 inches and a row spacing of 36 inches the ratio is:

$$C = \frac{12}{36} = 0.333 \qquad (3)$$

Using equation 1 above, the water needed for trees having canopies of 10×14 ft in spacings of 15×18 ft, the Ep = 0.4 inches per day and the factor = 0.75 will be:

$$Et = 0.4 \times 0.75 \times 0.518 \text{ or } 0.155 \text{ in/day} \qquad (4)$$

The water needed for row crops with a canopy 10 inches wide in rows 36 inches apart with an Et of 0.4 and a factor of 0.75 can also be calculated from equation 1. Entering the several components into the equation gives Et = 0.1 inches per day ($0.4 \times 0.75 \times 0.33$).

The Et can be converted to gallons per tree (gpt) for complete coverage by use of the following:

$$GPT = 0.623 \times Et \times \text{plant area (tree setting)} \qquad (5)$$

Using the data given for equation 4, the gpt is 26.07 ($0.63 \times 0.155 \times [15 \times 18]$).

The gallons per acre (gpa) for solid coverage can be calculated by multiplying the Et in inches per day $\times 27,154$ (gal/acre-inch). An Et of 0.145 will require 3,937 gal per acre each day. For less than solid coverage, 27,154 needs to be multiplied by the percentage cover (eq. 2 or 3).

Foliar Air Temperature Differentials

Regulating irrigation by foliar temperature differentials is based on the concept that leaf temperatures begin to rise as moisture necessary for maintaining the transpiration stream becomes limiting.

Leaf and air temperatures are measured by a small portable infrared thermometer on a daily basis at one hour past solar noon with the back to the sun and the instrument held at a 45° angle to the leaf canopy. Positive values when leaf temperatures are greater than air temperatures are recorded and accumulated.

The need for irrigation is signaled when the accumulated stress degrees exceed the threshold value. At just what point increases in leaf temperatures cause enough stress to warrant irrigation is still being determined for many crops at various stages of growth. The threshold value for wheat is a total of 15 accumulated stress degrees.

The method offers great promise for scheduling irrigation start in arid or semiarid climates, where leaf temperatures of a plant receiving ample water can be as much as 18°F (10°C) cooler than the air about it, if relative humidity is low. The difference value can be small where rain falls during the growing season or humidity is high. But even in the more arid climates, practical use for many crops awaits satisfactory cumulative thresholds.

Another difficulty with downward-pointing infrared instruments is that readings are not reliable on soil without cover, as bare soil may be 50°F (28°C) higher than the canopy temperatures of plants receiving ample water. The situations with bare soil may still give useful readings if a vegetative index derived from infrared and near-infrared images can be established. The vegetative index would be combined with paired radiant surface temperatures obtained with airborne sensors to give a trapezoidal two-dimensional index. While feasible, the system awaits automated process imaging from the data in order to give the fast response needed to be practical for the grower.

AquaProbe

A new type sensor for soil moisture, known as an AquaProbe, employs a wall coated with a gas plasma polymerized coating and a sensor body filled with osmotically active polyethylene glycol (PEG) or polyvinylpyrolidone (PVP). After establishing hydraulic contact in soil or other media, the sensor simulates a plant cell allowing water to move into and out of the sensor body based on the potential gradient existing between the sensor solution and soil matrix potential. The water potential in the sensor equals matrix plus osmotic potential and is reflected as internal pressure that can be measured by an electronic or gauge-type pressure transducer.

The Aquaprobe is a relatively new economical approach for measuring soil water in the field and could be highly useful since it enables effective measurement between 0 and -10 bar, compared to a tensiometer's effectiveness in the range 0 and -0.8 bar. Accurate knowledge of existing moisture conditions just beyond the tensiometer's range could be very helpful in supplying water to maintain the slight water stress conditions so necessary for producing satisfactory quality in several maturing fruits. Its suitability for this purpose awaits sufficient trials in the field.

WATER QUALITY

Water application by various irrigation systems can be compromised by poor quality. What is worse, poor quality can lead to poor plant response in the short run and soil degradation in the long run.

Poor plant response due to poor-quality water may result from foliar injury, uneven application of the water, or poor uptake of soil-applied water. Foliar injury, primarily from overhead irrigation, can result from excesses of several nutrients in the water. Poor response resulting from uneven application of water is associated with several materials that can clog the irrigation system or with inadequate or incorrect spacing of the irrigation nozzles. Poor water uptake may be due to excessive total salts or large amounts of sodium, carbonates, or bicarbonates that have deleterious effects on soil structure.

Soil degradation results as excess salts tend to accumulate or high concentrations of Na^+ or HCO_3^- in the water displace Ca^{2+} from soils, so that they are increasingly saturated with Na^+. Soils with high Na^+ (>10 percent base saturation) concentrations are difficult to handle, have poor water infiltration, and their exchange of air and water is slowed.

Chemical Tests

The suitability of irrigation water for various purposes can be ascertained by a series of chemical tests (see Table 15.4). Because water quality needs to be better for small-orifice emitters, amounts of suspended solids and number of bacteria also needs to be determined for drip irrigation and similar systems (see Table 15.5). To be meaningful, samples need to be

TABLE 15.4. Irrigation Water Quality Based on Chemical Tests

Water quality parameter		Severity of potential problems		
		None	Increasing	Severe
Acidity				
pH[a]		5.5-7.0	<5.5 or >7.0	<4.5 or >8.0
Salinity				
Conductivity EC	mmhos/cm	0.5-0.75	0.75-3.0[b]	>3.0[b]
Dissolved salts	ppm	320-480	480-1920[b]	>1920[b]
Permeability				
Caused by low salts				
Conductivity EC	mmhos/cm	>0.5	0.5-0.2	<0.2
Dissolved salts	ppm	>320	320-125	<125
Caused by excess sodium				
SAR[c]		<3.0	3.0-6.0	>6.0
Toxicity				
Bicarbonate[d]	ppm	<40	40-180	>180
Boron	ppm	<0.5	0.5-2.0	>2.0
Chloride[e]	ppm	<70	70-300	>300
Fluoride[f]	ppm	<0.25	0.25-1.0	>1.0
Sodium[g]	ppm	<70	70-180[b]	>180[b]
Clogging[h]				
Calcium[i]	ppm	20-100	100-200	>200
Carbonates	ppm	<35	35-75	>75
Magnesium[j]	ppm	<63	>63	
Iron[k]	ppm	<0.1	0.1-0.4	>0.4
Manganese	ppm	<0.2	0.2-0.4	>0.4
Sulfides	ppm	<0.1	0.1-0.2	>0.2
Nutritional				
Nitrogen[j]	ppm	<5	5-30	>30

[a] The water pH affects low-buffered media. High pH aggravates emitter clogging and foliar staining. Low pH increases corrosion of metal pipes and equipment.

[b] The conductivity and amount of dissolved salt causing problems are affected by soil class and crop tolerance to salts.

[c] Sodium adsorption ratio. See text.

[d] Bicarbonates cause problems by modifying soil calcium. If applied overhead, phytotoxicity also can be a problem.

[e] Amounts of chlorides that are phytotoxic to some plants when applied overhead may not be harmful if soil applied.

[f] These values are significant for fluoride-sensitive plants.

[g] Less severe if potassium is present in equal quantities or if plants are tolerant to sodium.

[h] Clogging is more serious in systems with small orifices.

[i] Calcium causes clogging as phosphates or sulfates are added in irrigation lines, particularly if the pH is high.

[j] Values >0.2 ppm increase clogging from bacterial slimes and can cause staining of plants as water is applied overhead. Concentrations >0.4 ppm can cause sludges if chlorine is used.

[k] Values >5 ppm can stimulate algae growth in ponds. Values >30 ppm can delay maturity of some crops and decrease sugar content of crops such as sugar beet. For nonsensitive plants the N content can be beneficial but should be considered in the fertilization program.

Source: Wolf, B. 1996. *Diagnostic Techniques for Improving Crop Production.* The Haworth Press, Inc., Binghamton, NY, p. 341.

TABLE 15.5. A System for Classifying Water for Drip Irrigation Systems

Rating	Suspended solids (ppm)	Dissolved solids (ppm)	Bacteria (no/ml)
0-ideal	<10	<100	<100
1	20	200	1000
2	30	300	2000
3	40	400	3000
4	50	500	4000
5	60	600	5000
6	80	800	10000
7	100	1000	20000
8	120	1200	30000
9	140	1400	40000
10-unsatisfactory	>160	>1600	>50000

Source: Bucks, D. A., E. S. Nakayama, and R. G. Gilbert. 1979. Trickle irrigation water quality and preventative maintenance. *Agricultural Water Management* 2: 149-162. Reprinted by permission of Elsevier Science, Amsterdam, Netherlands.

collected properly. About $1/2$ gal of water needs to be collected in a clean container that has been rinsed several times in the test water prior to filling. Water from a domestic tap line or well is allowed to run for at least 10 minutes before collecting samples. Samples from open bodies of water are collected beneath the surface and near the water intake. The importance of several tests and some means of modifying some of their undesirable effects are covered in some detail in the sections below.

pH

The pH value effects the solubility of various water compounds—those found naturally in the water, added as fertilizer salts to irrigation water, or formed as fertilizer salts react with each other or with the water components. The resulting precipitates affect the availability of added nutrients and can reduce the flow of irrigation by clogging small orifices or building up in pipes. High pH values favor and low pH values tend to prevent the formation of many precipitates. Very low or very high water pH values can affect poorly buffered media (sands, artificial media, and water) enough to have an impact on the availability of several nutrients.

High pH values in irrigation waters are usually due to the presence of considerable bicarbonates and/or carbonates. The amount of carbonates and/or bicarbonates can be lowered by acidifying the water to pH values

<6.0. The loss is accelerated as the pH falls below this value but corrosion of metal pipes and concrete ditches can be serious at lower values. The maximum safe amounts of sulfuric acid that can be used with varying amounts of bicarbonates and ammonia are given in Table 15.6.

Titrating the water with acids is the most accurate method of estimating acid needs for lowering pH. Lowering the pH to about 6.0 limits much of the carbonate and bicarbonate present. Lacking titration data, one of the following procedures for estimating acid needs can be used.

The amount of sulfuric acid required to neutralize 90 percent of bicarbonates in waters are given in Table 15.7.

The amount of acid needed to change a high pH to 6.5 can be calculated by multiplying the milliequivalents of bicarbonate plus carbonate by one of the following factors:

- 7.0 for 75 percent phosphoric acid
- 3.2 for 65 percent sulfuric acid
- 10.5 for 67 percent nitric acid

If data for meq are not given but amounts of bicarbonates and/or carbonates are given as ppm of hardness, meq can be obtained by dividing the ppm by 50. If data are given as grains of hardness, the ppm is obtained by multiplying the grains of hardness by a factor of 17.

TABLE 15.6. Maximum Amounts of 95 Percent Sulfuric Acid to Be Used for Irrigation Systems Containing Concrete Ditches or Metal Pipes

Water test	Acid rates			
ppm	ppm	ton/ac	lb/ac-ft	gal/ac-ft
Bicarbonate				
50	38	0.17	101	6.6
100	75	0.34	203	13.9
150	114	0.51	304	18.9
200	152	0.68	407	26.7
Ammonia				
25	76	0.34	303	13.4
50	152	0.68	407	26.7
100	304	1.37	814	53.4

Source: Miyomota, S. and J. L. Stroelein. 1975. Sulfuric acid. *Progressive Agriculture in Arizona* 27: 13-16.

TABLE 15.7. Quantities of Sulfuric Acid Required to Neutralize 90 Percent of Bicarbonates in an Acre-Foot of Water*

Bicarbonate	Acid required		
(ppm)	ppm	lb	gal
50	38	103	7
100	76	206	13
200	152	412	27
400	304	824	55

* 95 percent sulfuric acid.

Source: Stroehlein, J. L. and A. D. Halderman. 1975. *Sulfuric Acid for Soil and Water Treatment.* Arizona Agric-File Q-357.

If data on bicarbonates and carbonates are not available, acid should be limited to one of the following amounts per 1,000 gal of water:

- 1 lb of food-grade phosphoric acid
- 0.5 lb of sulfuric acid
- 1.5 lb 66 percent nitric acid

The addition of acid in irrigation lines will cause fewer problems if it is diluted to about 10 percent strength prior to injection. Dilution is always obtained by adding acid to water and not vice versa to minimize splashing. Using a piece of corrosion-resistant pipe (about 20 ft in length) just beyond the injection point mixes the acid with enough water so that corrosion of irrigation pipe, gate valves, emitters, etc. downstream is minimized. Resistant metal or plastics need to be used for the entire system if the anticipated pH will be less than 6.0. Also, handle all acids with proper precautions, using protective clothing, goggles, and rubber or plastic gloves.

Low pH values of water need to be raised if used for low-buffered media to avoid creating poor root environments. Modern hydroponic installations monitor the pH of irrigation water. Low pH values can be altered by the substitution of calcium nitrate or potassium nitrate for some of the ammonium compounds. Some installations will introduce alkalies such as ammonium hydroxide (aqua ammonia) or potassium hydroxide to raise the pH. Ammonium hydroxide should not be used in closed areas where plants are growing because free ammonia in very low concentrations can be extremely toxic to plants.

Conductivity and Dissolved Salts

Conductivity, measured by a simple test that can be run by a grower, is an indicator of the total dissolved salts. Its usefulness has been covered in Chapter 8. Methods of coping with high salts have been covered in Chapter 12.

Water with relatively high conductivity can be used for irrigation if the crop is salt-tolerant and/or there is sufficient rain to prevent the accumulation of harmful amounts of salts (see Table 12.3).

Low-conductivity water. Very low-conductivity water (<0.2 mmhos/cm) can cause corrosion of irrigation equipment and have an adverse effect on water infiltration. See "Low Salts" in Chapter 12 for several corrective measures.

Bicarbonates and Carbonates

Both bicarbonate and carbonate ions raise the pH and in so doing tend to precipitate Ca, Mg, P, and Fe. The precipitates can cause serious blockage problems in irrigation systems, produce unsightly deposits on leaves if overhead irrigation is used, and have serious adverse soil effects. The bicarbonates are more troublesome than the carbonates because: (1) they are more phytotoxic, and (2) their addition to the soil in large amounts tends to drive Ca from the soil, leading to soil dispersion, surface crusting, and compacted soil layers. Bypassing plant leaves reduces the problems of phytotoxicity but does not reduce soil damage.

Reducing problems from excess bicarbonates and carbonates. Both ions can be substantially reduced by lowering the irrigation water pH to 6.0 with constant acid injection as it is applied. If the water is not applied overhead or used for very low-buffered media, acid injection can be omitted and the soil treated periodically with gypsum.

The amount of acid needed to lower the pH of water to 6.0 is best determined by titration of the water with acid, but data in Table 16.7 indicate of the amounts of sulfuric acid that may be needed to neutralize most of the bicarbonates.

Calcium and Magnesium

Waters high in Ca and/or Mg can cause clogging of small orifices. The problem is greatly aggravated as pH values are >7.0 and if ammonia or phosphates are injected into the water. Waters containing Ca in amounts <20 ppm or 1.0 meq/L tend to disperse soils, thereby increasing water penetration problems. The problem can be corrected by the addition of a soluble calcium salt such as calcium nitrate.

Boron

The upper safe limits of B in irrigation waters depend on the tolerance of the crop to B, characteristics of the soil, and degree of leaching. Plants sensitive to B should not receive water containing more than 0.5 ppm but water with 2 ppm has been safely used for semitolerant plants and as much as 4 ppm for highly tolerant plants.

The amount of B that accumulates in the soil and can be toxic to plants depends on the amount of water applied, the type of soil, and the amount of leaching. Less B can be tolerated on coarse soils low in OM, but B can easily be leached from these soils and is seldom a problem if there is enough leaching. The heavier soils, especially with high OM, tend to tie up B allowing much greater amounts to be applied without causing injury. Such amounts can cause problems if soil pH is allowed to fall, making the B more available to plants.

Chlorides

As indicated in Chapter 8, chlorides in irrigation waters contribute to the salt problem. To avoid problems with chloride-sensitive plants, chlorides should be <70 ppm, although amounts up to about 300 ppm will not seriously affect yields for some tolerant plants.

Fluorides

The upper limit of F in irrigation water should be 0.25 ppm for sensitive plants but can be as much as 1 ppm for tolerant plants. A list of F-sensitive plants was provided in Table 3.1.

Slightly higher concentrations than those given above can be tolerated if the following conditions are met: (1) increase soil pH to 7.0, (2) reduce transpiration by the use of windbreaks and partial shade, (3) avoid overhead irrigation, and (4) limit the addition of F by using low-F fertilizers. Phosphates are the principle source of fertilizer F.

Fluorides can be removed from irrigation water by ion exchange, reverse osmosis or by treating the water with alumina or activated bauxite. Removal is an expensive process and probably is justified only for the production of some high-priced plants.

Iron

Iron present in water applied by overhead irrigation tends to be oxidized and precipitate on leaves, causing unsightly stains. The economic damage from these deposits is of minor consequence except for foliage or flowers, when the deposits detract from the plants' visual appeal. As little as 0.2 ppm of Fe can be troublesome if the water contains considerable bicarbonates or carbonates (pH >7.0). The problem is mitigated by acidifying the water to a pH of 6.0 or by the use of chelating agents, such as the hexametaphosphates (Quadrophos or Calgon) to keep the iron in solution. Unsightly Fe deposits on ornamental plants can be removed by spraying leaves with a 5 percent oxalic acid solution (6.7 oz or 16 tablespoons in 1 gal water) and then rinsing before the spray dries. The spray can be phytotoxic and should be tried on a few plants before using on a large scale.

Relatively small amounts of iron can cause serious blockage of micro-irrigation systems due to slimes caused by some bacteria. This problem is considered in greater detail in the section, "Microorganisms" later in this chapter.

Sulfides

Sulfides also can be responsible for the bacterial production of slimes that clog irrigation systems if present in amounts >0.1 ppm. Insoluble iron sulfides that can clog systems can be formed by the interaction of iron and sulfides naturally present in the water or by the introduction of Fe in added fertilizers.

Nitrogen

Nitrogen naturally occurring in irrigation water would appear to be a "freebie," which it often is, but its presence can cause problems at times. More than about 5 ppm can stimulate growth of algae in ponds, increasing the problems of orifice plugging. Amounts >30 ppm can delay maturity of some crops and decrease the sugar content of fruits. The presence of large quantities of N probably is beneficial for most crops during the early stages of growth but usually needs to be curtailed as the crop approaches maturity. Even when beneficial, the amount provided by irrigation needs to be considered.

CLOGGING PROBLEMS

Suspended Solids

Suspended solids can easily clog irrigation systems. Microirrigation systems with small orifices are most readily affected but nearly all pressure systems can be rendered ineffective in very short periods of time. The presence of large amounts of suspended solids usually requires some physical treatment prior to water use.

Dissolved Solids

Dissolved solids can be a source of clogging for microirrigation systems as these solids may precipitate out when various fertilizers are added to the lines. They can cause problems for all types of piped systems as some of these solids precipitate as scales, which can restrict the flow of water. The presence of large amounts of salts may reduce the effectiveness of irrigation water for salt-sensitive crops regardless of the irrigation system used, although it appears that there are fewer problems with high-salt waters when drip irrigation is used.

Microorganisms

The presence of large quantities of bacteria and/or algae can indicate potential problems, particularly with microirrigation systems. Two types of problems usually occur:

1. Bacteria may produce slimes or cause Fe or S to precipitate out of the water. Both processes readily restrict the flow of water in microirrigation systems, either by clogging emitters or binding with fine silt or clay to form aggregates.
2. Algae and other aquatic plants can multiply in reservoirs to seriously affect screening and filtration processes. Besides the physical problems of handling such waters, the presence of these organisms can block pipes and emitters. High levels require chemical treatment of reservoirs. Microirrigation systems require applications of bactericides on a regular basis, either as a constant injection or with large amounts applied at the terminus of each irrigation to supply a reservoir of active bactericide.

Water Quality Needed to Avoid Clogging

Most clogging is primarily due to particles of various types, bacterial precipitation of iron or sulfur, or precipitates of calcium and/or phosphates forming in the lines. Prevention is far superior to corrective measures. Treatment of the water as indicated below can prevent a great deal of the clogging problems.

Although clogging is potentially much more serious with drip irrigation, it can present problems with any system using pressurized pipes to move water. Water quality must be higher for these systems, often requiring special filtration and treatment of the water prior to use. The frequency of problems usually increases as the size of the aperture through which water moves decreases and pressure increases. Much of what is outlined below for drip irrigation can also be useful for systems using relatively small devices for finally spreading the water.

Particle clogging may result from materials introduced in the installation phase or present in the water as it is introduced into the lines. Materials introduced during installation should be flushed out just prior to completion. Particles of sand, rust, and organic debris present in the water needs to be removed by physical treatment of the water as outlined below. The presence of large numbers of bacteria can be troublesome, primarily for microirrigation systems (see Table 15.5). Problems can be reduced by treating water with chlorine.

Water Treatment

Clogging of lines, emitters, and nozzles can affect uniformity of application as well as restrict the flow. Early problems can be detected from a gradual decrease in the flow rate as measured by taking daily readings of flow meters installed on main pipelines, although such reduction may also indicate a problem with the pumping station. Treatment of the water as indicated below can prevent a great deal of the clogging problems.

Physical Treatment of Water

If the water contains suspended solids, it should be retained in settling basins to allow the solids to settle out. The settling basin is an economical primary treatment facility for cloudy water sources from streams or ditches. It also can function as a treatment area to remove iron and dissolved solids.

Sand, scale, or other particles heavier than water and no finer than 200 mesh can be removed by centrifugal sand separators. Pump wear can be reduced if they are installed on the suction side. It is important that separator size be fitted to flow rate.

Pressure screen filters that also remove these contaminants are available in various sizes and with screens varying from 20 to 200 mesh. The screen mesh should be between $^1/_7$ and $^1/_{10}$ of the opening of the emitter orifice. The screen filters need to be cleaned periodically by hosing particles from the surface. This cleaning process for most types involves disassembling, which makes most pressure screen filters impractical for waters containing considerable contaminants. The blowdown screen filter permits cleaning by opening a flush valve and therefore is much better suited for waters with moderate amounts of contaminants.

Gravity screen filters are simple, low cost, easy to install and maintain, and are effective in removing heavier-than-water particles. Some incorporate a self-cleaning water spray.

Media filters containing sharp-edged sand or crushed rock are especially suited to cleanse water for microirrigation because they effectively remove fine suspended solids. The sand or rock needs to be carefully graded, avoiding "fines." The filters are cleaned by backflushing. A screen filter should follow a media filter in order to remove any filter sand that might be dislodged and enter the irrigation system.

Screens are disassembled at regular intervals and hosed down to remove debris. Media filters need to be backflushed frequently to avoid excess buildup.

Submains and laterals need to be flushed periodically to remove sediments that may travel beyond screens and filters or enter as lines are being modified. The ends of the lateral lines are opened while the system is running and allowed to flow until the water appears clear as examined in a jar. Examination of sediment to determine its origin can be helpful in deciding treatment for correction (Boswell, 1990).

Chemical Treatment of Water

Several organisms can clog or seriously reduce the flow of water in various irrigation pipes, filters, and emitters. Algae and bacterial growth are the commonest problems. Large numbers of bacteria are troublesome primarily for microirrigation systems, where their presence can create various slimes that can block emitters. Treatment requires chlorination on a continuing basis.

Algae growth originating in storage ponds and lakes can clog screens and media filters. They can be controlled by adding copper sulfate or bluestone to the water at a rate to supply 0.5 ppm of copper sulfate or 4 lb per 1,000,000 gal of water. Three methods are used for application: (1) the copper sulfate is scattered on the water surface, (2) copper sulfate is placed in open mesh bags and dragged through the water in paths about 15 feet apart, or (3) the bagged copper sulfate is equipped with a float and anchored at several spots. In applying the copper sulfate by methods (1) and (2), only half of the area should be treated at one time to avoid fish kills. The remainder of the area can be treated about a week later. Copper levels in the water should be limited to 0.25 ppm or less.

Copper sulfate can seriously corrode aluminum pipe. Constant injection of chlorine to supply about 1 ppm can be substituted for the copper sulfate if aluminum piping is used.

Bacterial growth may yield a slime that can precipitate Fe or S. The slime can combine with silt or clay particles to effectively clog small orifices, which can be prevented by chlorination. Application of the chlorine at a 10 ppm rate for about the last 30 minutes will supply a residual chlorine content of 1 ppm. This approach or the constant injection of 1 ppm works well for most situations.

Bacterial contamination of a well source can greatly increase the problem of clogging by iron bacteria, but can be corrected by the injection of chlorine into the well to provide 200 to 500 ppm of chlorine. The volume of the water (in gal) to be treated is calculated by determining the cubic feet of water and multiplying it by 7.5 gal/cu ft. The volume is obtained from equation 6,

$$\text{Vol. (cu ft)} = \frac{r(2) \times D \times 3.1416}{27} \tag{6}$$

where r = the radius and D = water depth.

The use of chlorine to control emitter blockage is not satisfactory if the water contains more than about 0.2 ppm of Fe as the chlorine will precipitate the Fe. If the water contains too much Fe, problems can be avoided by (1) keeping the Fe in solution by injection of acid, (2) removing the Fe by aerating the water in the settling basin, or (3) removing the Fe before the water enters the lines by adding sufficient chlorine to precipitate it.

Injecting acid requires considerable expertise and is not without risk. This approach may be more appropriate if the water needs to be acidified because it contains high levels of carbonates or bicarbonates. Whereas a pH of about 5.5 is needed to eliminate most carbonates and bicarbonates, a pH of 4.0 is needed to control iron bacteria causing blockage. Any reduction of pH below 6.0 can cause corrosion of concrete and many metals, especially aluminum.

Method (2) is the simplest and cheapest if large amounts of iron are present. The water can be aerated by passing it over a number of baffles or by spraying it in the air. Sufficient time is needed to allow the precipitated iron to settle out. A media bed filter to remove any suspended iron will help avoid emitter clogging.

Iron is precipitated with chlorine if it is not possible to precipitate it by aeration and settling. It is necessary to inject 0.7 ppm of chlorine on a continuing basis for each ppm of Fe. Good mixing by turbulence is needed before the water passes through the filter, which will have to be backwashed frequently. This method is not suitable if the water also contains considerable Mn, as this will delay the Fe precipitation enough so that much of the Fe will precipitate in the lines and emitters.

Chlorine Use

Chlorine can be supplied as chlorine gas, liquid sodium hypochlorite, or dry calcium hypochlorite. Chlorine gas, which can be dissolved directly into the supply lines by a metering device (chlorinator) is the cheapest form of chlorine, but is practical only for installations employing very careful supervision because of its extremely toxic properties. Sodium or calcium hypochlorite can be metered into the system. The amounts of available chlorine in the various commercial sources, their equivalent values, and the amounts needed per acre-foot of water to supply 1 ppm of chlorine are presented in Table 15.8.

TABLE 15.8. Commercial Chlorine Sources and Their Equivalent Values

Chlorine form	1 lb equivalent	Amount to supply 1 ppm per acre-inch of water*
Chlorine gas 100% available Cl_2	1.0 lb	3.6 oz
Calcium hypochlorite 65-70% available Cl_2	1.5 lb	5.3 oz
Sodium hypochlorite		
15% available Cl_2	0.8 gal	23.5 fl oz
10% available Cl_2	1.2 gal	35.2 fl oz
5% available Cl_2	2.4 gal	57.8 fl oz

*Amount needed per acre-inch of water (27,154 gal) to supply 1 ppm of chlorine at the injection point.

Source: Boswell, M. J. 1990. Micro-Irrigation Design Manual. J.H. Hardie Irrigation, El Cajon, CA, pp. 4-13.

Calculating amount of chlorine for injection. The rate (R) of sodium hypochlorite injection in gal per hour (gph) can be calculated from equation 7,

$$\text{gph} = \frac{SR \times C \times 0.006}{\% \text{ NaOCl}} \tag{7}$$

where SR = system flow rate in gal/minute (gpm); C = desired Cl concentration in ppm. If the flow rate = 100 gpm, the desired chlorine concentration = 10 ppm and the percent active chlorine = 5.25 percent, then

$$\text{gph} = \frac{100 \times 10 \times 0.006}{5.25} \quad \text{or } 1.14$$

The same equation can be used for calcium hypochlorite but, since it is a solid, calculations will have to include the fact that it will take 12.8 lb of the 65 percent solid to produce a 1 percent solution.

Calculations for injection of gaseous chlorine can use equation 8,

$$\text{lb per day} = SR \times C \times 0.006 \tag{8}$$

where SR and C have the same meanings as in the previous equation (Boswell, 1990).

Correcting Clogging

Correction of clogging is indicated whenever flow rate or application uniformity is affected and must occur before there is total blockage of emitters. The type of correction will depend upon the cause. Blockage from calcium salts (bicarbonates, carbonates, and phosphate) can be corrected by injection of acid (commonly hydrochloric) at a rate to lower the pH below 4.0 for 30 to 60 minutes. Drippers partially clogged with calcium carbonate can be cleaned with 0.5 to 2.0 percent hydrochloric acid introduced under slight pressure for 10 minutes. Organic matter restricting emitter flow can be corrected by injecting 500 ppm of chlorine for a period of 24 hours. The system is shut down for this period, after which all submains and laterals are flushed thoroughly.

IRRIGATION SYSTEMS

Various irrigation systems affect the efficiency of water and thus impinge on the triangle sides. The effects of several systems on salt levels, which can affect all three sides of the triangle, have been noted in Chapter 12. Some of the other effects are briefly noted in the description of the various water delivery systems.

The simplest irrigation systems use water introduced at the head of a field, allowing gravity or hydrostatic pressure to move the water over the entire field. Use of inflow devices such as siphons, flumes and spills, and gated pipes allows for better control of incoming water and are still relatively cheap compared to pressurized systems for introducing water.

Sprinklers or tricklers allow more positive application and savings of water with fewer adverse effects on the aeration side of the triangle but require piped pressurized systems to move the water over the field, thereby increasing energy and maintenance costs.

Pressureless Systems

Some systems move water to the head of the field by pressurized pipe but use gravity and hydrostatic pressure to move water over the field. A brief description of the different irrigation systems follows.

Wild Flooding

This system introduces water by a ditch or pipe to the upper end of a field and allows it to flow downslope, the spread of the water following the topography. Unless the soil has been graded, there is usually tremendous variability in the distribution of the water. The system is commonly used only for the most primitive types of farming or for low-value crops such as pastures or hays.

Surface or Gravity Systems

With these systems, water is introduced at the head of the field and is spread by gravity and hydrostatic pressure. Nearly level land is best suited for this type of irrigation. Uneven fields or those with marked slopes will require extensive terracing or earthwork. If not already present, the necessary downward slope in the direction of the flow needs to be provided by land leveling and sloping. Modern laser technology has helped in accurately grading fields so that water application by the gravity system is more reliable than in the past.

Infiltration is usually greater at the upper end where water is introduced. The differences in upper- and lower-end infiltration can be partly overcome by reducing the length of run and improving the slope.

The type of soil and its compaction or layering will also affect infiltration. Coarse open soils will tend to take in more water close to the introduction point. The fine-textured or compacted soils allow larger amounts to move downslope. The variability of soils within a field tends to vary the effectiveness of these systems. In the furrow method, some of this variation can be reduced by the construction of strategically placed dams in the furrows that allow longer infiltration periods at key locations.

Basin Irrigation

By surrounding relatively small, nearly level areas with earth dikes or berms, water is more uniformly distributed than by wild flooding. Basins may surround only a tree, a few trees, or a garden patch. The size of the basin can be increased to about a couple of acres if the ground is nearly level. The system can be used on sloping land, but terraces are needed to supply nearly level areas. The system tends to be labor intensive, making costs appreciably higher than wild flooding.

Border Irrigation

Border irrigation uses earth banks to hold water in uniformly sloped (<0.5 percent), elongated areas (150 to 1250 ft) that are 10 to 30 times longer than wide. The inflow rate, which is in the range of 10,000 to 100,000 gal per hour, needs to be varied with plot size and variability of the soil as well as the crop and its stage of growth. Large plots require large inflow. The system works well for many crops but there are problems of design that may require a trial and error approach. The system is expensive to construct and is not readily adaptable for small applications.

Furrow Irrigation

Water is directed in a series of furrows separated by ridges of soil on which one or more rows of plants can be grown. Furrow lengths vary from about 60 to over 1,000 feet (18.3 to 305 m); furrow widths from 8 to about 25 inches (20 to 63 cm); height of ridges above the furrow bottoms from 12 to about 25 inches (30 to 63 cm); bed widths between furrows 25 to 72 inches (63 to 183 cm). The slope in the furrow usually ranges from 0.2 to 2.0 percent, but slopes as steep as 5 percent have been used at times. Land grading for slopes greater than 1 percent usually are not economical, and slopes greater than 2 percent are subject to erosion. The variations in dimensions of beds and furrows are needed to fit the system to different crops, soils, and infiltration capacities.

Water is usually brought to the upper portions of the field by ditches and is lifted over the bank by portable siphons or spiles inserted in the banks. Recently, gated pipe or "cablegation" has been used to more efficiently transfer water to the furrows. Gated pipe is large-diameter pipe with predrilled holes at intervals corresponding to the furrows. Cablegation uses a pipeline with holes corresponding to the furrows but with a plug inside the pipe that can be moved by cable to open the orifices furrow by furrow.

Furrow irrigation is a very satisfactory method of watering loams and heavier soil types. Startup costs relatively little but it is labor intensive and considerable expertise is needed for proper layout and operation. Another disadvantage is that water may reach the ends of the furrows at different times due to variations in slope and infiltration. This can lead to waterlogging, erosion, or pollution of underground water unless the excess or tail water is collected in a lower ditch to be reused.

Pressurized Systems

The water is moved by pump from various sources (lakes, ponds, canals, wells) to the field and spread by various pressurized devices.

Sprinkler or Overhead Irrigation

In this system, water delivered under pressure to the field is distributed by perforated pipe or rotating sprinklers. These rotating sprinklers may be set in: (1) overhead pipe permanently placed; (2) portable lightweight pipe that can be moved by hand or machinery; or (3) self-propelled movable pipe (traveling gun, center pivot, LEPA [low-energy precision application]). Sprinklers come in various sizes, shapes, and designs. Because of the many variations, it is possible to deliver the water in droplets ranging from a fine mist to large drops. Individual sprinklers may cover from about one square foot to over an acre in area.

The recent development of microsprinklers, also known as misters, mini or low-volume sprinklers, set about 6 feet (1.5 m) above ground can cover an area up to about 32 feet in diameter, (9.7 m) and deliver 40 to 50 gal of water per hour. Because these microsprinklers apply low rates of water and operate under low pressures, they will be considered in greater detail in the section, "Microirrigation Systems."

The macrosprinkler systems require a pump, a mainline pipe, lateral pipes, and sprinklers. Considerable expertise is needed to design the system, since the following should be taken into consideration: (1) water requirements of the crop, (2) area covered at any one time, (3) pump and line sizes to handle sufficient amounts of water, (4) performance of different types of sprinklers, and (5) placement of sprinklers to maximize uniformity of application.

Sprinkler irrigation can be used on soils of varying texture or topography without soil shaping or grading. To avoid runoff or erosion, the rate of application cannot exceed the expected infiltration rate. Principle disadvantages of sprinkler irrigation are: (1) its high initial cost, (2) high maintenance cost, (3) the need to operate at high pressures, (4) poor water application uniformity during windy weather, (5) increased fungal and bacterial disease problems due to wetting of the leaves, and (6) reduced efficiencies of spray programs by interfering with the application of sprays or washing them off after they are applied.

Solid Set Sprinklers

A series of overhead pipes with either a number of fixed nozzles or rotating sprinklers are set at distances so that the water from one can overlap that of the other (usually 45 feet or 13.7 m). Pipes containing fixed nozzles are rotated about 135 degrees by hand or water power to give the desired overlap, whereas rotating sprinklers are designed to provide sufficient overlap to give full coverage. Length of lines is subject to water pressure, pipe friction, and type of emitter. Because of labor requirements, the system with fixed nozzles has largely given way to the type with a number of rotating sprinklers.

Movable Systems

Because of high initial costs, the fixed sprinkler type is being superseded by portable systems. The portable types also have the added advantage that fields can be worked with less interference.

Hand-Moved Portable Systems

These systems use lightweight aluminum pipe sections that can easily be snapped in place to form long lines. Sprinklers mounted on or between sections of pipes or holes drilled in the pipe disperse the water. Disadvantages of the system are that considerable labor is required to move the pipe and considerable soil compaction can result from this movement on wet soils.

Automatic Linear Systems

A line of lightweight aluminum pipe with sprinklers or water outlets is moved across the field on a set of wheels activated by water pressure or a lightweight engine. The system needs a near-level field. Dimensions of the sprayed area are limited to the length of flexible pipe and the water pressure. Uniformity of application is usually better than solid-set or hand-moved systems, especially in windy weather.

Traveling Guns or Reel Hose Systems

The system consists of a drum or reel on which is wound a flexible hose attached to pulsating sprinkler. The sprinkler is slowly moved over the area to be irrigated.

Uniformity of application requires properly timed movement. Because the hose moves over the ground and is wound on the drum, it must be very strong and flexible. High-pressure guns delivering large droplets of water usually used with this system can cause soil crusting and compaction on fields without sod cover.

Center Pivot

A lightweight pipe with nozzles is centered over a water source and gradually moved in a circle. The problem of a circular pattern in rectangular fields can be overcome by special nozzles that are capable of covering what would otherwise be blank areas at the corners. Uniformity of application can be quite high with properly spaced sprinklers or boom-mounted nozzles rotated at an angle to the lateral. The system has become very popular for irrigating crops grown in large fields. Its use entails some destruction of plants as the system traverses the circle but this is of minor consequence. Principle disadvantages are its high initial cost and subsequent maintenance.

LEPA

LEPA is a modification of center pivot and linear movement systems. The system, guided by underground cables, moves linearly. Water can be applied by a set of sprinklers covering the entire area in a pattern similar to that of overhead sprinklers, or the water can be allowed to fall with a set of drop tubes into furrows between rows as in furrow irrigation. The water is held in the furrows on sloping ground by a series of small ditches made by a ditcher-diker. Advantages of the system are: (1) low energy use, (2) low labor requirements to operate, (3) versatility of water application, and (4) adaptation for large-field use. The disadvantages are its high initial cost and the expertise required to set up the system and keep it operating.

Subirrigation

A relatively constant head of water is maintained so that plants are continually supplied with capillary moisture. It requires soil capable of moving water readily from the water table laterally and upward through the root zone. Suitable soils are level peats or mucks, although large areas of sand overlying oolitic limestone in southern Florida are successfully irrigated by subirrigation.

Water is supplied by pumps from surrounding ditches and canals and introduced and maintained at a desired level by means of ditches spaced at regular intervals. The same ditches and/or mole drains in the peats and mucks serve as drainage ditches in case of rains. Additional surface or sprinkler irrigation will be needed to prevent the accumulation of salts in areas receiving too little rainfall for drainage.

Microirrigation

Microirrigation systems deliver relatively low rates of water at low pressure for long periods of time. As with other pressure systems, water is brought to the crop by a network of mainlines, submains, and lateral lines. They differ from other systems in that the lateral lines have emission points along their lengths and these emission points provide small but uniform amounts of water directly to the root zone. The emitting devices used are strip tubing, emitters or drippers, jets, and minisprinklers.

The low application rate of these systems require long application periods of 12 hours or more. This can be a disadvantage in that more equipment is needed but the advantages are lower capital and operating costs because the pumps, filters, and pipelines are sized for the lower flow rates.

Appreciably less water is used with these systems. By placing the water directly in the zone of active roots, water is not wasted on nonproductive soil. Less water is lost to evaporation, runoff, and deep percolation. Amounts can be varied readily to the meet the requirements of the developing root system.

Microirrigation can be used for difficult soils and terrain. By keeping a relatively constant wetted area, this type of irrigation has produced good crops on soils that would be too sandy, gravelly, rocky, or salty for normal production with other types of irrigation. Uniform application of water is possible on irregular or rolling terrain without costly leveling. The low application rate makes it possible to satisfactorily irrigate heavy soils or other soils with poor infiltration.

A large advantage of microirrigation is its capability of moving salts away from the plant. By properly placing emitters close to the plant, salts move away from the plant in the advancing front of the water to the outer edges of the root zone, permitting good growth of relatively sensitive crops with elevated salts in the water or soil.

Disadvantages of microirrigation are high initial cost, the need for expertly designed systems, constant maintenance and management, and the need for high-quality water to avoid clogging.

Drip Irrigation

The application of water by strip tubing or emitters is commonly referred to as drip irrigation. Water application by individual opening or emitter is limited to a relatively small area, producing characteristic wet zones extending sideways and downward. The dimensions of the wet zone are determined by the amount of water applied, soil infiltration capacity, and frequency of application. Usually the wet zone has a larger diameter at or near the surface that tapers to a narrow point with depth. The wet zone is appreciably wider but shallower in most fine-textured soils compared to the more open sandy soils.

Strip tubing, used extensively for row crops, consists of inexpensive plastic hose with a series of orifices spaced along its length. It is sometimes referred to as drip tape and comes in two different flow categories. The turbulent type controls emission by means of orifices or making the water follow tortuous paths. A common means of producing a tortuous path is to use a double hose. One chamber of the hose that is under high pressure delivers the water to another hose under lower pressure. The latter releases the water slowly along its entire length. The flow rate of the tortuous-path types changes slowly with changes in pressure, which allows for longer runs than the laminar flow. The laminar flow type, commonly made by overlapping two edges of plastic film, which produces a capillary passage for the water flow. The flow rate is directly proportional to the pressure and temperature of the water.

Small plastic flow-control devices are also installed at predetermined spacing in polyethylene hose of varying diameters. The flow-control devices or emitters can be placed in-line or at various points on the hose (point source). Capillary spirals with periodic openings for in-line systems have advantages. The in-line emitters work best on near-level soils and for runs less than 300 feet. Low pressures of 5 to 15 psi are satisfactory. The point source emitters (button), plugged into the tube at various places corresponding to plant location, operate at higher pressures (5 to 60 psi). The rate of flow is controlled by varying the pressure. Buttons are more expensive but tend to outlast the in-line emitters. The discharge rate of some point emitters remains fairly constant with considerable variation in pressure, thus making them much more suitable for uneven terrain. Emitters discharge water at rates varying from about 0.5 to 2 gal per hour.

Jets and Minisprinklers

Jets can apply water at higher rates than emitters and wet a greater area than emitters or strip tubing. Water is sprayed by the jets in either a fan-shaped pattern or as discrete streams. The lack of moving parts limit the throw distance.

Minisprinklers can apply water in a circular pattern extending in diameter from about 2 to 32 feet (0.6 to 9.8 m). Their low water flow for long periods has promoted their use for under-tree irrigation for frost control.

Bubbler Irrigation

A modified microirrigation system is very effective in greatly reducing clogging problems, while simplifying the system. Buried lateral irrigation tubes are attached to a series of standpipes 1 to 3 cm in diameter, at strategic locations. The standpipes are anchored to stakes and their height adjusted to deliver the water at a rate to supply an equal amount to small level basins. The hand-constructed basins are rimmed with soil. By removing drippers, clogging is practically eliminated and the pressure needed to operate the system is reduced, thereby reducing pipe and pump size and the need for pressure regulators.

The system is very well adapted for orchards or vineyards on relatively level land where a basin can be constructed for each tree or small group of vines. The cost of installing the system may be only slightly less than that of a drip system but costs of maintenance and pump energy are appreciably less. Disadvantages are the high cost of labor to maintain the basins and its poor adaptability to adverse terrain and different crops.

Comparative Efficiency of Different Irrigation Systems

The efficiency of an irrigation system depends on such factors as the type of installation, the size of motors, piping, etc. Efficiency and uniformity of application affect the amount of water needed. Some applied water does not recharge the soil but is lost as runoff or percolation, and as wind drift for some irrigation systems. Much of the efficiency and uniformity of irrigation is determined by the type of system, the comparative efficiencies of which are given in Table 15.9.

TABLE 15.9. Efficiency of Irrigation Delivery Systems

System	Efficiency
Sprinkler	
Center pivot	0.80-0.85
Skid-tow	0.75
Solid-set	0.75
Side roll	0.75
Big gun traveler	0.70
Surface	
Gated pipe and reuse	0.85
Gated pipe	0.60
Siphon tube and reuse	0.75
Siphon tube	0.30-0.50
Drip or trickle	0.75-0.90
Border	0.65-0.80
Basin	0.75-0.90

Sources: A compilation of data from *Irrigation Scheduling.* 1977. Coop. Ext. Serv., Univ. of Nebraska, Lincoln, NE; Goldhammer, D. A. and R. L. Snyder (Eds.). 1989. *Irrigation Scheduling.* Division of Agriculture and Natural Resources Publication 21454, Univ. of California, Oakland, CA.

TIMING THE APPLICATION

In addition to knowing whether to irrigate, such factors as water availability, irrigation capacity, uniformity of the irrigation, and type of system need to be considered for properly timing the application.

An efficient scheduling program requires water availability on demand. This requires sufficient well or other source capacity to handle maximum water needs. In situations where water is available only on certain days or is subject to use upstream, scheduling is often meaningless. The irrigation capacity should be designed to fulfill maximum needs considering type of soil, infiltration, type of crop and its acreage, sensitive crop stages, maximum climatic stress, and the limitations of the irrigation system.

The timing of irrigation using tensiometers (one of the more popular methods for determining soil moisture status) can be used as an example of scheduling water application. The centibar readings indicate tension or water stress, with a zero reading signifying saturation or free water, but 10 to 30 cb equals field capacity depending upon the soil texture. The zero reading signifies ample water and severe lack of air. Field capacity will exhibit no water stress and little or no air stress. Values above the field capacity indicate increasing water stress, with serious stress for nearly all crops as tensions increase above 75 cb.

The tensiometer readings at which irrigation should start varies with different crops, their stages of growth, and climatic conditions (see Tables 2.3, 2.4, 2.8, and 2.11).

Irrigation needs to be started at relatively low soil moisture tension for young or shallow-rooted crops or for critical periods of growth. Very low tensions (10 to 20 cb) as starting points for irrigation are desirable for crops grown on soils of low MHC. Tensions of 12 to 15 cb for citrus and only 10 cb for plastic-mulched vegetables are considered ideal starting points for irrigating the flatwood sands of Florida—soils with very low MHC. Frequent small applications are more suitable for shallow-rooted crops grown on light soils than heavy waterings applied less frequently.

Starting at lower soil moisture tensions is also advisable under conditions of high evaporation (high temperatures with ample solar radiation and/or strong winds).

Appreciably higher tensions (30+ cb) can be used as a starting point for crops grown on heavier soils. Greater tensions (40-50 cb) can be used as starting points for crops with deeper roots grown on heavier soils, especially as these crops are approaching maturity. An estimate of the rooting depth of several crops can be obtained from Table 2.11. More accurate values can be obtained by examining the root systems, since soil conditions often prevent roots from attaining their normal depth.

Irrigation scheduling is often dictated by the capacity of the system. If all the areas cannot be watered at the same time, irrigation may need to be started earlier on a portion of the farm in order to irrigate all of it before harm is done to some of the crop.

Greater accuracy in timing the start and duration of the irrigation is possible if a curve is drawn for each field showing the moisture contents at different tension readings. Samples of soil taken at the same depth as cup placement are collected over a period of time from full capacity to about 80 cb readings. Moisture in the samples is determined gravimetrically and converted to volume percent. These values are graphed against centibar readings existing at time of sample collection (see Figure 15.2).

These readings can also be helpful in determining amounts of water to be added. For example, if at field capacity the cb reading is 8 and the moisture held equals 25 percent by volume, and the current cb reading is 41 and only 12 percent of the water remains, the amount of water to be added to bring the soil to field capacity can be calculated as follows:

a. 25% (moisture at field capacity) − 12% (moisture held now) = 13%
b. 13% × 12 (in/ft) = 1.56 in of water per ft of soil

FIGURE 15.2. Relationship of Soil Moisture Contents of a Sandy Soil to Different Tensiometer Readings Taken at a Depth of 6 Inches

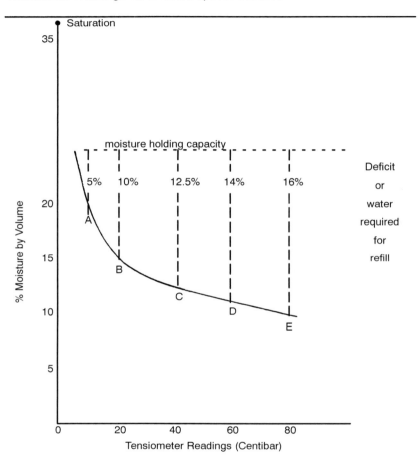

* Percent by volume.

Source: Rhoads, F. M. 1981. Plow layer soil water management and program fertilization on Florida utisols. *Proceedings Soil and Crop Scientific Society of Florida* 40: 12-16.

The amount of water needed at various tensions can be read as the difference between soil moisture present and that needed to bring it to full capacity. For example, in Figure 15.2, showing typical moisture percentages at different tensiometer readings, 10 percent of the water volume is deficient at point B. If the cup was placed at a 6-inch depth, it would be necessary to add 0.60 inches of water (0.10×6). Since there are 27,154 gal of water in an acre-inch, the 0.6 inches of water would require 16,292 gal per acre.

APPLICATION RATES

To maximize the water leg of the triangle without unduly upsetting aeration or adversely affecting the soil or medium, the water must be added at a rate suitable for the particular conditions. Application of water must never exceed the infiltration rate. A slow rate can be effective in reducing runoff and puddling on slowly permeable soils or in avoiding excessive percolation in highly permeable soils. Drip irrigation is an ideal method of applying water at slow rates. The normally slow rate of 0.15 to 0.25 in/hr can be reduced to 0.1 in/hr if high-pressure emitters delivering 0.5 gal/hr are used. Care must be taken that the use of such low rates with low-pressure emitters provide enough volume to fully pressurize the lines. If lines are not fully pressurized, water distribution may be uneven.

Intermittent irrigation cycles are a means of reducing runoff and puddling in soils of low permeability. Intermittent cycles of 15 minutes on and 15 minutes off will cut the normally slow rate of drip irrigation in half, which is usually satisfactory for light soils of poor permeability. Even longer on-off periods may be desirable for the heavier soils with poor infiltration rates.

Water Delivery Measurement

It is important for the grower to know how much water is being applied in a given period in order to plan the application of fertilizer or other chemicals through the system and also to effectively recharge the soil with water without overdoing it.

Meters

Flow meters can be installed in pipelines to measure the volume of water applied. They can be placed at the head of the farm to show total volume consumed or at the head of individual fields. Metering devices that shut off the flow of water after a predetermined amount is delivered are an effective means of measuring the volume of water used.

Sprinkler Delivery Rate

The delivery rate of nozzles, as supplied by the manufacturer or determined by measurement, can be a means of calculating delivered volumes. Determination of discharge rates in the field is a more accurate method of determining the volume delivered. The average discharge rate of several sprinklers is determined by directing each nozzle flow into a 10 to 50 gal tank for a period of 10 minutes and then measuring the volume of water collected. The rate (R) of water applied can be calculated from equation 9,

$$R = \frac{SR \times 96.3}{DS \times DL} \tag{9}$$

where R = inches per hour (in/hr); SR = average sprinkler rate in gpm; DS = distance between sprinklers in feet; and DL = distance between lines in feet.

Drip Irrigation Delivery Rate

The delivery rate of drip systems can be estimated in a similar manner except it is desirable to make the collections in small containers for point emitters or troughs for porous tubing. The contents are transferred to 100 ml graduates to measure volume in milliliters. It may be necessary to remove some soil under each emitter or a length of porous tubing equal to the trough length so that the collection containers can be placed underneath without lifting the pipe. The ml/min can be readily converted to gpm with Table 15.10. The above equation can be used for estimating inches per hour delivered by point emitters; DS = the distance between point emitters or the length of the trough used to collect delivery from porous tubing.

Open Channel Delivery Rate

The volume of water passing through open channels can be calculated from equation 10,

$$V = A \times MV \tag{7}$$

where V = volume in cubic meters (m^3 = 264.17 gal); A = cross section area of flow in sq ft; and MV = mean velocity of flow through the cross section.

TABLE 15.10. Conversion of Milliliters per Minute to Gallons per Minute

ml/min	gpm	ml/m	gpm
45	0.0118	75	0.0198
50	0.0132	80	0.0212
55	0.0145	85	0.0225
60	0.0158	90	0.0238
65	0.0172	95	0.0250
70	0.0185	100	0.0263

The cross section area (A) of smooth-sided channels can be calculated, but several points along channels with nonuniform sides must be averaged to obtain their area. The mean velocity (MV) can be taken from a meter measuring the current but this may not be available for most growers. A simple method of obtaining MV is to measure the movement of a float over a standard straight length of the channel for 150 feet. The average velocity in feet per second for smooth-lined ditches will be 0.8 of that obtained with the float, or only 0.6 for rough ditches.

Measurement from Horizontal Pipes (Trajectory Method)

Two different measurements of the water trajectory need to be made to determine flow rate from full pipes: the horizontal distance X in inches parallel to the centerline of the pipe needed for the trajectory to fall a vertical distance Y of either 6 or 12 inches. If the pipe is full, the measurements are made on the upper surface of the jet; if partially full, at the center of the jet. Well yield equals the horizontal distance of the trajectory (X) multiplied by a factor from Table 15.11. For example, if the length of

TABLE 15.11. Factors for Converting Trajectory Height of Water Flowing from Pipes of Several Dimensions to Gallons Delivered Per Minute*

Pipe diameter (inches)	Factors for height trajectories (Y)	
	6 inches	12 inches
2	5.02	3.52
3	11.13	7.77
4	17.18	13.4
6	43.7	30.6
8	76.0	52.9
10	120.0	83.5
12	173.0	120.0

* For directions of measuring X and Y and also other methods of calculating water flow, see also V. H. Scott and C. E. Houston. 1981. *Measuring Irrigation Water.* Leaflet 2956, Univ. of California, Berkeley, CA.

Source: Schwab, D. *Irrigation Water Management.* OSU Extension Facts #1502. Coop. Ext. Serv., Div. of Agric., Oklahoma State University, Stillwater, OK.

the trajectory is 19 inches when the height of the trajectory is 12 inches for an 8-inch pipe, the factor = 52.9. This factor × 19 (the trajectory distance) = 1,005 gpm (Scott and Houston, 1981).

SUMMARY

Irrigation water affects the aeration and nutrition sides of the fertile triangle as well as the more obvious water side. To obtain maximum benefits from the three sides of the triangle, the grower must apply water only when it is needed, using amounts and quality suitable for the crop and soil. But the amounts and quality of water are closely connected to the irrigation system used.

Some of the more important methods for determining need for irrigation are: (1) gravimetric, (2) feel and appearance, (3) tensiometers, (4) gypsum blocks, (5) thermal change sensors, (6) neutron scattering, (7) capacitance, (8) time domain reflectometry, (8) evapotranspiration, (9) foliar air differentials, and (10) AquaProbe.

The quality of water, which can be measured by a series of chemical, biological, and physical tests, is closely related to effectiveness of irrigation, but the quality required varies with different crops and irrigation systems. For example, water used for crops sensitive to chlorides will be satisfactory if it is low in this element, but much more chloride can be tolerated if the water is applied to the soil rather than to the leaves. Microirrigation systems are easily rendered useless by clogging, thereby requiring water low in suspended solids and ions that might cause clogging by biological or chemical means (Fe^{2+} or Fe^{3+}, HCO_3^-, $Ca^{2+} + PO_4^{2-}$, $Ca^{2+} + SO_4^{2-}$).

Irrigation systems can be essentially pressureless (wild flooding, basin, border, or furrow) or may require some pressure to move the water through the final distributor (sprinkler types, or drip). Sprinklers can be fixed overhead or portable. The portable types are moved by hand, water pressure (both straight line or center pivot), small motors, LEPA (guided systems), and reels. The drip systems employ emitters or strip tubing for dispensing the water.

The different systems have various advantages and disadvantages. The pressureless systems are generally cheaper to install but may be more costly to operate and usually lack precision in delivering water. The pressure systems are more useful in providing precision application, but may be subject to clogging and breakdowns. The drip system is one of the more expensive systems to install and is very subject to clogging, but uses the smallest amount of water.

Measurement of water delivery is needed to avoid adding it faster than the soil's ability to infiltrate it. Methods for measuring water delivery of several irrigation systems have been presented. Drip irrigation is especially suited for soils with low infiltration capacity.

Chapter 16

Adding Nutrients

NUTRIENT MANAGEMENT

Profitable crop production usually cannot be attained, and certainly not maintained, without proper nutrient management. Nutrient management involves not only the replacement of needed nutrients but adding them in such a manner that sufficient quantities are available as the plants need them. This is complicated by variable requirements of different plants, by changing requirements at different stages of the same plant, and varying losses of nutrients due to erosion, leaching, fixation, and volatilization.

For optimum crop production, the essential nutrients, as described in Chapter 3, must be available in solution upon demand. After much cropping, few soils left in the world have sufficient natural content of nutrients to raise satisfactory crops without some additions.

Although some artificial soils are prepared with large amounts of nutrients requiring little or no addition, many have almost none. The media with little or no nutrients will need continuous feeding from the very beginning. Even those supplied with large amounts of slow-release materials will in all probability also need some nutrients with time, if production is to be maintained.

By altering certain parameters, the efficiency of added nutrients in both soils and artificial media can be increased considerably, thereby reducing the amounts that need to be applied to make good crops. Some of the more obvious conditions that are amenable to adjustment and can greatly influence the availability of both added and native nutrients are: pH, organic matter, cation exchange capacity, percentage base saturation, water level, and salt concentration, discussed in Section II. These conditions need to be optimized regardless of whether nutrients are added or how they are applied.

In addition to what is already in the soil or will be made available to the crop, the kinds and amounts of nutrients that need to be added for optimum yields depend not only on the demands of the crop but also on the (1) release rate of nutrients added, (2) salt content of the fertilizers, and (3) when and where the nutrients are placed.

To optimize the nutrient side of the fertile triangle so that air and water can be fully utilized, it is usually necessary to add the maximum amount of nutrients that does not exceed the salt tolerance of the crop.

In supplying large amounts of nutrients, care must be taken to avoid "burning" caused by excess salts, which has been covered in Chapter 12. Basically, salt damage is reduced by limiting the the application of extraneous ions sparingly used by the plant (Na^+, $Cl-$, SO_4^{2+}) to small amounts, and applying the large quantities of needed nutrients (N, P, K, Ca, Mg) either as slow-release materials or, if quick-release materials are used, adding them in ways that will not harm the crop. By limiting extraneous ions, and better managing the necessary ions, much larger quantities of the latter can be safely utilized.

Large amounts of quick-release materials can be safely used if (1) they are added as several small amounts, (2) the soil naturally holds or can be conditioned to hold large amounts of fertilizer without causing harm (see Chapters 4, 5, and 7), (3) they are added as low-salt materials, and (4) if they are placed to minimize contact with existing roots.

Supplying Needed Nutrients

Adding insufficient or excess nutrients for a crop can be costly to the grower and society. To the grower, a shortage or an excess of nutrients leads to reduced crops of inadequate quality; the excess has the additional onus of increased costs. Society pays for inadequate fertilization by spending more for food and fiber, and in the case of excesses, by covering the costs of aggravated pollution problems.

Before attempting to replace needed nutrients, it is important to determine what elements are needed and the amounts necessary. Whereas some reliable general figures are available for the amounts of the elements needed to produce different yields of some of the important crops, we still lack precise knowledge of the minimal amount required by a particular soil or medium in any given period to produce the desired crop. Some good estimates of nutrients available at planting time are possible from soil tests, but their usefulness is reduced at times due to the different amounts of available nutrients caused by climate variability.

SOIL AND PLANT ANALYSES TO GUIDE
THE FERTILIZER PROGRAM

Some idea of the amounts of the various nutrients needed to produce good yields of several crops can be gained from amounts removed as presented in Table 16.1, but these values are very rough estimates since

TABLE 16.1. Common Inorganic Fertilizer Materials

Material	Analysis (%)						Solubility[a] g/100 ml
	N	P_2O_5	K_2O	Ca	Mg	S	
Ammonia-ammonium nitrate	8-25	0	0	0	0	0	liquid[b]
Ammonia-ammonium nitrate-urea	36-44	0	0	0	0	0	liquid[b]
Ammonia-sulfur	74	0	0	0	0	10	liquid[b]
Ammonia-urea	21	0	0	0	0	0	liquid[b]
Ammonium bisulfite	14	0	0	0	0	32	380
Ammonium bisulfite	8	0	0	0	0	17	liquid
Ammonium chloride	26	0	0	0	0	0	30
Ammonium nitrate	33.5	0	0	0	0	0	116
Ammonium nitrate-lime (ANL, Cal-Nitro)	21-27	0	0	7-15	0-4	0	ss
Ammonium phosphate mono (MAP)	11	48	0	1	0	2	38
di (DAP)	16-21	48-53	0	0	0	0	70
ortho	8	24	0	0	0	0	liquid
poly (48)	10	34	0	0	0	0	liquid
poly (78%)	11	37	0	0	0	0	liquid
poly (APP)	10-15	35-62	0	0	0	0	
solutions	10	34	0	0	0	0	liquid
	11	37	0	0	0	0	liquid
Ammonium polysulfate	20	0	0	0	0	40	liquid
Ammonium sulfate	21	0	0	0	0	23	72
Ammonium thiosulfate	19	0	0	0	0	43	148
Ammonium thiosulfate	12	0	0	0	0	26	liquid
Ammoniated super-phosphate (normal)	2-5	14-21	0	0	0	9-11	ss
Ammoniated super-phosphate (triple)	4-6	44-53	0	0	0	1	ss
Anhydrous ammonia	82	0	0	0	0	0	liquid[c]
Aqua ammonia	20-24	0	0	0	0	0	liquids[b]
Calcium chelates	0	0	0	3-6	0	0	s
Calcium chloride monohydrate	0	0	0	31	0	0	110
dihydrate	0	0	0	27	0	0	40

TABLE 16.1 (*continued*)

| Material | Analysis (%) | | | | | | Solubility[a] g/100 ml |
	N	P$_2$O$_5$	K$_2$O	Ca	Mg	S	
Calcium nitrate	15.5	0	0	19	1	0	550
Epsom salts	0	0	0	0	9.9	13	90
Lime sulfur	0	0	0	9	0	23	liquid
Magnesium chelates	0	0	0	0	4-9	0	s
Magnesium chloride	0	0	0	0	8	0	55
Magnesium nitrate	0	0	0	0	6	0	liquid
Nitric acid	12	0	0	0	0	0	liquid
Nitric phosphates	14-29	14-28	0-20	0	0	0	ss
Nitrogen solutions							
low-pressure	37-41	0	0	0	0	0	liquid
no pressure	28-32	0	0	0	0	0	liquid
Phosphoric acids							
glycero	0	41	0	0	0	0	liquid[d]
food grade 75%	0	54	0	0	0	0	liquid[d]
ortho-wet process	0	30-52	0	0	0	0	liquids[d]
ortho-furnace	0	56	0	0	0	0	liquid[d]
super	0	70-83	0	0	0	0	liquids[d]
Potassium chloride (muriate of potash)	0	0	60-62	0	0	0	35
Potassium hydroxide	0	0	41	0	0	0	liquid[e]
Potassium nitrate	13	0	44	0	0	0	40
Potassium phosphate							
di	0	10	14	0	0	0	liquid
meta	0	60	40	0	0	0	liquid
meta	0	53	47	0	0	0	100
ortho	0	20-47	11-31	0	0	0	liquids
poly (tri)	0	47	52	0	0	0	
Potassium sulfate	0	0	50-53	0	1	17	12
Potassium magnesium sulfate (Sul-Po-Mag, K-Mag)	0	0	22	0	22	11	ss
Sodium nitrate (nitrate of soda)	16	0	0	0	0	0	95
Sodium-potassium nitrate (nitrate of soda-potash)	15	0	14	0	0	0	s

Material	Analysis (%)						Solubility[a] g/100 ml
	N	P₂O₅	K₂O	Ca	Mg	S	
Sulfuric acid	0	0	0	0	0	31	liquid[c]
Superphosphate							
ordinary	0	18-20	0	20	0	12	ss
treble (TSP)	0	44-53	0	13	0	1	ss
Urea*	46	0	0	0	0	0	50[f]
Urea-ammonium-	28,30,						
nitrate (UAN)	32	0	0	0	0	0	liquids
Urea-sulfuric acid	28	0	0	0	0	8	liquid[d]

[a] Solubility. s = soluble. Figures given are in grams per 100 cc at 20°C. Multiply g/100 cc by 8.5 to obtain lb/100 gal). ss = slightly soluble; not suitable for preparing solutions, but OK for suspensions or dry fertilizers.

[b] Pressure solutions. The free ammonia contained can injure plants and personnel. High-pressure solutions require special handling. All require use of goggles and protective clothing when preparing mixed fertilizers. Not to be used in greenhouses or for overhead irrigation, although suitable for production of soluble fertilizers or furrow irrigation.

[c] Anhydrous ammonia requires special handling and equipment for application.

[d] Acids or alkalies can be injurious to plants, personnel, and equipment. Handle with care, using protective clothing and goggles.

[e] Highly corrosive! Reacts with acids to neutralize them, creating considerable heat. Add after acid has been diluted. Handle with care as for acids.

[f] Urea is an organic compound but analysis and solubility more appropriately places it with the inorganic materials. Although sold in some states as an organic source of N, losses of N are often equivalent to highly soluble inorganic sources.

they do not include nutrients required for roots or those normally lost by leaching, volatilization, or fixation.

The problem of estimating the net amount of nutrients available to the crop during the growing season can be lessened considerably by using later soil and plant tests. (The net amount of nutrients available to the crop = amount added + that derived from soil or medium – that which is leached, fixed, or volatilized.)

Some of the methods of soil and plant testing and their use in crop production have been outlined in my book, *Diagnostic Techniques for Improving Crop Production* (Wolf, 1996). Most growers can get some very good information about the nutrients and amounts needed for a particular crop by submitting soil samples before the crop is planted and soil plus leaf samples during the crop season to one of the many commercial or governmental laboratories.

Most laboratories will make recommendations for liming materials and fertilizers based on soil test results. Unfortunately, many laboratories provide the recommendations in pounds of nutrients per acre but often do not consider the source of the nutrients, the type of fertilizer, nor the method or time of their application. These factors, often unconsidered, alter the efficiency of applied materials. Growers, by working closely with a laboratory, can obtain more reliable recommendations that are more apt to be consistent with yield goals, better fit the grower's application procedures, and include various inputs (manure, legumes, etc., irrigation) which pertain to the the particular operation.

Rapid soil tests obtained shortly before planting a crop allow adjustment of pH and salts and provides data allowing the grower to limit application to amounts needed to bring levels into optimum ranges.

A soil sample collected before the crop is planted should supply sufficient information about the amounts of nutrients needed and the soil conditions that affect fertilizer efficiency for both surface soil (0 to 8 inches) and subsoil samples (at least 8 to 16 inches). The following information should be helpful for most situations:

- pH
- Texture
- Lime requirement
- Organic matter
- Cation exchange capacity
- Percent saturation of Ca, Mg, K, Na
- Soluble salts (conductivity)
- Available nutrients and harmful elements

N (NO_3-N and NH_4-N)	B
P	Cu
K	Fe
Ca	Mn
Mg	Zn
Sulfate-S	Al

In addition to providing the above results, the collection of the soil sample should also provide information about depth of the effective soil, compaction, drainage problems, and adequacy of water. The interpretation of the results is greatly improved by accompanying information about the crop to be grown, previous problems and yields, yield goals, method of tillage and fertilizer application, irrigation, and the quality and amount of water normally applied.

Soil tests coupled with leaf analysis of samples collected after the plants are well established can refine the fertilizer program by indicating whether additional nutrients are needed. Such later tests are especially useful in defining contributions made by humus and organic materials or other slow-release materials.

Soil Tests for Nitrogen

Recommendations for N are fraught with more problems than those for any other element, because its availability is subject to greater changes. These changes are difficult to predict because they are largely dependent on (1) amount of leaching, which can vary from season to season; and (2) microbial action affecting OM decomposition, which in turn is influenced by climate, soil pH, aeration, and energy sources. Nevertheless, soil OM or the additions of various organic materials is used as a basis for applying N or modifying normal N applications.

Our own experiences suggest that using soil OM tests or various organic additions as a basis for recommending N, while helpful, can be greatly improved by making soil analyses for available N (NO_3-N+ NH_4-N) several times during the year.

In the northern states, serious N deficiencies can arise in the spring when soils are too cold for N release from OM. Eliminating or greatly reducing the N application because of large amounts of soil OM can seriously limit crop yields. A better approach for these and other areas is to measure the available NO_3 and NH_4-N and eliminate or greatly reduce the N application only if high levels of available N are present. If soils contain good to high levels of OM or have received manures, cover crops, or other organic matter additions, only about 1/3 of the recommended N application is added at planting. In the Morgan-Wolf method (Wolf, 1982), the amount added is enough to bring available soil N from a "poor" (0-14 ppm) or a "fair" (15-24 ppm) to a "good" level (25-75 ppm). For example, if only 10 ppm of NH_4-N + NO_3-N were present, it would be desirable to raise the available N to midway in the "good" range, and so 40 ppm (80 lb/ac) would be needed. The addition of manures or legumes as potential sources of relatively available N permits using the much lower "fair" level as guide. In this case, an application of 10 ppm (201 lb/ac) is enough to bring available N midway into the "fair" range. This method works well if combined with later tests to determine the need for side dressings.

The proponents of a test now being adapted in the northeastern and to some extent in the central states suggest only starter amounts of N be applied for corn at planting. The NO_3-N content of a sample collected from the upper 12 inches when corn plants are 6 to 12 inches tall is used to determine whether N will be applied as a side dressing. An NO_3-N content of 40 to 60 lb/ac at that time is considered ample for the remainder of the season and no N need be applied.

Analyses for available N for early crops (northern hemisphere) to be planted in soils subject to considerable leaching can be omitted. Generally, the amounts of available N in these cold soils will be too low to materially affect crop production. But tests should not be omitted for soils receiving little or moderate rainfall in autumn or winter. Frequently, such soils may have accumulated enough available N to affect crops, but this N may be at deep levels sometimes requiring samples collected to a depth of about 3 feet.

For such drier areas, use of available N tests can be helpful even if tests are made only prior to planting the crop. Kansas suggests the following reductions in N rates if NO_3-N+ NH_4-N in the first two feet of soil is at varying levels:

N Rating	Adjustment in Fertilizer Recommendation
low	no reduction
medium	reduce by $1/3$
high	reduce by $2/3$
very high	no N recommended

Although evaluation of available soil nitrogen for spring plantings in humid northern climates has little value, its measurement in the summer prior to replanting a crop in the northern states or anytime when replanting in warmer climates has been very helpful. Often, there is enough residual available N that the recommended application can be reduced or even eliminated. Failure to measure this N can cause serious economic losses and pollution of groundwaters.

Ideally, several tests for available N will be made on many soils as the crop is growing to more fully evaluate the changing situation. The frequency of testing needs to be varied with the crop, type of soil, and rainfall. I have recommended tests every two weeks for high-valued short-season annual crops grown on light soils subject to extensive leaching, but several tests or even only one test during the growing season can suffice for less exacting growing conditions.

Soil tests coupled with leaf analysis of samples collected after the plants are well established can refine the fertilizer program by indicating whether additional nutrients are needed. Assuming that a plant sample will be run at the same time, the analyses of postplanting samples can be limited to pH, conductivity, available N, P, K, Mg, and sulfate-S. The tests for OM and CEC are not rerun because these change so slowly; tests for Ca and Al are not rerun because these should have been corrected by addition of liming materials, most of which leach or are fixed rather slowly; and the micronutrients are not only better evaluated by leaf analyses but their correction is possible by foliar application, making the amounts present in the soil less important. If the micronutrients cannot be added foliarly, then they also need to be determined in the postplanting sample.

Availability of Other Elements

Soil analyses for recommending nutrients other than N are much less subject to errors. Nevertheless, there are problems that make recommendations based solely on a single sample collected before planting less reliable. Leaching losses from coarse or light soils can make it difficult to accurately prescribe amounts of K and Mg as well as N. Also, unless fertilizers have been applied in bands, fixation of P and K in heavy soils can seriously compromise recommendations for these elements. Both situations can be corrected by tests of samples collected while the crop is growing. But the later tests are of little value unless there are facilities for applying additional fertilizer.

Leaf Analysis

The leaf analysis can help decide whether the basic program is adequately providing necessary nutrients. It never should be run without an accompanying soil test and an evaluation of soil and roots, which are needed to help explain leaf test results. For example, low values in the leaf, which are usually due to shortages of the element in the soil, may also be due to high salts, low or high pH values, high values of an antagonistic element (Ca shortages due to excess NH_4-N or K, Zn shortages due to excess P, etc.), or to such nonnutrient problems as compaction, too little or excess water, or nematode and disease attack. In the case of excess salts, the soil may already have an excess of the element that is low in the leaf. Blindly attempting to correct shortages by soil application without considering the soil can be wasteful or even harmful to the crop, as well as increase pollution problems.

Correction of low leaf levels of macronutrients by foliar applications is much less subject to problems but it is usually difficult to supply necessary amounts of N, P, and K. Correction of Ca and Mg shortages by foliar application is more practical because of the relatively small amounts needed. Also, foliar applications for correction of Ca shortages in fruits are much more efficient than similarly timed soil applications.

The correction of most micronutrient shortages by foliar application is relatively easier also because very small quantities are needed. Foliar applications are often more efficient than soil applications, especially if pH values or antagonistic ions in the soil are high.

Timing the Nutrient Application

One of the major problems in nutrient management is to have enough available nutrients present at various plant growth stages. As was pointed out in Chapter 3, most of this demand (in total amount of nutrients) is usually very small in the first third of growth, but becomes very large in the second third and remains almost as large in the final third. The concentration of nutrients needed to supply this demand, though, is quite different—with the highest concentration required soon after germination, declining as the plant ages. But because the totals removed are greatest in the second third and large in the final third, much more fertilizer or other nutrients will be needed to replenish the soil or nutrient solutions during these periods. The high concentrations needed soon after germination can be supplied by strategic placement rather than adding large amounts. The second and third segments can be taken care of by large applications through (1) timely fertigation (the application of fertilizer by irrigation) during these periods, (2) broadcast fertilizer before planting, or (3) properly placed bands before planting or during the growth period.

The timing of nutrient applications for perennial plants that are well established and have considerable reserves of nutrients is somewhat different than for annuals. The large reserves of N and other elements carried in the plant makes it somewhat less dependent on daily levels of nutrients in soil or solution. Large amounts of nutrients are needed primarily for renewed growth after dormancy or prior to new flushes of growth.

SLOW- OR QUICK-RELEASE NUTRIENTS

Quick-release nutrients can provide the answer if they are added periodically as they can be quickly taken up by plants providing they are positioned properly. Their use is ideal for hydroponics or if the grower has an irrigation system capable of quickly adding readily available nutrients. With these systems, needed nutrients as measured by soil tests can be added cheaply and quickly.

Their use in conventional agriculture or where irrigation is not available is also highly practical if leaching or fixation of nutrients is not serious. In these situations, preplant broadcast applications of large quantities of quick-release fertilizers by placing them in multiple bands can provide adequate nutrition. In such case, the primary objective is to provide sufficient nutrients without adding an excess.

If however, leaching or fixation of nutrients is a problem, quick-release nutrients may not adequately nourish the crop as it advances. But the losses from leaching are less pronounced in the heavier soils and placing the nutrients in a series of bands can eliminate much of the fixation (primarily P and K), making quick-release nutrients still satisfactory. Some losses from fixation can still be tolerated with many quick-release materials because they are relatively cheaper than many slow-release materials. Some losses from leaching also can be tolerated from quick-release materials but only if such losses do not cause pollution problems.

Where losses from fixation or leaching are greater than the cost differential the grower can use slow-release materials. Besides having less loss by leaching and fixation, slow-release materials reduce the chances of toxic salt accumulation. This allows much larger applications with greatly reduced risks of "burning" the crop.

There are several different types of slow-release materials. Most of them fall in the category of (1) natural organics, such as cover crops, hays, and plant residues; manures, sludges, and composts; animal and fish byproducts (see Chapter 5); or (2) the slowly soluble synthetic materials; or (3) soluble chemicals coated with resins or plastics, which slow the release of nutrients. Many of the natural organic materials have been used for centuries, but inorganic materials that are relatively insoluble or soluble fertilizer salts coated with different materials to slow releases are a more recent development.

There is considerable variability in the effectiveness of different slow-release materials. Because many of the natural organics (hays, cover crops, crop residues) have to undergo decomposition by microorganisms before releasing nutrients, their effectiveness at times can be slow and erratic. The legumes or materials relatively high in N will decompose faster but N release, especially if soils are cold, may also be too slow for many crops. Manures, composts, and activated sludges, the analyses of which are given in Table 5.9, generally release their nutrients much quicker than the hays, etc. The nutrients in several fish and animal byproducts (Table 5.8), high in N, are usually rapid enough for many crops, but their costs have escalated, making them almost unavailable for most crop use. Several high-analysis but slowly soluble synthetic organic materials may provide fairly active release if particles are ground very fine. Performance of these at times has been less than satisfactory. The release of nutrients from coated high-analysis synthetic materials does not depend on microbial decomposition but rather on diffusion, which primarily depends on temperature. The release time can be varied by using coatings of various thickness. (Analyses of slowly soluble inorganics are given in Table 16.2; slowly soluble synthetics and coated inorganics in Table 16.3.)

TABLE 16.2. Major Nutrient Contents of Plant and Animal Byproducts and Rock Phosphates Used for Organic Mixes

Product	Composition		
	N	P_2O_5	K_2O
Animal tankage	7	9	0
Ashes, cotton hull	0	4-7	22-30
Ashes, hardwood	0	1	5
Beet sugar residue	3-4	0	8-10
Bone meal	4	22	0
Bone, precipitated	—	35-46	—
Bone tankage	3-10	7-20	3-9
Castor pomace	5-6	2	1
Cocoa shell meal	3	2	3
Cocoa tankage	4	1.5	2
Cottonseed meal	6-9	2-3	1.5-2
Crab scrap	3	3	0
Distillery waste	1	0.5	14
Dried blood	8-14	0.3-1.5	0.5-0.8
Dried king crab	9-12	—	—
Fish, acid	6	6	
Fish meal	5-10	5-13	—
Fish scraps, fresh	2-8	2-6	—
Garbage tankage	3	2	3
Hoof and horn meal	13	0	0
Linseed meal	5	1	1
Olive pomace	1	1	0.5
Rapeseed meal	5-6	—	—
Rock phosphate	—	26-40	—
Peanut hulls	1.5	0.2	1
Seaweed kelp	2	1	4-13
Shrimp scrap, dried	7	4	0
Soybean meal	6	1	2
Steamed bone meal	3	20	0
Tobacco dust and stems	0	2	5
Winery pomace	1.5	1.5	1
Wool wastes	7	0	0

Sources: Collings, G. H. 1955. *Commercial Fertilizers,* Fifth Edition. McGraw-Hill Book Co., Inc., New York; Lorenz, O. A. and D. N. Maynard. 1988. *Knott's Handbook for Vegetable Growers,* Third Edition. John Wiley and Sons, New York; Follett, R. A., L. S. Murphy, and R. L. Donahue. 1981. *Fertilizers and Soil Amendments.* Prentice-Hall Inc., Upper Saddle River, NJ.

TABLE 16.3. Relatively Insoluble Synthetic Materials Used As Slow-Release Fertilizers

Type of material	Trade or generic name	N	P_2O_5	K_2O	Mg
		— %			
Uncoated Materials:					
Guanylurea	G. sulfate[a]	37	0	0	0
Magnesium ammonium-phosphates	Mag-Amp	8	40	0	14
Oxalic acid diamide	Oxamide	31.8	0	0	0
Potassium calcium phosphate[b]	KCP	0	17-22	21-22	0
Potassium polyphosphate[c]	KPP	29-32	24-25	0	0
Urea-aldehyde	IBDU (pure)	32	0	0	0
	(fert. grade)	30	0	0	0
	CDU	32	0	0	0
	Crotadur	32	0	0	0
	Floranid	28	0	0	0
	Gllyccluril	39	0	0	0
	Ureaform	38	0	0	0
	Agriform[d]	28	18	4.8	0
	Urea-Z	33-38	0	0	0
Coated Materials:					
Resin coated	Osmocote[e]	14	14	14	0
		18	9	9	0
		18	6	12	0
		24	4	8	
	Sierrablen[f]	19	6	10	0
	Polyon	25	4	12	
	Procote	20	3	10	
	Nutricote[g]	13	13	13	1.2
		18	6	18	1.2
		14	14	14	
		16	10	10	
		20	7	10	
		18	6	8	
	Woodace	20	4	11	
Sulfur coated	SCU[h]	37	0	0	0

[a] The material is soluble in water but is tightly held on soil colloids and only slowly mineralized.

[b] A fairly large group of compounds that vary in solubility depending on composition of aluminum and iron oxide.

[c] The solubility of these polymerized mixtures depends on proportion of dipotassium phosphate in the mix.

[d] This material has a ureaform base. The slow-release characteristics are enhanced by the formation of pellets by compression.

TABLE 16.3 (*continued*)

[e] The rate of availability is regulated by the extent of the resin coating. The 14-14-14 and the 18-9-9 are rated for 3 to 4 months and the 18-6-12 for 8 to 9 months, but the availability period is affected by temperature, with high temperatures reducing the length of time that an application will support plants.

[f] Contains a small amount of Fe.

[g] The 13-13-13 and 18-6-8 also contain small amounts of B, Cu, Fe, Mn, Mo, and Zn. Rates to release 80 percent of its N at constant 77°F vary from 40 days for 20-7-10, 16-10-10, and 14-14 to 100 days for 13-13-13 and 180 days for the 18-6-8.

[h] Urea coated with S. A sulfur coated urea-potassium chloride product testing 32-0-16 can be made but has not been used extensively.

Sources: Dinauer, R. C. (Ed.). 1971. *Fertilizer Technology and Use*. Soil Science Society of America, Inc., Madison, WI; Hignett, T. P., E. Frederick, and B. Halder (Eds.). 1979. *Fertilizer Manual*. International Fertilizer Development Center, Muscle Shoals, AL.

High-analysis slow-release chemicals, although expensive, are justified for growing high-valued annual crops. The need for slow-release N is less for perennial plants that are well established and have considerable reserves of plant N.

The long-lasting effects of slow-release high-analysis nitrogen with the more uniform growth obtained and less supervision required usually pay for high-priced annual crops. Because of simplicity, better predictability of N release, and the advantage of having a more complete fertilizer there has been a tendency to utilize coated slow-release NPK fertilizers rather than natural or synthetic slow-release N for potted plants, production of seedlings, and various nursery items.

The cost of slow-release materials is also justifiable if the crop will be subjected to leaching rains, particularly during the early periods when canopies are small. The presence of slow-release N close to the plant can make considerable difference in early growth. In south Florida, with its open soils subjected to considerable leaching, including at least 25 percent of the N in the planting fertilizer as slow-release materials has helped ensure early development of many flower and vegetable crops. Early use of animal and plant wastes has given way in recent years to synthetic materials or to the cheaper sulfur-coated urea (SCU).

ORGANIC VERSUS INORGANIC NUTRIENTS

Both inorganic and organic materials can supply needed nutrients. Despite the arguments of a vocal uninformed group, both forms are entirely satisfactory and safe for plants and humans. The organic forms are less apt to burn crops, but the inorganic materials can be used so that injury is largely avoided. The relatively low-analysis organic materials that are used in large amounts (tons per acre) have an advantage of better soil conditioning, but unless grown in place their cost, largely due to moving the huge amounts needed, is prohibitive for many crops. The organics generally are much slower in releasing nutrients, which can be an advantage under warm wet growing conditions, but often is a detriment if soils are cold. The primary advantage of the inorganics, and the reason that they are used so extensively, is their ease of handling and relatively low cost.

For the majority of farming operations, proper management should include both organic and inorganic nutrient sources. While the slow release of nutrients is advantageous, the major benefit of organics lies in their effects on soil properties, although very large amounts are needed.

Nutrient Release from Organic Sources

The decomposition of organic material requires suitable moisture, temperature, soil pH, and soil aeration. Release of nutrients from low-N materials (wide C/N ratio) is much slower than from the high-N materials. But even with the high-N organic materials, there may not be enough N released to adequately maintain growth during very cool or dry periods, although quite adequate for warmer, moister periods.

Cost of Nitrogen

Nitrogen derived from animal waste organics is too expensive for general agricultural use. The relatively small applications of these higher-analysis materials are insufficient to provide the beneficial physical effects associated with organic matter. To obtain improved water infiltration, MHC, increased CEC, and reduced erosion and compaction requires large quantities, which eliminates most of the animal waste materials but does include composts, manures, sludges, and organic matter grown in place.

Amounts of Inorganic Sources Needed

The amount of inorganic nutrients necessary will vary not only with the crop but the amount, kind, and quality of organic materials used. Inorganic nutrients needed with large inclusions of sludges, composts, and manures may be very small or none at all, but very large for most other organic additions. A better understanding of the additional inorganic nutrients needed can be obtained by examining some of the organic sources.

Organic Sources of Nutrients

Nutrients from Soil Organic Matter

A substantial amount of nutrients can be derived from the soil OM and needs to be considered in planning a fertilizer program. As indicated in Chapter 5, amounts of nutrients released from organic matter can be quite variable depending on amounts present, climatic conditions, soil aeration, and moisture, but recommendations as given in Tables 5.3 and 5.4 or made possible with the NPK Predictor can be helpful in estimating these amounts.

Manures. The release of N and other elements from manures has been covered in Chapter 5. Essentially only about $^1/_3$ to $^1/_2$ of the N in manures, sludges, and composts can be expected to be released for the first crop year, or only about 5 lb N being released per ton of manure.

Other Organic Materials

Applications of various animal waste or good leguminous crops can contribute to much of the N needs of the succeeding crop. Amounts of N, P_2O_5 and K_2O of some of the more common legumes and animal products are given in Table 5.8. But as with OM and manures, the major release of these nutrients is affected by microorganism activity and only about $^1/_3$ to $^1/_2$ of the N can be expected in the first crop year.

Nonleguminous cover crops will provide little N or may even rob N from the succeeding crop. As pointed out in Chapter 5, mature nonlegumes have such wide C/N ratios that decomposition will tie up available N. No initial release of N can be expected if the cover crop analyzes less than 1.75 percent N; usually extra N needs to be added at planting time if the N content of the cover crop (or any other organic material) on a dry weight basis is less than 1.5 percent.

Relatively immature nonlegumes can be expected to release N and several other elements for the succeeding crop, especially if they followed a cash crop that had been fertilized heavily. The amounts of nutrients released by such nonlegume crops are difficult to estimate. Nutrient value of such crops per acre can be determined by multiplying the average dry weight of a square foot of cover crop × percent of the different nutrients × 43,560. A release of 33 to 50 percent of the total nutrients can be expected in the first year.

Using Inorganic Fertilizers

It is possible to produce high-yielding crops economically with inorganic fertilizers. Such production is more likely if the grower in conjunction with his supplier selects materials or mixed fertilizers that are suitable for the crop and soil conditions.

Cost of Inorganic Fertilizers

The costs of supplying nutrients can be reduced considerably by using cheaper inorganic materials, a list of which is given in Table 16.1. Some of these materials, such as the superphosphates, have relatively low solubility and can be used in rather large amounts with little danger of burning. Many of them are soluble and may cause burning but relatively large amounts can be used safely if (1) unneeded elements, such as Na, are eliminated, (2) the essential elements needed in small amounts (Cl and S) are applied at minimal rates; (3) low-salt materials are used to supply the macronutrients, and (4) fertilizer additions are placed and timed to minimize damage to roots.

Minimizing Salt Effects of Quick-Release Inorganic Fertilizers

In addition to proper placement, some reduction of salt effects can be made by choosing the proper fertilizer material. Often, a lower-salt material can be substituted for one of higher salts. For example, in choosing the same amount of N from inorganic sources, the amount of salts added will be in the following order: anhydrous ammonia < urea < most fertilizer solutions < ammonium nitrate < ammonium sulfate. Less salt will be added in the same amount of K_2O by using materials as follows: potassium hydroxide < potassium sulfate < potassium nitrate < potassium chloride (muriate of potash). Supplying Ca from either limestones or gypsum will add far less salt than calcium nitrate.

FLUID VERSUS DRY

Fertilizers can be applied in either liquid or dry forms. The fluids can be true solutions or suspensions. The dry mixtures can be (1) bulk blends (mixtures of compatible ingredients), (2) ammonium granulated (ammonia nitrogen solutions and acids to produce a chemical mix), or (3) coated with various materials.

Each type has advantages and disadvantages. The fluids use some very economical materials and also have some advantages in application. Although liquids usually can be dispensed via irrigation systems, only the true solutions are readily used for drip irrigation or other systems with small apertures. High-analysis solutions have a disadvantage of salting out (precipitating) as temperatures fall.

Both kinds of fertilizers can be formulated for quick availability, but the solutions have an edge, being instantaneously available, whereas dry fertilizer and some ingredients of suspensions may have to dissolve before being taken up by plants. Instant availability can be a disadvantage on light soils in humid climates, especially when plants have limited root systems, since several nutrients can be readily leached. The dry fertilizers can be formulated to have resistance to leaching. The true solutions still can be effective under leaching conditions if they can be applied frequently in low doses, which is possible with drip irrigation.

FERTILIZER INGREDIENTS

Various materials are used for fertilizers. Whereas at one time the fertilizer trade was a scavenger industry, depending largely on remains of animals, today it is a highly sophisticated chemical industry capable of producing high-analysis goods suitable for many different types of conditions. The material chosen depends not only on nutrients required and costs, but also the kind of fertilizer to be prepared, the crop and the climate in which it is grown, and how the fertilizer is applied.

Liquid Fertilizers

True Solutions

The ingredients of true solutions are more limited than those of suspensions or dry fertilizers. Some of the more important sources of inorganic nutrients, along with their solubilities, are listed in Table 16.1. Materials for true solutions must be in liquid form or highly soluble. The degree of solubility needed is closely related to temperatures to which the solution will be exposed and also to other ingredients with which it is mixed. The pH also has an important bearing on the resulting solution. The effects of temperature and pH on the solubilities of some important inorganic materials and the amounts of N, P_2O_5 and K_2O that can be maintained without salting out are given in some detail in Wolf, Fleming, and Batchelor, 1985.

Suspensions

Suspensions can be made from a much larger group of chemicals. The solutions and highly soluble materials useful for true solutions can also be used for suspensions, but most of the relatively insoluble materials listed in Table 16.1 also can be used providing suitable suspension agents are included and sufficient agitation is provided in their application. Common suspending agents used are attapulgite clay, seipolite clay, sodium bentonite clay, and xanthum gum.

Dry Fertilizers

A large list of materials is also available for dry fertilizers. Besides price, availability, and ability to supply needed elements, the choice will depend on whether the fertilizer needs to contain slow-release nutrients, be suitable for organic farming, be highly soluble (fertigation through small orifices), and the degree of residual acidity that can be tolerated.

Organic Fertilizers

Although not an organic material, rock phosphates are used combined with various plant and animal byproducts listed in Table 5.8 to produce organic fertilizers.

Slow-Release Fertilizers

Relatively insoluble synthetic materials and common mixes used as slow release fertilizers are presented in Table 16.3. Slow-release fertilizers are also prepared from animal and plant byproducts and rock phosphates, or in combinations with the synthetic materials. Sulfur-coated urea is the material of choice for mixes with some slow-release N used for vegetables and field crops. IBDU has had considerable use for golf course, lawn, and garden fertilizers. The coated fertilizers (Osmocote, Nutricote, Procote, Polyon, and Woodace) have primarily been used for container nursery plants.

Acid Fertilizers

Fertilizers can influence soil or medium pH and thus affect liming or acidification programs. Most of the ammoniacal sources of N tend to have an acid reaction whereas the nitrates are basic. The equivalent acidity or basicity of a number of fertilizer ingredients are given in Table 6.3.

METHODS OF APPLICATION

The method of application can have an important effect on the economy and efficiency of applied fertilizer. Several methods of application with advantages and disadvantages are described in the following sections.

Fertigation

Fertigation has become increasingly popular because: (1) it is simple, requiring little labor or equipment; (2) the application is usually uniform; (3) it can be a quick and efficient way of adding nutrients while the crop is growing; (4) there is less soil compaction than with most other methods; (5) the application can readily be combined with pesticides; and (6) timely applications of small amounts of fertilizer can improve efficiency while reducing leaching losses. Nearly all methods of irrigation can be used to apply fertilizer.

The periodic application of nutrients with water can increase yields, especially on sandy or gravelly soils. The increase in yields apparently is due in part to reduced nutrient loss by erosion and leaching. Moving the fertilizer into the soil with water reduces erosion losses of surface-applied fertilizer. The split application of N made readily possible by fertigation reduces losses from leaching.

Both dry and liquid materials or mixtures can be used for fertigation but the choice must be tailored to the injector, the type of irrigation, and water composition as well as the crop and soil.

Fertigation of inert media such as sands will require complete solutions. Several complete solutions suitable for fertilizing inert substrates are given in Tables 17.7, 17.9, 17.10, and 17.11. Low-chloride complete mixes with micronutrients are commercially available in several analyses as soluble powders or as liquids in lower analyses. These prepared complete mixes, while expensive, may be the correct approach for small growers of high-priced crops because of simplicity in handling and nutritional suitability.

Bulk fluid fertilizers can be purchased in many parts of the United States and Canada to meet most required formulations. These fertilizers are competitive with dry fertilizers and entirely suitable for large-scale production.

Growers can prepare their own mixes with the soluble or liquid ingredients listed in Table 16.1. The analyses of other combinations can be calculated by adding the formulas of the different ingredients in columns multiplied by the pounds used and dividing that amount by the sum of fertilizer units used.

In preparing mixes, combinations noted below that can cause precipitates need to be avoided. Also, avoiding pressure solutions, strong acids, or alkalies will simplify mixing and handling.

Mixing can be avoided entirely by introducing individual materials into the stock tank and moving them directly into the lines. This is a common approach as much of the fertilizer applied by fertigation supplies only one or at most two elements. Multiple injectors capable of handling two or more ingredients simultaneously are also available.

Materials for Microirrigation Systems

Only soluble or liquid materials can be used for microirrigation systems using injectors or emitters easily clogged by small particles. Soluble chemicals coated with clay, limestone, or waxes designed to reduce their absorption of water from air and to limit their caking are not suitable for these systems.

Not only must these chemicals be soluble as introduced, they must remain soluble in the system. This often limits the introduction of some soluble chemicals or the simultaneous introduction of materials that can react with each other. Of prime concern is the introduction of phosphates in hard water (containing appreciable amounts of Ca and Mg). The polyphosphates cause fewer problems than the orthophosphates. For example, use of a 10-34-0 grade fertilizer containing 70 percent polyphosphate is not recommended for use with very hard water, but is satisfactory for soft water. Phosphoric acid may be used with hard water if sufficient acid is used to lower the pH to <4.5. This pH will keep most phosphates soluble. Use of glycerophosphates, urea phosphate, or glycerophosphoric acid as sources of P also causes few or no problems. But even with the best of situations, it is desirable not to inject phosphates into drip systems unless jar tests with phosphates and water as treated show no precipitates upon standing.

Introduction of Ca salts can also cause problems, particularly with water containing appreciable phosphates or sulfates. Low pH, especially as it falls below 4.5, can eliminate this problem, but it is desirable to evaluate the feasibility of calcium introduction by means of jar tests. (Lines and equipment handling the liquid fertilizers need to be resistant to corrosion, if low-pH solutions are used.)

The reaction of chemicals that form precipitates with each other eliminates the simultaneous application of Ca salts with sulfates (calcium nitrate or chloride with Epsom salts or potassium sulfate) or with phosphates (calcium nitrate or chloride with ammonium phosphates).

Jar Tests

Because fertilizers may react with the water or each other to produce precipitates capable of plugging the lines, it is best to run a test of any newly injected material with the irrigation water before introducing the fertilizer. A gallon of irrigation water should be treated with a proportionate amount of the chemical and allowed to stand for some time to observe whether a precipitate is formed or contents separate. The amounts of chemical per gallon of water for the test varies with the rate of the chemical and water to be applied. Formation of even a slight precipitate or a cloudy mix should exclude a chemical from use in small-orifice systems. Small amounts of fine precipitates can be tolerated in more open systems.

It may be possible to inject the material into microirrigation systems even though the jar test shows precipitate formation. Lowering the pH can prevent precipitate formation, which also can be evaluated in a jar test. Various amounts of acid should be tested to determine the suitability of the combinations.

Materials for Nonmicroirrigation Systems

There are far fewer restrictions on materials or mixed fertilizers that can be used for systems without small orifices. Almost any fertilizer or combination of materials that are soluble or capable of forming stable suspensions can be used in these systems, providing proper positive displacement pumps are used. It is best to check with the manufacturer as to the suitability of the injector for the intended materials.

Fertilizer suspensions made with a suitable dispersant can be used with these fertigation systems. Attapulgite clay is commonly used as a dispersant but the sepiolite and sodium bentonite clays as well as xanthum gums also have been used for this purpose. Some agitation may be necessary before introducing fertilizer into the irrigation lines, and furrows may have to be shortened to provide uniform application.

Despite the greater freedom of material use in these systems, it is desirable to avoid undue deposits of precipitates in lines as these can eventually restrict flow rates and sprinkler operation. Suitable flushing following introduction of suspensions or chemicals that may cause precipitates reduces this problem.

Ammonia Injection into Hard Water

Ammonia and other high-pH materials can precipitate Ca and Mg in hard waters. The precipitation of Ca plays havoc with nearly all systems as it deposits carbonates in siphons, pipes, and behind sprinklers. The high pH may also lead to losses of free ammonia and can increase the sodium hazard in the soil. The rate of ammonia to most water should not exceed about 100 ppm and should be less than 50 ppm for alkaline water. Acidification of water to eliminate carbonates and bicarbonates can solve many of these problems.

Adding inhibitors helps prevent precipitate formation as anhydrous ammonia is injected into hard water.

Phytotoxic Materials

As was pointed out in Chapter 12, potassium or other chlorides should not be used for overhead irrigation of plants sensitive to foliarly applied chlorides unless the chlorides can be washed off the foliage soon after application. To avoid chloride foliar damage, potassium can be introduced as hydroxide, nitrate, or sulfate. These materials increase the cost of the fertilizer with the hydroxide being the most expensive and difficult to handle. Both nitrate and sulfate forms also tend to reduce the maximum amount of nutrients carried. Sulfate causes the greatest restriction; the nitrate solubility, which is not much better than sulfate at very low temperatures, approaches that of the chloride at high temperatures (>75°F).

Other Application Methods

Most fertilizers are applied by means other than irrigation, and the method of application, particularly the placement of fertilizer, can have an effect on its availability. A wide variety of equipment is used for such purposes.

Applicator trucks capable of moving fertilizer onto a spinner or dual spinners mounted on the back are used for quickly applying dry bulk blended materials or single ingredients as a broadcast. Similar trucks that direct the fertilizers into chutes distributing it by air pressure tend to provide more uniform application. Trucks mounted with sprayers quickly apply liquid fertilizers either as broadcast or as surface bands. Those equipped with spargers can handle suspension fertilizers. Fertilizer applicators mounted on various farm equipment can apply either dry or liquid fertilizers while planting or cultivating crops. Fertilizer is also applied by helicopters or fixed-wing airplanes, usually for land that is too wet or rough to use ground equipment. Most fertilizers applied by airplane are used for forests, but some application has been used for rice production.

Placement of Fertilizers

Much of the difference noted in agronomic response to different application methods is due to their placement of fertilizer. We have seen in Chapter 12 that placement can have an effect on salt damage, but in selecting ideal methods, the fixing power of the soil, climate, and cost of application must also be considered.

Larger amounts of soluble fertilizers (relatively high salt) can be used if not placed in contact with roots. Only very small amounts of fertilizer (about 10 lb/ac of N and K_2O) can be placed with seeds of most crops without causing injury. Slightly larger amounts (15 to 20 lb of N and K_2O) may be harmful if placed in the row directly under the emerging seed. Mixing the fertilizer with a lot of soil before planting may reduce its toxicity, but it is safer to apply it in bands away from the emerging roots. Applications of 50 lb of N + 50 to 75 lb of K_2O per acre have been safely made in bands about 2 inches from the seedline and 2 to 3 inches deeper than the seedline. Roots developing as the seeds germinate do not immediately reach the concentrated band, allowing the roots to make contact with the fertilizer as needed. Much larger amounts (100 lb/ac or more of N plus similar quantities of K_2O) can be placed in bands if moved 4 to 6 inches away from the seedline, but some nutrients need to be placed close to the seeds or early growth will be stunted.

Broadcast

Broadcasting fertilizer is a rapid, cost-effective method of applying high rates to large acreage of existing pastures and hayfields or before a crop is planted. Large amounts of fertilizer can be applied at one time, yet initial concentrations close to the developing root can be low enough to ensure good germination and development. Negative aspects of broadcasting fertilizers are that crop effectiveness may be reduced due to leaching and fixation losses while the pollution potential is increased.

Leaching is of less concern if low-solubility materials are used or if soluble materials are applied to soils of adequate CEC, to hayfields and sods, or to various crops in arid climates, but it can be serious in humid climates, especially for light soils.

Fixation is usually of minor concern on soils of low CEC, but that of P and K can be a problem on the heavier soils. Other situations where broadcast applications can reduce the effectiveness of fertilizer, making it mandatory to band them, are: (1) fixation of N on soils containing large amounts of OM with wide C/N ratios; (2) fixation of P on soils with either low or high pH values; (3) fixation of Cu, Fe, Mn, and Zn on soils of high pH; and (4) fixation of Cu on organic soils.

Banded Fertilizer

The nature of some materials, such as anhydrous ammonia and pressure nitrogen solutions, dictate band placement at some depth to avoid ammonia (NH_3) loss. But band placement of other fertilizers is preferred to broadcast applications if: (1) soil pH values are high or low; (2) soil is cold, compacted, or excessively wet; (3) soil contains crop residues having wide C/N ratios (small grain straw); and (4) fertility is low to medium. Banding fertilizer or concentrating the application in a small volume of soil has the following additional benefits: (1) minimal adverse effects from salts if the band is placed ahead of roots; (2) ideal placement can be made relative to developing roots; (3) fertilizer can be placed in soil with sufficient moisture for rapid uptake; and (4) there is much less fixation of P and K on heavy soils.

Dropping the fertilizer in bands about 6 to 8 inches apart before the soil is turned by disk or plow limits the amount of P and K_2O fixation, yet reduces the salts enough that large amounts of fertilizer can also be applied to the heavier soils without danger of undue fixation.

Some estimates of comparative efficiency of P and K fertilizers in band and broadcast have suggested a three- to fourfold advantage in favor of the band method for soils of low fertility and/or those with strong fixation for P or K. Considerable savings in amounts of applied Mn and Zn for high-pH soils and Cu for organic soils also can be made if these elements are banded as compared to broadcast applications.

Band placement may mean increased costs, varying with different methods. Dribbling fertilizer in surface bands should cost no more than similarly applied fertilizer broadcast. But bands placed deeply or close to the seed increase costs by requiring more labor, slowing planting, or increasing energy costs.

Placement in direct contact with seeds or developing roots can cause injury. Applying a 2-inch band over seed in rows spaced 36 inches apart concentrates the fertilizer eighteenfold. A 100 lb/ac application carries the salt equivalent of an 1800 lb/ac broadcast application. Injury is avoided by reducing the amounts, placing the fertilizer slightly ahead of developing roots, increasing the number of bands, or widening the band.

When calculating rates of banded fertilizer per acre, it is necessary to consider width and spacing of the bands and the seedbed utilization ratio (SBU). The SBU is calculated by the following equation (Roberts and Harapiak, 1997):

$$\% \text{ SBU} = \frac{\text{width of seed row}}{\text{row spacing}} \times 100$$

Utilizing SBU and width of fertilizer band for calculating suitable N rates for direct seeding of cereals on the Canadian plains is presented in Table 16.4.

TABLE 16.4. Effect of N Application Method Upon Yields of Corn Grown on a Sandy Soil

Method of N application	Yield		
	Silage (tons/ac)	Dry matter (tons/ac)	Grain (bu/ac)
Band	23.4	10.1	173
Broadcast	24.4	10.2	166
Fertigation	26.2	10.9	183

Source: Gascho, G. C. 1985. Use of fluid fertilizer in irrigated multicropping systems. *Fluid Fertilizer Foundation Symposium Proceedings,* p. 242. Reprinted by permission of Agricultural Retailers Association, St. Louis, MO.

Mixed with Seed ("Pop-Up")

Although not strictly a band method, mixing fertilizers with or adding them to seed is a means of concentrating nutrients and placing them so that they are immediately available for developing roots. "Pop-up" fertilizers are easy and cheap to apply because no additional openers are needed, saving fuel costs. In some cases, fertilizer can be mixed with seed beforehand or it is dropped by tube onto the seed at planting time.

There are limits to the amounts of fertilizer that can be used effectively without injuring the seedling. There is no problem in supplying the few ounces of needed sodium molybdate per acre in this manner, which is applied by spraying a 1 percent solution on the seed or dusting the moist seed. It becomes much more of a problem with the greater amounts of macroelements—especially if these materials add large amounts of salts. Nevertheless, there have been successful plantings of bahia, bromegrass, fescue, orchardgrass, ryegrass, or wheat seed mixtures with suspension fertilizers. Rates up to 100 lb/ac N + K_2O from either 15-10-10 or 9-18-18 suspensions produce satisfactory crops of ryegrass and wheat.

Such high rates of nutrients in contact with the seed are only possible if the seed-fertilizer suspension contains 20 to 30 percent water and it is planted in well-prepared moist soil soon after the mix is prepared. Good mixing of seed and fluid fertilizer is necessary but excessive mixing or any practices that can physically damage the seed should be avoided. Delaying planting for more than a few hours after preparation can also cause damage to emerging seedlings. Such high rates of fertilizer with the seed may also be a problem on soils of low CEC.

Usually, much smaller rates of fertilizer are recommended for pop-up applications. Safe amounts appear to be related to type of soil, its moisture, and the fertilizer's makeup. Recommendations for corn have suggested a maximum total of 10 lb/ac of N and K_2O. As much as 45 lb/ac of $N + K_2O$ have been satisfactory for corn in some years but only on the heavier soils. On the Canadian prairies, as much as 40 lb of N derived from ammonium nitrate but only 25 lb from urea has been satisfactorily used for the cereal grains (Roberts and Harapiak, 1997). These must be soils of good CEC, as a combination of $N + K_2O$ of as little as 5 lb/ac have injured corn and a total of 10 lb injured soybeans on light soils in the United States.

Moisture is an important factor in determining safe rates, because it dilutes salts. Larger amounts of high-salt fertilizer can be placed with or near the seed if the soil contains ample but not excessive moisture. The higher moisture content of fluid fertilizers may account for reduced damage by these fertilizers compared to equivalent amounts of dry, high-salt materials placed with the seed.

The source of fertilizer has an important bearing on the amount of nutrients that can be placed in contact with the seed. Low-salt N sources from natural organics and other slow-release forms can be used in larger amounts than most soluble sources, but their slow release of nutrients may fail to give the quick start required from pop-up sources. Super or rock phosphates, because of their low salt content, can be used in much greater amounts than equivalent amounts from soluble phosphates. Diammonium phosphate (DAP) can be much more toxic to germinating seeds than equivalent amounts of P_2O_5 from MAP (monoammonium phosphate). The increased toxicity of DAP appears to be due to the release of ammonia, small amounts of which can be extremely toxic. Eliminating free ammonia or materials that can release ammonia from mixed fertilizers helps reduce damage from pop-up fertilizers. Safe usage of pop-up fertilizers is also ensured by eliminating or greatly reducing the content of biuret, B, Cl, and Na.

Larger amounts of direct-seeded fertilizer can be used with the newer seeders that cause some spread between seed and fertilizer, or if they have the capacity to place fertilizer away from the seed.

Surface Bands

Bands of fertilizer applied to the surface as a preplant are a cheap, effective means of adding nutrients for many crops if the fertilizer is incorporated soon after application. Surface bands also work well if followed soon afterward by rain or irrigation, providing that much of the fertilizer is soluble. Failure to incorporate the bands can lead to considerable loss of N and leave the other elements positionally unavailable. Nitrogen losses from surface bands of fertilizer containing urea can be quite severe in reduced-tillage systems.

The optimum width of surface bands that will be incorporated varies with the amount and type of fertilizer and type of soil. Maximum uptake of P for some of the heavier soils appears to occur when the fertilizer is mixed with about 10 percent of the soil volume, but for lighter soils it is slightly over 20 percent. Applying fertilizer in 3-inch bands on 30-inch centers gives about a 10 percent mix of fertilizer to soil; 6-inch bands give 20 percent and 9-inch bands supply 30 percent mixing when incorporated.

Spacing between bands needs to be no more than about 30 inches apart for row crops and closer spacing of 12 to 15 inches may be more desirable for solid-seeded crops, small grains, or those planted in narrow-spaced rows. With a 2-inch band placed 12 inches apart applied in the same direction as the planting, no roots will be more than 5 inches from fertilizer. Incorporation of fertilizer by disk tends to place the band at a slant so that developing roots of plants in narrow-spaced rows would soon be in contact with fertilizer.

Surface bands can also be applied to growing row crops if the concentrated band is not applied too close to the row. Ideally the band should be positioned so that it will move into the soil slightly ahead of the furthest roots. Such fertilizer is of little value unless soil has good moisture or application is soon followed by rain or irrigation.

Application of surface bands to actively growing hayfields, pastures, or small grains may cause burning, particularly if rates are high, fertilizer contains free ammonia, temperature is high, droplets are fine, or foliage is dense. Damage from fluid fertilizers is reduced by eliminating free ammonia and increasing droplet size. Damage from dry fertilizers is reduced as dust is eliminated, particle size is increased, and when leaf surfaces are dry. Much of the damage, if it occurs, may be cosmetic. Often, plants will outgrow the damage, yielding more than those not receiving fertilizer or receiving the same fertilizer applied as a broadcast.

Bands Placed Near Seed (2 × 2)

Placing fertilizer close to seeds but not in contact with them avoids much of the harmful effects of salts, yet nutrients are close enough for rapid uptake by developing seedlings. By placing the fertilizer in 2-inch bands, 2 inches to the sides of and 2 to 3 inches deeper than the seed line, much larger amounts can be used and elements such as B can be included. Such fertilizers help crops get off to a fast start despite cold soil, poor soil, or other inhibiting factors.

With this type of placement, it has been possible to apply 600 to 1000 lb/ac of 5-10-10 fertilizer containing considerable chloride and micronutrients on sandy soils for a wide variety of crops with no apparent injury.

Fertilizers very low in salts, made from materials such as superphosphate and/or slow-release materials, can be used in relatively large amounts placed in bands under the row. I have effectively used 300 to 400 lb/ac of treble superphosphate (138-184 P_2O_5) in this manner on a wide variety of soils for several crops. Even including a small amount of N and K_2O (3 lb/ac of each) has been satisfactory and at times improved early growth as compared to P_2O_5 alone. But high-salt fertilizers should not be placed under the row even if fertilizer is mixed with the soil prior to planting.

Deep Bands

There may be certain advantages in placing bands deeply rather than on or near the surface. Much of the advantage appears to be related to better moisture conditions at lower depths, especially as the season advances, and the response to improved fertility of very poor subsoils.

Deep band placement is usually 3 to 6 inches deep for reduced-tillage systems. For conventional tillage, depth varies from 4 to 6 inches for fertilizer disked or knifed in to 8 to 10 inches for fertilizers applied with chisel plows. Some fertilizer has been placed with subsoilers at 16 to 24 inches deep. Fertilizers placed deeper than about 8 inches may be in a zone of reduced fertility and often in one of much lower pH.

Deep placement is desirable for areas of low rainfall that do not receive irrigation. In many areas, placement at 4 to 6 inches would help ensure enough water to provide good uptake. It has given us good results for side dressings where moisture drops as the season advances. Very good results have also been obtained in Europe as well as the United States when mixed fertilizers containing both N and P_2O_5 have been banded before planting winter wheat.

Placement at depths greater than 6 inches has not given consistent results. It is suspected that poor uptake in some soils may be due to lower pH, high aluminum or manganese content, and/or reduced oxygen at lower levels.

Comparative Value of Different Application Methods

The response to different application methods varies depending to some extent on the climate, soil, amount of nutrients supplied, and the tillage method. Fertigation methods, especially fertilizer applied through drip irrigation, is very versatile and may have special advantages for coarse or salty soils. (It was highly useful for producing corn on sandy soil, Table 16.4.) Broadcast works well for heavy applications of fertilizer on light soils in nonhumid climates, but can be wasteful if soils are subject to extensive leaching or fixation. Broadcasting micronutrients, some of which are subject to extensive fixation, requires much larger amounts than if these materials are applied by fertigation or foliar sprays (see Table 16.5). The efficiency of macronutrients applied by broadcast, dribble, or soil injection, is influenced by soil class as well as method of tillage (see Table 16.6).

TABLE 16.5. Recommended Micronutrient Rates As Influenced by Method of Application

Micronutrient	Recommended rate (lb/ac)		
	Broadcast	**Band**	**Foliar**
B	0.5-2.0	0.2-1.0	0.1-0.5[a]
Cu	2.0-10.0	1.0-4.0	0.1-0.5[b]
Fe		0.5-2.0	0.5-3.0[a]
Mn	5.0-25.0	2.0-10.0	0.1-1.0[a]
Mo	—	0.1-0.2	0.05-0.1[a]
Zn	2.0-10.0	0.1-5.0	0.1-0.5[b]

[a] Upper rates need to be divided into several applications.
[b] All but very low rates need to be applied as sparingly soluble compounds or chelates. The high rates need to be divided into several applications.

Source: Micronutrients: Unlocking agronomic potential. Adapted from Mortvedt, J. J. (Ed.). 1990. *Solutions* 64(6): 30-31. Reprinted by permission of Agricultural Retailers Association, St. Louis, MO.

TABLE 16.6. Efficiency of Broadcast N As Influenced by Soil Type and Tillage Method

Tillage	Application method	
	Sand	Silt loam
Moldboard	Broadcast = dribble = injected	————
Chisel	Broadcast < dribble < injected	Broadcast = dribble = injected
Ridge-till	Broadcast < dribble < injected	Broadcast = dribble = injected
No-till	Broadcast < dribble < injected	Broadcast < dribble < injected

SUMMARY

The addition of nutrients to nearly all soils is necessary to obtain satisfactory crops. Soil tests before planting a crop, combined with those made after planting and later leaf analyses can help guide the kinds and amounts of added nutrients.

Nutrients can be organic or inorganic. Inorganic sources tend to be quicker, cheaper, and easier to handle than organics not produced on the farm, but are usually more subject to leaching and more apt to burn crops if improperly placed. Ideally, crop production will combine enough organic materials to supply optimum soil physical conditions and inorganic sources for quick availability.

Organic materials provide nutrients less vulnerable to leaching than most inorganic materials. Nitrogen is the element most subject to leaching, but K and Mg can also suffer considerable losses. Most organic sources of N suitable for fertilizers are more expensive than the inorganic sources. Several slow-release inorganic sources of N are available but these also tend to be expensive. Inorganic sources of N coated with sulfur to produce a slow-release form of N (SCU) are cheap enough to use for a wide variety of crops. Cheap inorganic soluble fertilizers are also coated with plastics or resins to produce slow-release mixed fertilizers. While not cheap enough to be used for general crops, these fertilizers have found special niches for container-grown floral and nursery crops, and for starting small fruit trees.

Fertilizers can be in dry or liquid forms, the latter being either true solutions or suspensions. The suspensions permit use of a wider range of materials but need to be agitated as they are dispensed. The true solutions use soluble or liquid sources, often more expensive than the suspensions, but they have the advantage of being instantaneously available to plants and they are ideally suited for application through microirrigation systems.

Fertigation is economical and an ideal way to provide quick nutrients for a growing crop. Suspension or true solution fertilizers can be used for such purposes but only true solutions or highly soluble materials are suitable for micro- irrigation systems. The introduction of anhydrous ammonia or new ingredients need to be evaluated by jar tests to see whether they are compatible with irrigation water before use.

Most fertilizers are applied not through irrigation but by various types of equipment directly to the soil. The agronomic response to these methods often is related to the placement of the fertilizer, which can have an effect on salt damage to crop, fixation of applied P and K, and leaching of nutrients, particularly N. Banding fertilizers reduces fixation and can lower losses of N, especially if bands applied at planting are supplemented with later deeply placed bands. Such banding may be more costly than broadcast operations.

SECTION V:
MAXIMUM LENGTHS
OF THE TRIANGLE SIDES

The various ways of improving the sides of the fertile triangle—proper tillage, organic matter additions, regulation of pH and salts, limiting damage from excess water and the beneficial inputs of air, water, and nutrients—are being used on millions of acres to produce better crops. Considerable increases in production are still possible by extending these inputs to full advantage on all agricultural land.

Although these measures have added greatly to our production of food and fiber, there are conditions that warrant additional extension of the triangle sides if we are to obtain maximum returns. The extra measures are justified because the monetary value of the crop permits much greater investment in its production to meet the exacting requirements of some crops and/or to overcome the undue stress to which the crop is exposed.

Thus far, these approaches have been rather expensive, limiting them to highly specialized situations or high-priced crops. There is a good possibility that this may change in the future, either as these approaches are improved or if population increases place great pressure on land resources, making the cost of producing many crops much greater.

Extraordinary measures to lengthen the sides of the triangle are covered in three parts of Chapter 17. "Modification of Field Soils" deals with special additions for field soils enabling large yields of high-value specialty crops or the growing of grasses subject to heavy traffic. "Modifying Soil or Using Amendments for Potting Mixes" outlines methods of modifying soil or use of soil substitutes that enable good growth of many plants in containers. The third section is "Aeroponics and Hydroponics." Aeroponics allows the growing of plants in a maximum air side with periodic additions of water and nutrients to provide ideal ranges of both. Hydroponics permits growing plants either (1) in water, which provides a maximum water side with continuous additions of air and nutrients in optimum ranges; or (2) in various aggregates that allow excellent air sides with periodic addition of water plus nutrients at most favorable levels.

Chapter 17

Maximizing the Sides

MODIFICATION OF FIELD SOILS

The modification of field soils to provide for maximum yields of sensitive plants or to allow plants to thrive under undue stress requires considerable additions of materials. The amount of amendment required depends on the properties of the amendment and the type of soil. Larger amounts of substances that improve water-holding capacity (primarily organic materials) are needed for the sands than the loams, and it would take much more fir bark than peat moss to provide the same amount of improvement. Likewise, larger amounts of materials capable of improving drainage and porosity (both organic and inorganic materials) are needed for silts and clays than for loams or sands. Average suggested amounts of amendments needed for the various soil classes are presented in Table 17.1. Some of the more common materials available for soil improvement and their effect on several properties are given in Tables 17.2, 17.3, and 17.4.

Organic materials are effectively used on a wide variety of soils. Adding about 50 percent by volume of humus materials to sandy soils will usually suffice for bedding plants, cut flower beds, golf greens, and landscaping, but the fibrous types would be more suitable for silty and clay soils.

The choice of materials may also be affected by the longevity needed. Some materials, such as grass clippings, manures, and mushroom composts last a very short time—usually less than a few weeks. The humus-like materials (leaf mold and composts) can last up to about six months but the fibrous materials (cedar, cypress, fir bark, redwood, and rice hulls) up to several years. Much longer periods can be obtained with the inorganic materials, although vermiculite will often lose much of its porosity effect (collapse) within a few months.

Most materials need to be broadcast uniformly and then worked into the soil to give intimate mixing. Care must be used to avoid pockets at the turns during application and mixing. After mixing, grading allows positive uniform drainage.

TABLE 17.1. Amendments Needed for Different Soil Textural Classes to Markedly Improve Soil Aeration, Water-Holding Capacity, and CEC

Soil class	Addition of amendment %*
Sand	35
Loamy sand, sandy loam	30
Sandy clay loam	25
Sandy clay	25
Loam	25
Silt loam	30
Silty clay loam	30
Clay loam	30
Silt	35
Silty clay	35
Clay	35

* Percent of total volume. The author has effectively modified several Colombian clay soils so that good crops of chrysanthemums or carnations may be grown by adding 3 inches of rice hulls and mixing it to a depth of 7 inches. The rate is equivalent to a 43 percent addition.

Source: Western Fertilizer Handbook. 1990. Horticultural Edition. The Interstate Printers and Publishers, Danville, IL, p. 161. Reprinted by permission of the California Fertilizer Association, Sacramento, CA.

TABLE 17.2. Some Common Amendments Used for Improving Soil or Other Media and Their Effect on Density, Permeability, and Water Retention

Type of material	Amendment	Density	Permeability	Water retention
Fibrous	sphagnum peat	low	low-medium	very high
	sedge peat	low	low	high
	hypnum peat	low	low	high
	wood residues	low	high	low-medium
	ground fir bark	low	high	low-medium
	rice hulls	v. low	high	low-medium
	peanut hulls	v. low	high	low-medium
Nonfibrous	composts	medium	low-medium	medium-high
	compost-sludge	medium	low	high
	rotted manures	medium	low	medium
Inorganic	calcined clay	high	high	high
	perlite	v. low	high	low
	pumice	medium	high	low-medium
	sand, builder's	v. high	high	low
	sand, fine	v. high	low	low

Source: Modification of a table in *Western Fertilizer Handbook.* 1990. Seventh Edition. Interstate Printers and Publishers, Danville, IL, p. 233. Reprinted by permission of the California Fertilizer Association, Sacramento, CA.

TABLE 17.3. Physical and Chemical Characteristics of Some Common Materials Used for Soil and Soilless Mixes

Material	Bulk density	MHC	Porosity	CEC	C/N ratio
Bagasse	L	H	L	M	H
Sawdust	L	H	M	H	H
Rice hulls	L	L	H	M	M-H
Shavings	L	M	H	M	M-H
Vermiculite	L	H	M	H	L
Peat moss	L	H	H	H	M
Bark	L	M	M	M	M
Sand	H	L	M	L	L

Low =	<0.25 gm/cm^3	<20%	<5%	<10 meq/100 cm	<1:200
Medium =	0.25-0.75	20-60%	5-30%	10-100	1:200-500
High =	<0.75gm/cm^3	>60%	>30%	>100 meq/100 cm	>1/500

Sources: Johnson Jr., H. *Soilless Culture of Greenhouse Vegetables.* Univ. of Calif. Coop Ext., Div. of Agric. and Natural Resources Leaflet 21402. Originally adapted from R.T. Poole, C.A. Conover, and J. Joiner. 1981. *Foliage Plant Production.* Prentice-Hall, Englewood Cliffs, NJ.

Growing Crops for Transfer to Pots

Satisfactory modification of a relatively level, slightly acid loam soil so that ericaceous (acid-loving) plants can be grown for one to two years before lifting is made possible with the addition of the following materials per 1000 sq ft:

Sphagnum peat	20 6-cu ft bales
Sawdust or wood chips	6 cu yd
Perlite (supercoarse)	30 4-cu ft bags
Dolomitic limestone	25 lb*
Superphosphate, treble	50 lb
Long-lasting fertilizer (coated 18-6-12)	30 lb

* Omit if pH > 5.5

The first three amendments that are added in about 2-inch layers have to be mixed with about 2 inches of the topsoil to be effective. The materials need to be applied uniformly on the surface and then mixed thoroughly

TABLE 17.4. Physical Properties of Several Materials Used As Soil Amendments

Material	Bulk density		Water retention		Air % volume	
100% materials	Dry (lb/yd³)	Wet (lb/yd³)	Vol (%)	Wt (%)	Total porosity	Air space
Bark, fir 0-1/8"	387	1027	38	165	69.5	31.5
Bark, fir 1/8-5/8"	312	565	15	81	69.7	54.7
Bark, redwood (3/8" fiber)	210	725	30.8	245	80.3	49.5
Loam, clay	1585	2510	54.9	59	59.6	4.7
Loam, sandy	2640	3247	35.7	23	37.5	1.8
Peat, sedge, AP	353	1234	52.3	253	69.3	17.0
Peat, sedge, BD	430	1585	68.6	361	77.0	8.4
Peat, sedge, BP	403	1208	47.6	200	68.1	20.5
Peat, sedge, MP	374	1279	53.7	242	73.5	19.8
Peat moss, hypnum	313	1313	59.3	320	71.7	12.4
Peat moss, sphagnum	176	1166	58.8	560	84.2	25.4
Perlite, 1/50-1/16"	155	873	42.6	463	75.8	33.2
Perlite, 1/16-3/16"	161	663	47.3	312	77.1	29.8
Perlite, 3/16-1/4"	150	470	19	213	75.3	56.3
Perlite, 1/4-5/16"	162	487	19.5	200	73.6	53.9
Pumice, 1/50-1/16"	774	1456	40.5	88	62.2	31.7
Pumice, 1/16-1/8"	722	1277	33	77	65.2	32.2
Pumice, 1/8-5/16"	905	1342	25.9	48	60.3	34.4
Pumice, 5/16-5/8"	800	1230	25.5	54	70.5	45.0
Rice hulls	172	385	12.3	124	81.0	68.7
Sand, builder's	2800	3245	26.6	16	36.0	9.4
Sand, fine A	2525	3093	33.7	22	36.2	2.5
Sand, fine B	2400	3050	38.7	27	44.6	5.9
Sawdust, cedar	353	997	38.2	183	80.8	42.6
Sawdust, redwood	296	1126	49.3	280	77.2	27.9
Vermiculite, 0-3/16"	182	1077	53	492	80.5	27.5
Manure, dairy	580	1630	66.7	182	74.3	7.6

Source: Johnson, P. and A. Johnson. 1968. *Horticultural and Agricultural Uses of Sawdust and Soil Amendments.* Technical Bull. Compton, CA.

(two times) with a rototiller. When settled, the resulting 6-inch lightweight mix produces plants suitable for transfer to an 8-inch container. Plants scheduled for larger containers with greater depth should be grown in modified soil that is prepared to greater depth and that has correspondingly greater amounts of amendments.

Rather similar treatments can be used for nonericaceous plants, but sufficient limestone needs to be added to bring the pH in the 6.0 to 6.5 range.

High-Traffic Areas

Special amendments are needed for growing grasses in high-traffic areas (playing fields, golf greens, etc.) Some of the more common amendments have limitations for this use. Coarse sand (particles $>1/8$-inch diameter) give good drainage but hold very little water. Vermiculite also has high water retention but has low mechanical strength, being reduced to a pasty mass if kept wet and packed. Perlite increases aeration but has poor moisture holding capacity and tends to be too fragile to stand up to the traffic. Ordinary clays hold large quantities of water but tend to increase compaction problems with time. The calcined clays tend to retain a stable granular form and adequate pore space despite traffic and they have excellent water absorption and retention. Disadvantages are their poor release of water and poor retention of nutrients.

Organic materials that have been used extensively for this purpose improve water holding capacity and aeration but also have limitations. They are subject to decomposition, and the use of peat moss beyond about 15 percent by volume tends to keep the areas too wet. Nevertheless, organic materials have been used in sand-based root zones to assist in the establishment and management of turfgrass. Besides increasing the water-holding capacities of sand mixtures, the organic materials substantially lower bulk density, help retain nutrients, and improve soil strength which help provide stable pore space capable of supporting extensive root penetration space despite extensive traffic.

Certain minimal amounts of organic materials are needed to supply satisfactory bulk densities and water retention. In a study of five different organic matter sources (Canadian sphagnum peat, Michigan sphagnum peat, reed sedge peat, compost, and muck) mixed with sand at various rates, it was found that to attain a satisfactory bulk density of less than 1.45 mg m^3, required a mixture containing more than 3 percent by weight of organic matter. To obtain available water greater than 0.2 m^{3}m^{-3}, it was necessary to incorporate more than 3.5 percent by weight organic matter.

Organic materials may vary in their effects on bulk density and availability of water held. In the above study it was found that all materials tended to decrease bulk densities of sand with little differences at the 10 percent volume incorporation rate, but compost and muck gave smaller decreases at 20 percent and 30 percent incorporation. The Michigan sphagnum, reed sedge, and muck contained water that was more available to the plant than the Canadian peat or compost, indicating that organic sources with fiber contents greater than about 45 percent could be too coarse, reducing the availability of retained water (McCoy, 1992).

Building High-Traffic Areas

The building of suitable greens or playing fields usually requires special treatments for particular conditions that are best handled by architects familiar with such projects. In addition to providing suitable surface and internal drainage with tiles and slopes for putting greens, best mixtures of soil, peat moss, sand, and calcined clay are often evaluated by trials. Tentative mixtures are made up, saturated with water, drained, uniformly compacted at field capacity, and the rate of water movement through the compacted sample is evaluated. Common topsoil mixtures consist of (1) 3 parts sand, 1 part peatmoss; (2) 2 parts sand, 1 part soil, and 1 part peat moss; (3) 5 parts sand, 1 part peat, 1 part soil. Various amounts (1-4 parts) of calcined clay are added to these mixtures to give the desired permeability.

Calcined clay can be added to existing soil, the amounts needed increasing with the percent clay in the soil. Whereas a 15 percent mix (by volume) may be sufficient for putting greens if the soil is sandy, a 60 percent mix is more suitable for clay soils.

A common approach for rebuilding athletic fields is to add a $^3/_4$-inch layer of calcined clay to prepared soil that has been properly subsoiled, drained and graded. The calcined clay is worked into the soil to a depth of 3 inches, and the treated areas shaped, smoothed and seeded. The surface can be lightly top-dressed with calcined clay and kept moist for adequate germination.

Maintaining Playing Fields

Various athletic fields are often in need of repair due to traffic or the abrasive action of the sport. Much of the damage occurs at the height of the season. Rebuilding these fields at such times is usually untenable because the rebuilding process takes them out of use for long periods. Continued use when the grass has been largely destroyed can lead to unnecessary injuries to sports players.

Topdressing of sand, or mixtures of sand and organic matter, are often used to repair damage to such fields. Some benefits accrue due to smoothing of areas and thatch modification, but often the effects are short lived. In case of sand or sand and organic mixtures, for certain sports the sand acts as an abrasive, scarifying the crowns and further debilitating weak turf.

The relatively poor performance of these topdressings has prompted the search for better alternatives. Trials at Michigan State University have indicated that the topdressing of crumb rubber probably is a much better approach to the problem. Crumb rubber is made from discarded tires.

Particle size is $1/4$ inch or less. The materials are broadcast on existing healthy turf and dragged in for uniform coverage. The applications, made three times per growing season, supplying a total of $3/8$ to $3/4$ inch have improved density and "wearability" of the grass.

MODIFYING SOIL OR USING AMENDMENTS FOR POTTING MIXES

Soil placed in containers (pots, cans, flats, benches, and planters) reacts differently than the same soil in the field. The small volume limits root development, thereby necessitating more frequent replenishment of nutrients and water. The base of the container tends to restrict the drainage of water and this, coupled with the adhesion of water to the sides of the container, tends to produce a perched (raised) water table. Perched water table problems of excessively wet conditions and poor aeration are common as mixes having few large pores are placed in shallow containers.

Some of the limitations of volume can be corrected by more frequent feeding and irrigation with relatively small amounts of each, but the latter may increase the problems of aeration.

A better approach is to modify the soil with coarse-textured materials to increase the number of large pores, allowing good drainage and satisfactory aeration despite conditions favoring a perched water table. In addition to improving porosity and drainage, it is often desirable that the amendments also (1) increase the water-holding capacity, (2) provide a reservoir of water and nutrients, (3) allow the slow release of nutrients, (4) increase the buffering capacity, (5) add permanence to the media with little or no subsidence, (6) provide stability to plants, and (7) maintain satisfactory weight. All of these changes need to be accomplished without introduction of pests or toxic concentrations of elements or salts. In addition, the materials need to be readily available, easy to handle and mix, and cost effective.

Materials for Soil Modification

Numerous materials can modify soils for container use. The more common ones are arcillate, bagasse, bark, calcined clay (Sorbalite, Terra-Green, Turface), coconut fiber, Cofuna, expanded polystyrene flakes (Styromull), gravel, hadite, leaf mold, composts, manures, peat, peanut hulls, pecan shells, perlite, polyurethane foam (Nutriform), rice hulls, sawdust, shavings, urea formaldehyde foam (Hygromull, Floramull), vermiculite, and wood chips.

The characteristics of some of these materials were given in Tables 17.2, 17.3, and 17.4. Some mixes that can provide special requirements for various plants are presented in Table 17.5.

Mixing particles of different sizes often yields volumes smaller than the sum of the ingredients' volumes. The loss of volume upon mixing is greater as the difference in size or structure of the individual ingredients increases. This is a result of amendments filling in large pores. Sufficient coarse material has to be added beyond a critical proportion to improve aeration. The extra pore space introduced by the material needs to be greater in volume than the reduction by filling in existing voids. The critical proportion varies with different combinations and is determined primarily by the natural packing structure of the amendment and total porosity. The amounts needed to meet this critical proportion varies with different soils. Whereas it may be necessary to add coarse organic materials up to 40 percent by volume of a sandy soil to provide for satisfactory oxygen diffusion rate, it may be necessary to add as much as 90 percent of a loam soil volume to provide for equivalent pore space. Figure 17.1 illustrates the the critical proportion and changes in pore space as perlite is added to soil.

TABLE 17.5. Basic Soil Mixes: Parts on a Volume Basis

	Loam[a]	Peat	Sand	Leaf mold	Osmunda fiber	Fir bark
General potting	1	1	1[b]			
Cacti and succulents[c]	2		1			
Tropical plants	1	1	1	1		
Ferns	2	1	1[b]	2		
Orchids and bromeliads[d]						
a. epipytes					X	X
b. terrestial orchids					X	X
c. terrestrial bromeliads					X	X

[a] A suitable loam can be prepared from heavier soil by mixing 1 part leaf mold and 1 part peat with 2 parts soil. All loam and soil mixes should be sterilized before using.

[b] Perlite can be substituted for coarse sand.

[c] Christmas cacti to be grown in general potting mix.

[d] Can use either osmunda fiber or fir bark. The fir bark used for terrestial orchids should be finely chopped. Any mix that drains well but retains good moisture should be suitable.

Source: Kaufman, P. B., T. L. Mellichamp, J. Glimn-Lacy, and J. D. LaCroix. 1985. *Practical Botany.* Reston Publishing, Reston, VA.

FIGURE 17.1. Critical Proportion and Change in Pore Space As Perlite Is Added to Soil

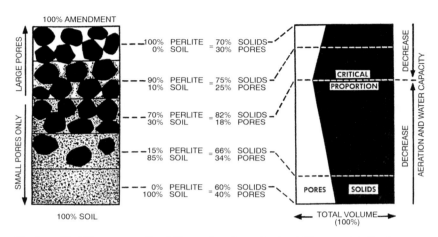

Starting with 100 percent soil, additions of perlite up to about 70 percent (critical proportion), reduce total pore space without increasing the number of large pores. Starting with 100 percent perlite, additions of soil up to 30 percent reduce total pore space by filling in the large pores. The critical proportion of amendment to soil varies with the kind of amendment and kind of soil.

Source: Spomer, L. A. 1973. *Soils in Plant Containers: Soil Amendment, Air and Water Relationships.* Illinois Research, Illinois Agricultural Experiment Station 15:3, 16-17. University of Illinois, Urban, IL.

Peat moss, derived from the decomposition of sphagnum moss, is perhaps one of the most satisfactory ingredients used for amending soils. It increases pore space, provides a reservoir of nutrients and water, and helps buffer the medium. Combined with soil or by itself, it can be a satisfactory medium for acid-loving plants, such as azaleas, gardenias, and rhododendrons. Limestone needs to be added to most peat moss mixtures to make them suitable for the nonericaceous plants.

Substitutes for Peat Moss

Because of costs, various substitutes have been sought to replace peat moss. Reed and sedge peats, which can be cheaper in some areas, are not as satisfactory as peat moss derived from sphagnum moss, as they are usually finer textured and thus yield fewer large pores. They also are not as long lasting as the sphagnum moss. Leaf mold can substitute at least partly for peat moss but it tends to be more variable and it also may be difficult to obtain in satisfactory quantities at consistent prices. Composts can be substituted in part for peat moss. They tend to decompose somewhat faster: the mushroom composts in several weeks but leaf mold and humus-type composts can last up to about six months. Composted sewage sludge may contain heavy metals, which may be a problem, especially for food production.

Manures can be used only in a limited way as substitutes for peat moss because of the salts they contain. Manures also are highly variable, can introduce weeds, and tend to decompose rather rapidly. Well-rotted manures are much more desirable because of greatly reduced problems of salts and weeds but still have the problem of fairly rapid decomposition. Bagasse, coconut fiber, peanut hulls, pecan shells, rice hulls, and various barks—cypress, fir, pine, and redwood—have also have been substituted for peat moss. They usually are not as efficient in holding water. Also, the C/N ratios of these materials are wider, making it necessary to add more N although this need is greatly reduced if they have been composted.

The lower MHC of substitute materials often can be compensated by the addition of hydrogels or polymers that expand 1000-fold their non-hydrated size as they absorb water. These gels, with such names as Super Slurper, Terra-Sorb, Broadleaf P4, Viterra 2, and Liqua-Gel, added to soil mixes at rate of 1.5-5 lb/cu yd, have greatly improved the MHC of many mixes. The gels have given poor results at times, some of which have been associated with high fertilizer rates. Also, failure to allow for expansion of the gel can give poor results if containers are filled with the mix.

Soil Mixes

For most pot production, there has to be a balance of permeability and water-holding capacity, with the ideal MHC varying with the plant to be grown and its duration of growth. At least five different types of mixes are utilized for different plants and situations: (1) a high-MHC mix for seed germination and short-term production, (2) lower MHC but with more permeability for longer-term plantings, (3) good permeability with low MHC for cacti and succulents, (4) mixes with good water retention and good permeability for ferns and a number of tropical plants, and (5) a mix with high permeability and little MHC for epiphytes (air plants). The make-up of mixes providing suitable MHC and permeability for different plants is given in Table 17.6.

TABLE 17.6. Ingredients of Several Fortified Potting Mixes

Ingredient	Western[a]		John Innes[b]		Ohio State[c]		Cornell[d]			
	#1	#2	#1	#2	#1	#2	#1	#2	#3	#4
				cu ft						
Sandy loam	13.5	—	—	—	—	—	—	—	—	—
Loam	—	—	13.5	15.75	13.5	—	—	—	—	—
Sand[e]	—	9	6.75	4.5	6.75	—	—	—	—	—
Peatmoss[f]	—	9	6.75	5.4	6.75	13.5	13.5	9	13.5	13.5
Milled bark[g]	—	—	—	—	—	—	—	9	—	—
Redwood	13.5	9*	—	—	—	—	—	—	—	—
Perlite[h]	—	—	—	—	—	—	6.75	9	—	—
Vermiculite[h]	—	—	—	—	—	13.5	6.75	—	13.5	13.5
				lb						
Dolomitic limestone[i]	.5	—	—	—	—	5	8.25	7	5	5
High-cal limestone[j]	1.5	—	1	1	—	—	—	—	—	—
Gypsum	1.25	—	—	—	—	—	—	—	—	—
Hoof and horn meal	—	5	2	2	—	—	—	—	—	—
Superphosphate[j]	2.5	2	2	2	2	2	2	4.5	1.15	2.3
5-10-5 fertilizer	—	—	—	—	—	—	—	—	—	—
10-10-10 fertilizer	—	—	—	—	—	—	2.75	2.5	—	—
Pot. nitrate 13-0-44	.375	.25	—	—	—	—	1	1	1	1
Pot. sulfate	.25	.25	—	1	—	—	—	—	—	—
				oz						
Borax	—	—	—	—	—	—	—	—	—	—
Iron sulfate	—	—	—	—	—	—	0.75	0.5	—	—
Osmocote 14-14-14	—	—	—	—	—	—	—	—	—	5
Magamp 7-40-6	—	—	—	—	—	—	—	—	—	5
Trace elem.[k]	—	—	—	—	2	2	2	2	3	3
Wetting agent[l]	—	—	—	—	3	3	24	24	3	3

TABLE 17.6 (continued)

a Western #1 = landscape container mix; #2 = greenhouse or foliage mix.

b John Innes #1 = seed compost; #2 = potting compost. The loam is taken from pastures and meadowland. Sandy, heavy, or calcareous soils are to be avoided. The loam is composted and steamed before adding to the peat and the fertilizer and limestone are added to the mix. The mix may require further steaming and tends to be quite variable, making it rather expensive because of labor requirements.

c Ohio State University #1 = adjusted to pH 6.5 and used as a bedding plant mix. The plants are fed soon after germination is started with a complete fertilizer such as 20-20-20; #2 = a peatlite mixture for growing plants.

d Cornell #1 = foliage mix or for plants requiring high moisture retention; #2 = epiphytic mix; #3 = a mix for seedlings and bedding plants; #4 = for potted plants except poinsettias and lilies. For poinsettias, the superphosphate is increased to 4 lb triple superphosphate or 9.2 lb ordinary superphosphate per cu yd and liquid feed is used. For lilies, the mix consists of 2 parts peat moss, 1 part vermiculite, and 1 part perlite, to which is added 10 lb dolomitic limestone, 1.5 lb calcium or potassium nitrate per cu yd but superphosphate is omitted. For plants requiring large amounts of slow-release fertilizers, the Osmocote and Magamp are each increased to 7.5 lb and potassium nitrate to 1.5 lb per cu yd.

e Medium or coarse sand. Builder's sand is satisfactory.

f Sphagnum peat moss. Screening through 1/2-inch mesh ensures uniformity.

g Cornell recommends Douglas red or white fir bark, 1/8 to 1/4 inch size. Other barks, such as redwood, pine, and cypress have proven useful in many different mixes. To avoid problems from toxicities, it is best to compost the bark.

h Horticultural grade perlite and vermiculite are used.

i Finely ground limestones should be used to give quick reaction. At least 90% should pass through 100-mesh screen.

j All the entries are based on ordinary superphosphate (0-20-0) although Cornell recommendations call for triple superphosphate (0-46-0). The 0-20-0 because of its gypsum content supplies considerable sulfur. Use of the 0-46-0 may require extra sulfur for some mixes.

k The trace element requirements can be met by use of one of several proprietary mixes; 3 oz of FTE #503; 4 oz. of FTE #555; 4 lb of ESMIGRAM; or 4 lb PERK per cu yd of mix. A previous dilution with 10-20 lb of material such as limestone and superphosphate before adding to the pile helps ensure uniformity in distribution.

l Dry peat moss is difficult to rewet once it is in the pile. The addition of a wetting agent, such as Aqua-Gro, Hydro-Wet, Triton B 1956, Tetronic 908, or Ethomid 0/15 greatly aid in rewetting the peat and makes for more uniform growth. The liquid wetting agent is diluted with about 10 gal of water and the dry wetting agent is mixed with 1 qt of vermiculite before sprinkling on a cu yd of mix.

Sources: Bunt, A. C. 1976. *Modern Potting Composts.* George Allen & Unwin Ltd., London; *Western Fertilizer Handbook.* 1990. Horticultural Edition. Interstate Publishers, Danville, IL.

Soilless Mixes

As clean soil free of pests or pesticides has become more difficult to obtain, it has been gradually omitted from the mixture, and sand, a more predictable material, has usually been substituted for the soil. Sand, added primarily to improve porosity, also adds to the weight of the mix. Coarse or builder's sand, having a particle size capable of completely passing through a 10-mesh screen but only 30 percent through a 40-mesh screen, should be limited to about 25 percent of the total volume, lest weight become a problem. Other coarse materials (both organic and inorganic) can be substituted for or used with sand to improve porosity.

The substitution of sand or several other coarse materials to potting mixes devoid of loam soil, while improving porosity and permeability, fails to provide adequate MHC. The elimination of loam soil increases the need for materials with considerable MHC (peat moss, compost) to compensate for that supplied by loam soil.

Moisture-holding capacity can also be increased by increasing the fines in a mix, since MHC and permeability are affected by particle size as well as the nature of the material. Most of the particles in a fine mix, suitable for short-term production, will pass through an NBS sieve #20; most of the particles in a coarse mix suitable for long-term production will not pass through an NBS sieve #8. The fine materials are <0.59 mm in diameter; the coarse particles >2.38 mm. To maintain permeability and avoid water-logging, the fine particles in a long-term mix should not exceed 20 to 30 percent of the total volume. To keep the MHC high in short-term mixes, all coarse materials >2.38 mm should be eliminated.

Several soilless mixes have been proposed. The University of California has recommended five different mixes varying in sand and peat moss as follows (Matkin and Chandler, 1957):

a. all sand;
b. 3 parts sand, 1 part peat moss;
c. 1 part sand, 1 part peat moss;
d. 1 part sand, 3 parts peat moss;
e. all peat moss (redwood shavings or rice hulls can
   be substituted for some or all of the peat moss).

Mix (e) is to be used for the acid-loving plants—azaleas, gardenias, and camellias; the others are used for plants requiring different degrees of aeration and water retention—with the best aeration and water-holding capacity in mixture (d), decreasing as the mixture contains less peat moss, with the poorest in the all-sand mixture (a). Varying fertilizer and limestone formulas are used with the different mixtures, depending whether the mix will be used within a week of preparation or stored indefinitely and also the type of plant and speed of growth desired.

In Canada, a mix of 11 bushels sphagnum moss and 11 bushels horticultural vermiculite for cucumbers and tomatoes has been used. To this mix are added 10 lb limestone, 2 lb 0-20-0 superphosphate, 1.5 lb potassium nitrate, 0.5 lb magnesium sulfate, 1 oz chelated iron, 0.5 oz borax, and 3 oz of fritted trace elements (FTE) per cu yd (Ward, 1975).

Researchers in Florida have recommended 5 parts Florida peat, 3 parts builder's sand, and 3 parts vermiculite for growing a number of ornamental plants. Several mixes have been recommended for foliage plants:

a. A 1-1-1 medium-weight mix consisting of 1 part peat moss and 1 part builder's sand or calcined clay plus 1 part perlite, shavings, vermiculite or bark.
b. A 1-1-1 medium-weight mix consisting of 1 part peat moss, 1 part perlite, bark, or vermiculite, plus 1 part shavings.
c. A 1-1 potting mix consisting of 1 part peat moss plus 1 part shavings, bark, or calcined clay. The shavings, bark, or calcined clay can be replaced (second choice) by 1 part perlite, vermiculite, or sand (Poole and Waters, 1972).
d. Two other combinations used for growing tomatoes consist of: (1) equal parts of milled pine bark and vermiculite, to which is added 10 lb limestone, 10 lb 10-10-10 fertilizer, 3 lb magnesium sulfate, and 5 lb calcium sulfate per cu yd of mix; or (2) 1 part perlite, 2¼ parts vermiculite and 2¼ parts peat moss, to which are added 8 lb limestone, 2 lb 0-20-0 superphosphate, 1 lb calcium nitrate, 10 g borax, and 35 g chelated iron per cu yd.

Lightweight Mixes

Recently, there has been a tendency to use the lighter mixtures and much of the sand or other dense materials have been omitted. Typical examples are the Canadian and tomato mixes described previously and the Cornell mixes in Table 17.6. These mixes employ sphagnum peat or wood compost with varying amounts of perlite or vermiculite to provide sufficient MHC with adequate permeability. Ideally suited for short-term production or for small plants, these light weight mixes often require special plant supports or restraints for the larger plants.

Modification of pH and Nutrient Content

To compensate for unfavorable pH values and nutrient status of both soil and soilless mixes, varying amounts of limestone, gypsum, superphosphate, calcium or potassium nitrate, potassium sulfate, hoof and horn meal, 5-10-10 or 10-10-10 fertilizer, and micronutrients are added to the artificial mix. Most of these items are included to give plants a satisfactory start. Later fertilization consisting of frequent dilute liquid fertilizers or small applications of dry fertilizers are used to grow and maintain plants.

Various slow-release materials have been added to the mixes to give longer-lasting nutrition and reduce the need for later fertilization. Organic materials such as hoof and horn meal were used, but these have given way to Magamp, and more recently to the resin-coated fertilizers (Osmocote, Nutricote) to supply macronutrients. Materials such as FTE 503, Esmigran, and Perk are used to supply the micronutrients. Some of the resin-coated fertilizers also provide micronutrients, reducing the need for separate application. Typical examples of fortified soil mixes (Western, John Innes, and Ohio State #2) and fertilization of the Cornell lightweight soilless mixes are presented in Table 17.7.

AEROPONICS AND HYDROPONICS

There has been a consistent search over the years to develop systems that do a better job than soil in growing plants. Early trials using water as a medium probably grew out of the development of greenhouses about 1700 and the desirability of avoiding hauling heavy soil in and out of them. Later developments using artificial soil moistened periodically with nutrient solutions also developed as a replacement for soils in greenhouses, which in the 1920s were used extensively for the production of off-season flowers and vegetables. The use of water or artificial soils as growth media was accelerated by the development of nutrient solution culture along scientific lines (Jensen, 1997).

The use of water or artificial soil and the later development of aeroponics allows the maximization of triangle sides. It is virtually impossible for any one of the sides (air, water, or nutrients) to support crop growth for any length of time without the others. But excluding one or two legs for a short period makes it possible to maximize the remaining side(s) to provide optimum levels.

Nutrients cannot be utilized without water or air, making it impossible for a medium to consist solely or primarily of nutrients. But it is possible to have media that are primarily water or air—at least for short periods—

TABLE 17.7. Complete Nutrient Solution Used for Aeroponics and Both Closed and Open Hydroponic Systems in Arizona

Chemical		Tomato		Cucumber	
		A	B	A	B
		(g/1000 L)			
Magnesium sulfate	$MgSO_4.7H_2O$	500	500	500	500
Monopotassium phosphate	KH_2PO_4	270	270	270	270
Potassium nitrate	KNO_3	200	200	200	200
Potassium sulfate	K_2SO_4	100	100	—	—
Calcium nitrate	$Ca(NO_3)_2$	500	680	680	1357
Chelated iron*	Fe 330	25	25	25	25
		(ml/1000 L)			
Micronutrients**		150	150	150	150

Note: Micronutrient formula for Table 17.7. Amounts are for 15.95 g packet. Dissolve one packet to make 450 ml of solution.[a]

A = seedlings to first fruit set.
B = plants setting fruit to end of crop.
*Double amount for calcareous medium.
**Made up per table below:

Chemical supplied	Element	ppm of element	g per packet
Boric acid (H_3BO_3)	B	0.44	7.30
Manganese chloride ($MnCl.2H_2O$)	Mn	0.62	6.75
Cupric chloride ($CuCl_2.2H_2O$)	Cu	0.05	0.37
Molybdenum trioxide (MoO_3)	Mo	0.03	0.15
Zinc sulfate ($ZnSO_4.7H_2O$)	Zn	0.09	1.18

[a]Heat to dissolve.

Source: Jensen, M. H. *New Developments in Hydroponic Systems: Descriptions, Operating Characteristics, Evaluations.* Environmental Research Laboratory, University of Arizona, Tucson, AZ.

because plants can survive providing water is supplied intermittently with the air medium, and air is supplied intermittently with the water medium. This ability has made it possible to extend the air and water sides to maximum lengths and obtain large yields of many different crops.

Nutrient application in water is very flexible. It can be maintained on a constant basis if water is the medium or applied intermittently with irrigation for soils or other media. If soils are very coarse (coarse sands) or the medium has high porosity, the air side of the fertile triangle will be at a maximum, and it is a simple matter to keep nutrient and water supply optimum by frequent irrigations of nutrient solutions.

Aeroponics

The term "aeroponics" describes the method of growing plants with roots suspended in air. The plant is maintained by frequent misting of the roots with a nutrient solution, the surplus solution draining back into a closed reservoir. The reservoir and roots are shielded from light to prevent the growth of algae. The lower section is enclosed to minimize evaporation of the solution and loss of moisture from the roots.

Plants are usually started in Jiffy 7's, small pots, or cubes. When roots have reached the bottom of the container, the entire Jiffy 7 or the ball is lifted and transferred to openings drilled 6 to 8 inches apart in Styroform boards, or placed in open-mesh rigid cloth positioned several feet above the nutrient solution. The height of the plant support is adjusted to provide maximum growth of the root system above the nutrient solution. Frequent mists from nozzles spaced about 18 inches apart are directed to the bottom of the root system to supply sufficient water and nutrients.

Little research has been done with the system so that ideal nutrient solutions and timing of sprays are not known. A complete nutrient solution, such as listed in Table 17.7, misted for 5 to 10 seconds at intervals of 1.5 to 3 minutes should be satisfactory for many plants but probably should be changed every 30 to 45 days.

Claims have been made that recycling water and solution is economical and the method allows maximum use of greenhouse space, but the procedure has not been readily adopted for commercial use. It probably has little to offer over standard hydroponic cultures, except for the evaluation of root systems and the use of small volumes of water. The evaluation of root systems may have some advantages for special research or teaching studies; the use of smaller water volumes may have some advantages for space travel or exploration.

Hydroponics

Technically, hydroponics is the growing of plants in water, but common usage of the term has given it a slightly different meaning. Our present understanding is that hydroponics consists of growing plants without soil and feeding them by fertilizer solutions. Plants may be grown in various relatively inert media or in water itself. I would like to extend the meaning even further to include the frequent fertigation of coarse soils that are rather inert and have little MHC.

Hydroponics, often used as a research tool, has had limited commercial use. It received a great deal of attention in the 1930s and there was limited production of specialty crops on barren islands during World War II. But many commercial enterprises started in the mainland United States during the 1930s and the 1940s, as well as in the 1950s during another period of renewed interest, failed to compete with more conventional production methods. Its success was largely confined to the propagation of rooted cuttings, bedding plants, and the production of some high-priced flowers and vegetables such as carnations, roses, lettuce, and tomatoes.

More recently hydroponics has had a boost from the development of the nutrient film technique (NFT) in England and establishment of large-scale production in the deserts of Arizona, California, Abu Dhabi, Saudi Arabia, and Iran. The NFT has the economic advantages of using small volumes of water and simple support construction; the desert production is favored by the extent of solar radiation and in the Arab countries, a shortage of fresh vegetables (Jensen, 1997).

With recent large-scale off-season production of vegetables, fruits, and flowers in Mexico, Central America, and the Caribbean and their importation into the United States, the successful hydroponic production of even the high-priced crops has been limited to a few situations or locations that have special advantages: (1) islands with little suitable soil, (2) little out-of-season production of high-quality items, and (3) the presence of cheap sources of warm water from power plants or thermal springs or the abundance of solar radiation with little need for heating.

The chances of economic success may improve with time as suitable agricultural land becomes scarce, forcing higher yields from smaller areas. The increased populations that reduce the availability of the land require ever-increasing amounts of food. Hydroponics has the advantage of greater yield returns per area and this production can be maintained in many kinds of locations—even in blighted areas or the rooftops of inner cities. Substantial increased use of hydroponics is possible if and when the costs of producing crops on the present agricultural land rise dramatically so that the currently more costly hydroponic system can compete. Hydroponics may become more competitive if one or more of the following should occur: (1) there is a scarcity of good agricultural soil resulting from population pressures, (2) present field growing of many crops is subjected to laws requiring no leaching of fertilizers or pesticides into ground waters, (3) water becomes scarce and its costs rise dramatically, or (4) transportation costs rise appreciably.

Hydroponic production may also increase if greater numbers of people are willing to pay a premium for vine-ripened fruit or very fresh produce. The current system of producing fruits and vegetables offshore or at great distances from the markets necessitates harvesting some produce before it is fully mature, with some loss of quality. Local production of these crops by hydroponics offers better quality but at a higher price.

Basic Systems

There are two basic hydroponic systems: (1) liquid culture, in which the medium is water plus nutrients (nutrient solution); and (2) aggregate culture, in which the medium is a nonsoil aggregate substrate or a coarse soil to which a nutrient solution is added on a regular basis.

In liquid culture, the nutrient solution can be (a) noncirculating or (b) constantly recirculated as a continuously flowing liquid or mist. Liquid cultures provide the maximum amount of water but as such have very little air (only that which is dissolved in the water). The noncirculating solution needs to be constantly aerated and frequently replaced. The aeration of a circulating solution is less difficult since air can be introduced as it circulates, making it possible to use the solution for longer periods if properly maintained. Liquid cultures are being used commercially to produce some high-priced vegetables with the nutrient film technique, which is described in further detail below.

Inorganic materials (gravel or sand) or organic materials (sphagnum moss, pine bark) were some of the media used for aggregate culture in the early development of hydroponics. More recently, the inert materials rock wool, perlite, and the commercial mixes of inert and organic materials have been used extensively.

In both liquid and aggregate culture, the nutrient solution can be handled as: (1) an open system, whereby the solution is used for a very limited time as a stationary liquid system or aggregate culture before being discarded or (2) closed systems, whereby the nutrient solution after bathing the plant roots is collected and circulated many times before discarding. The closed system has the advantage of using less water and nutrients but has the disadvantages of increased costs of monitoring the solution and the potential of increasing disease spread.

Liquid Cultures

The main advantage of liquid culture is the low cost of the medium. Various solid substrates used for hydroponics need to be sterilized before planting or, at least, before replanting. Many will have to be replaced before planting another crop due to decomposition of organic materials or settling of the mix. Replacement of solid substrates can be a lengthy and costly procedure that is avoided with the use of liquid culture.

Some of the disadvantages of liquid culture are: (1) the relatively high cost of building and maintaining watertight containers and the need for plant supports (the cost of building and maintaining the containers is appreciably less for nutrient film technique than for other types of liquid cultures, as it uses cheap plastic sheets to contain small amounts of nutrient solution); (2) greater problems in maintaining sufficient oxygen for plant roots, although aeration can also be lacking in aggregate systems; (3) breakdown of equipment can be disastrous because liquid cultures depend on continuous aeration; (4) rapid changes in pH due to the poor buffering capacity of the liquid cultures. Generally, more rigid controls have to be maintained with liquid culture than the aggregate systems so that problems (2), (3), and (4) do not adversely affect the crop.

Noncirculating liquid cultures. In this system, plants are grown for relatively short periods (7 to 14 days) in a nutrient solution, through which air is constantly being bubbled. Water losses are replaced as needed, but the entire solution is usually discarded after the preset period. Nutrient concentrations have to be appreciably high at the beginning to last the 7 to 14 days. This type of culture is seldom used commercially because of its great waste of water and nutrients and generally poorer yields.

Circulating systems. Constant circulating systems have the following advantages: (1) air is introduced by the circulation, although this may need to be augmented by other means; (2) nutrients are constantly supplied, allowing optimum growth; (3) the solutions can be automatically monitored for their pH, salt content, and some ions; and (4) water or specific solutions can be automatically added as needed.

Because of these advantages, there has been considerable interest and some new developments with better commercial application. Two basic systems are being used. (1) The liquid culture system has a header tank at a level above the plant area, in which nutrient solutions are prepared and allowed to move by gravity into the growing channels. The excess solution that drains from the channels is caught in a catchment tank and after a given length of time is pumped back to the header tank. (2) The second method differs from the first in that the catchment tank located at a lower level than the plants is also used to prepare the solutions. The solution is pumped from the catchment tank to an upper level from where it drains through the growing area back to the catchment tank.

Nutrient Film Technique

Various commercial operations of liquid circulating systems employ NFT or some modification of the system developed by A. J. Cooper and Associates of the Glasshouse Crops Research Institute of Littlehampton, England. The system can effectively produce several kinds of crops in a shallow, flowing film of nutrient solutions. The system has become popular because of the following advantages: (1) it uses relatively cheap containers for growing the plants; (2) minimum water is used and lost; (3) the contents of the solution can easily be monitored and controlled; (4) the nutrient solution can easily be heated; and (5) the application of water, nutrients, or pesticides as needed can be automated.

Essentially, plant roots grow and are confined to small gullies or channels in which a nutrient solution is circulated. Plants are started in various containers, such as cubes, Jiffy 7's, or small pots before placing them in the center of the gully. Different media have been used in the pots for starting the plants. Once established, the plants are transferred to gullies made of "white-on-black" or black polyethylene film (1.5 mm). A shallow stream of nutrient solution, about 1 cm deep, is circulated in the gullies by gravity or a pump. The gullies are sloped at least 1 foot in 100 feet, permitting the nutrient solution to flow from the top of the bed into a catchment tank. From there, it is handled in one of two ways: (1) it is pumped back into the upper end of the gullies to be discharged through plastic flow pipes, which have holes approximately $1/16$ inch in diameter opposite each gully, permitting the nutrient solution to pass readily in the gutters; or (2) the nutrient solution is constantly fed from the catchment tank to the header tank. The header tank has an overflow permitting the excess to return to the catchment tank. A large-capacity pump is used to provide continuous overflow, ensuring a steady supply of the nutrient solution above the valve controlling the water inlet in the header tank. Gravity moves the solution from the header tank into the gullies.

Forming the gullies. The gullies are formed by placing the plastic film on presloped ground or supports, and stapling or clipping the edges of the film directly to the various types of pots or media containing plants placed in rows in the middle of the film. The edges of the plastic film can also be attached by clothespins to two wires that are stretched taut between wooden stakes about 4 inches high. The edges of the plastic film must be drawn close together to limit light falling on the nutrient solution but not so close as to restrict ventilation. Light is restricted to avoid algae growth in the solution.

There are two very important considerations in constructing the gullies or channels: (1) they must be self-supporting even when plants have reached their maturity; and (2) there must be a uniform flat base that allows continuous flow of a shallow film of solution. Diagrams of a typical installation are presented in Figure 17.2. Details for constructing an NFT system are available in the book *Commercial Applications of NFT* (Cooper, 1978). An audiovisual course on the subject is offered by Soil-less Cultivation Systems, Aldershot, Hampshire, England.

Liquid flow control. Proper flow of the liquid in the gullies is regulated by the slope, which varies from about 1:75 to 1:100. This helps maintain a desirable depth (<10 mm) of liquid in runs <65 feet at a rate of about 2.1 qts per minute per gully. Flow rates as high as or even <4.2 qt per minute for longer gullies are satisfactory, providing depth of the liquid remains <10 mm. Addition of water is varied with plant age and solar radiation. Once roots are established in the gullies, flow rates need to provide a continuous film of water or root death can occur, seriously damaging the crop.

Nutrients and their maintenance. The composition of the solution must supply all needed elements in concentrations suitable for various stages of growth. The maintenance of the desired concentration coupled with adequate flow rates helps provide required amounts. The composition of the original solution and methods of maintaining it are described in greater detail in the section, "Nutrient Solutions."

Hygro-flo. A modification of NFT employs flexible plastic tubing to contain plants. The tubes are placed on wooden platforms sloped 1 inch per 8 feet, permitting ready flow of the nutrient solution into PVC return gutters to catchment tanks from which it is recirculated. The nutrient solution is recirculated by pump and gravity, 24 hours per day, using 10-minute on and 5-minute off periods.

The plants are grown in BR-8 blocks, which provide initial support. The blocks are placed in slits cut in the tubing. Dike sticks are placed crosswise under the plastic to prevent the solution from bypassing young plants. The sticks are removed several weeks later and placed over the tubing to help support the plants.

Cascade system. In this modification of NFT, a series of plastic pipes, one above the other, rather than plastic film, are used to support the plants and the nutrient film. The solution flows from the top pipe into consecutive pipes below it. The tilting of the pipes allows continuous flow. Plants are placed in holes or slits cut into the pipe. The header tank containing the nutrient solution is placed above the uppermost pipe. A catchment tank is

FIGURE 17.2. Diagram of a Nutrient Film Technique (NFT) System.

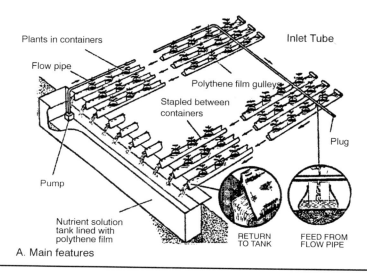

A. Main features

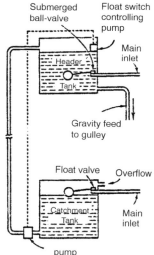

B. Diagram of Header and Catchment Tanks

Source: Schippers, P. A. 1979. *The Nutrient Flow Technique.* Vegetable Crops Mimeo 212. P-4, New York State College of Agriculture, Cornell University, Ithaca, NY.

placed below the lowest pipe. The solution draining into the catchment tank is pumped to the header tank to restart the process.

Pipe system. In another modification of the NFT system, a series of vertical pipes in a receiving tank carry the nutrient solution from a trough above the pipes. Started plants are placed in a series of openings cut or burned in the pipe. The solution flows by gravity through the pipes into the catchment tank, from which it is pumped back into the container at the top of the pipes. Float switches in both containers help control the amount of solution flowing through the pipes.

Aggregate or Solid Material Cultures

Aggregate culture employs various nonsoil materials or coarse soil to support plants fed with nutrient solutions. While commonly referred to as hydroponics, it actually is a cross between aeroponics and hydroponics. Coarse aggregates allow maximum aeration while water and nutrients are added periodically to maintain high levels of both.

The nutrients and water are applied (1) in, on or over the soil, or (2) from below (subirrigation) nonsoil materials to provide needed water and nutrients at frequent intervals. Unlike liquid cultures, which need a constant flow of solution about the plant roots, aggregate culture allows periodic wetting because of its ability to retain some water. Wetting frequency varies not only with the crop and its stage of growth but also with the nature and the fineness of the medium and evaporation rates. Appreciably lower frequencies are needed with organic materials.

In most operations, the solution is allowed to fill the bed. It is then drained back into a subtank, from which it is pumped back into a header tank to start the cycle all over again. In some systems, including soils, the nutrient solution is used only once. It is either discarded after it percolates through the medium, or in the case of soils, the excess is either leached from the soil or remains after evaporation. The reuse of the solution is less costly but presents some problems of maintaining a suitable pH and nutrient concentration.

Beds. Typical beds to hold various materials are constructed with slopes to the center and the length of the bed sloped to one end. Sides made of wood, building blocks, or metal support the substrate. A black plastic liner, PVC, or polyethylene, 0.15 to 2 mm, is placed on the base and supported by the sides. The base can be tamped earth or made from concrete or building blocks. A PVC pipe, 1.5-inch diameter that has been sliced in two is placed on the plastic at the bottom of the entire length of the bed to ensure drainage and improve aeration. Inverted barrel tiles can be substituted for the cut plastic pipe. The height of the sides will vary with the plant to be grown and the media used but ranges from about 6 to 18 inches. A depth of 1.5 to 2 times the diameter of maximum plant canopy size is recommended but extra space between plants in beds can compensate for less depth. Bed size depends on the crop but usually is 2 to 3 feet wide with an aisle of 2.5 feet. The substrate is placed on the plastic liner to a height that will provide adequate rooting. (Details of constructing suitable beds are given in Schwarz, 1977.)

Coarse open soils instead of prepared beds have great economic advantages over most types of hydroponic systems because of the elimination of construction costs. They are not entirely cost free, since there may be a need to install tiles or provide other forms of suitable drainage, as well as possible nematode and soil disease control. Also, use of these soils calls for drip irrigation to maintain sufficient moisture for most plants. These irrigation systems are expensive, but so are the methods of providing water for nonsoil aggregates. The latter, however, have the advantage of requiring less care in maintenance.

In prepared beds, nutrient solutions or water are supplied either by: (1) irrigation of the surface of the medium that is allowed to drain back into a catchment tank, from which it will be pumped back to repeat the process; or (2) the liquids, applied by subirrigation, are allowed to stand for some time before draining back to the catchment tank for later reuse.

Bed substrates. Various substances have been used in beds as substrates or media. Ideally the material should be readily available, cheap, lightweight, support plants effectively, drain readily but still retain sufficient water, have good aeration properties, and not readily subside or pack. Some of the more common materials used are listed in the following sections.

Sand. Washed sharp sand, about 0.5 to 2 mm in diameter has been used for some time as an effective medium for hydroponics. Beach sand should be avoided unless it can be effectively washed free of salts. Coarse sand is suitable for closed systems that apply the nutrient over the sand. It is less suitable for closed systems that are subirrigated because the sand is seldom coarse enough to supply rapid drainage, so essential for proper aeration. The sand is placed in beds to a depth of about 1 foot.

Sand may be very cheap in some areas, but it is quite heavy and tends to pack, giving poor aeration. Sand mixed with peat or another organic material will give better aeration until the organic materials are broken down.

Peat. Sphagnum peat, by itself or more commonly mixed with perlite and/or vermiculite, can be used for hydroponics. The sphagnum peat alone tends to pack and present some aeration problems. The pH of peat is satisfactory for azaleas and other acid-loving plants but can be too low for most others. Adding dolomitic limestone to the mix can help raise the pH and provide needed Ca and Mg. The pH of an organic mix for hydroponics needs to be kept close to 5.5 and that of a nonorganic mix close to 6.5 to avoid micronutrient shortages.

Woody materials. The high cost of peat has favored the introduction of cheaper wood by-products such as pine bark. Bark chips, sawdust, or sawdust mixed with sand have also been used and can be economical media in certain areas. Moisture moves better laterally with a sawdust-sand mixture than with sand alone. A mixture of fine sawdust with shavings also gives better moisture movement than sand. The sawdust or sawdust/sand mixtures should be placed about 8 inches deep. All of these substrates need to be thoroughly wetted before placing plants. Wood products have presented some problems. Good results have been obtained with sawdust from fir, pine, and hemlock but rather poor growth with sawdust or bark from western red cedar. The rather wide C/N ratios of most of these materials can result in N shortages unless extra N is added or the material is composted prior to use.

Gravels and cinders. Various gravels or cinders can be used effectively. Suitable materials are about 0.75 cm in diameter and are free of fines and calcareous materials. Cinders may have a high pH and contain toxic substances and therefore, after sieving to remove the fines, need to be thoroughly washed prior to use.

Foam plastics. Particulate foam plastic, produced as a by-product, can be cost effective because of its light weight. Polystyrene is especially useful because it has some water-holding capacity. If used, it is desirable to cover the foam plastic with several inches of gravel to avoid losses to wind. Also, beds need to be watered for a couple of weeks before planting to remove potentially toxic formaldehyde that may emanate from the plastic.

Modules. Several materials that have sufficient rigidity or are packed in bags can be placed on a concrete floor or isolated from the greenhouse soil by a layer of plastic. If plastic is used, a white on black with the white on the upper side to reflect light is preferred. The plastic >500 gauge is laid with at least a 6-inch overlap on preleveled soil that has paths outlined and slightly higher than the module area. The polyethylene sheets need to be well butted to the walls and carefully cut to fit around any pipes to avoid contamination from the soil. Plastic also has to be handled to avoid contamination of the plastic surface with soil. Once the plastic is laid, modules can be placed. The distance between modules depends on the length of the module and spacing of the plants. A typical placement for tomatoes is given in Table 17.8.

TABLE 17.8. Suitable Distance Between Modules for Growing Tomatoes

Length of modules (inches)	No. of plants per module	Plant spacing in row (inches)						
		13	14	15	16	17	18	20
36	3	3	6	9	12	15	18	24
37		2	5	8	11	14	17	23
38		1	4	7	10	13	16	22
39		0	3	6	9	12	15	21
40	4	12	2	5	8	11	14	20
41		11	1	4	7	10	13	19
42		10	0	3	6	9	12	18
43		9	13	2	5	8	11	17
44		8	12	1	4	7	10	16
45		7	11	0	3	6	9	15
46		6	10	14	2	5	8	14
47		5	9	13	1	4	7	13
48		4	8	12	0	3	6	12
49		3	7	11	15	2	5	11
50		2	6	10	14	1	4	10
51		1	5	9	13	0	3	9
52		0	4	8	12	16	2	9

Note: For example, distance of gaps between modules should be 8 inches for 46-inch modules and planting at a distance of 18 inches.

Source: Tomatoes in Peat. Growers guide #8. 1980. Growers Books, London.

Rock wool mats and blocks. Rock wool (Grodan) absorbs water and has a great deal of air space (about 97 percent). It is a long lasting, lightweight, sterile product that can be used to produce several crops.

The rock wool mats, $3 \times 12 \times 35$ inches wide, are commonly placed end to end on a plastic liner that is about 12 inches wider than the mats. The ends of the mats and sides of the plastic liner are folded over to create a basin. The folds are held in place with wooden pins inserted in the mats.

Nutrients and water are supplied by drip tubes placed on the blocks or water held by the basin. Mats need to be thoroughly saturated before placing transplants. Transplants are grown on small rock wool sections and are placed directly on the mats when roots are emerging from the sections. Care must be taken to keep the mats wet at all times. This can be monitored by observing water emerging from the mats if drip tubes are used or surface moisture if a liquid is introduced in the basin.

Bags of peat or peatlite mixes. Plants can be grown directly in bags of peat or peatlite mixes. Plants grown in pots are planted no more than 1-inch deep in an opening made in the bag. A single tomato plant or three pepper plants can be grown in a bag containing $^1/_2$ cu ft of peat or peatlite mix. The plant and water delivered by drip irrigation are introduced about the same time. With tomatoes, setting out the pot and plant are delayed until the first cluster of fruit is set to avoid losing the first fruit. Water is allowed to wet the entire substrate before slits are made in the bottom of the bag to ensure proper drainage. Nutrients are delivered by the drip tube, usually starting a few weeks after the plant and pot have been set out (Wittwer and Honma, 1979).

An advantage of the system is the rapid setup of large areas of medium with relatively little expense. The disadvantages are: (1) the bags cannot be reused, making the medium more expensive; and (2) a relatively high standard of management and skill is needed to produce satisfactory crops because of the small margin of error allowed in fertilizing and watering.

Peat bolsters. Bolsters of peat are prepared by placing peat on 2-foot wide polyethylene sheets. Enough peat is added ($^1/_2$ cu ft of peat for every tomato plant) to form a bolster as the sides of the sheet are brought together over the peat and stapled. Slits are made in the tube of peat to insert pots and drip tubes. Slits for drainage on the bottom sides of the tube are made only after the peat is thoroughly wetted. Plants in containers are set in the peat and handled much as for peat bags.

Straw bale culture. Bales of straw, placed on 4-mil polyethylene sheets, are wetted and allowed to heat up by the fermentation of the straw. The fermentation is hastened by the addition of 15 lb N per ton of straw. Temperatures will rise to about 140°F, which is high enough to destroy a number of disease organisms. The fermentation is slowed by thoroughly wetting the bales. They are allowed to cool to below 100°F and plants, established in containers, are transplanted into the straw. Nutrients and water are supplied by drip hose. Straw may be relatively cheap in some areas. The release of heat in the fermentation process can also be an advantage. Disadvantages are that the bales can only be used for a single crop and labor costs of handling the straw are high.

Troughs. Troughs, 20-80 feet in length, 8 inches deep, and 8 to 30 inches wide, are simply constructed to hold sphagnum peat or peatlite mix as a growing medium. They are sloped for drainage and lined with polyethylene. The plastic liner should be 12 mil or heavier for hydroponic use. A $^1/_4$-inch drainage bed of gravel (2 inches in center) is laid or 3-inch minimum tile (clay or plastic) is placed in the V groove at the base. The peat or peatlite mix is laid over the gravel bed or tile to a depth of 3 to 5 inches. Slits cut periodically 2 inches above the base of gravel can be helpful in maintaining adequate drainage. Tile has an added advantage that it can be used for watering and feeding if a T or L from thetile rises above the mix. A slope of

1-inch per 100 feet is sufficient for drainage if a continuous tile is used with peatlite mixes, but a drop of 1 inch per 10 feet is more suitable for gravel beds with peat.

Plants can be grown in pots and transferred to the peatlite after they reach suitable size. Plants in 7.5-inch whalehide pots may be left directly on the peat or planted at about a 2-inch depth. Disadvantages of the trough system are its initial cost and some problems with inadequate drainage. Occasionally, there have been plant toxicity problems in which galvanized materials were used for sides. Drainage can be improved with tiles, increasing slope, and reducing bed length. Galvanized sides should not be a problem if lined with plastic or if painted with a nonphytotoxic paint.

Construction details for troughs are presented in (1) *Tomatoes in Peat* (1980); and (2) *Peat-Vermiculite Mix for Growing Transplants and Vegetables in Trough Culture* (Larsen, 1980).

Aeration

Aeration often is insufficient unless air is positively introduced for liquid hydroponic cultures or irrigation is frequently applied for aggregate cultures. These procedures were covered in some detail in Chapter 14.

Nutrient Solutions

More than a hundred different formulas have been developed since 1861, when Sachs made his first studies with hydroponics. Most of the early formulas dealt with various nutritional properties. They advanced our knowledge of hydroponics, making possible current commercial solutions, as outlined in Tables 17.7 and 17.9, 17.10, and 17.11

The composition of the solution should reflect the fertility of the substrate. Since water and many inert materials usually cannot be expected to supply significant quantities of any nutrients, solutions for these media need to be complete in all known nutrient elements. If waters are high in salts or calcium, nutrient solutions need to be modified to compensate for these ingredients. Solutions used for modified media that may contain one or more essential nutrients in ample amounts also need to be prepared minus the known additions.

Most commercial operations use concentrated solutions that are diluted prior to their application. The addition of all the nutrients as a concentrated solution in a single tank can present serious problems of Ca precipitation, with phosphates and sulfates.

TABLE 17.9. Preparation of Complete Nutrient Solution Used in the ADAS System. Solutions #1 and #2 to be Diluted 1 in 100

Material	Rate per 100 liters[a]
Solution #1[b]	
Calcium nitrate	5.0 kg
Solution #2[c]	
Potassium nitrate	9.0 kg
Potassium dihydrogen phosphate	3.0
Magnesium sulfate	6.0
Iron sequesterene EDTA	300 g
Manganese sulfate	40
Boric acid	24
Copper sulfate	8
Zinc sulfate	4
Ammonium molybdate	1
Solution #3[d]	
Nitric acid	6 liters
Phosphoric acid	3 liters

[a] Solutions #1 and #2 to be diluted 1 in 100 before circulation; Solution #3 to be added to maintain a pH of 6.0 to 6.2 and conductivities between 3000 and 4000 ms from planting to at least 8 weeks after picking starts.

[b] This amount is for hard water and needs to be increased to 7.5 kg for soft water. The calcium nitrate can be omitted entirely if the Ca content of the water is >100 ppm. In such cases use 12 kg of potassium nitrate and omit potassium sulfate.

[c] Omit magnesium sulfate if water contains >50 ppm Mg. If the Mg content is 25 to 50 ppm reduce magnesium sulfate proportionally. Remove boric acid if water B >0.4 ppm. Reduce by 1/2 if water B >0.2 ppm.; Omit zinc sulfate if water Zn >0.5 ppm. Omit copper sulfate if water Cu >0.2 ppm.

[d] The nitric acid + phosphoric aicds are for hard water. A mixture of 4 liters of phosphoric acid and 4 liters nitric acid in the beginning (picking start) is preferred to avoid P deficiency. Use 10 liters of nitric acid for soft water. Various combinations of the two acids may be necessary to change N or P levels in the solution.

Source: Guide Notes on the Nutrient Film Technique for Tomatoes. 1981. Booklet G1, Ministry of Agriculture, Fisheries, and Food, London. © Crown copyright. Reprinted by permission of the Ministry of Agriculture, Fisheries, and Food, London.

TABLE 17.10. Modified Hoagland's Solution

Material	Rate per 50 gal	
Stock solution #1[a]		
Potassium nitrate	21	lb
Potassium phosphate, dihydrogen	12	
Epsom salts	21	
Micronutrient concentrate	5	gal
Stock solution #2[b]		
Calcium nitrate	45	lb
Sequesterene 330 Fe	2	
Micronutrient concentrate[c]		
Boric acid	54	g
Manganese sulfate	28	g
Zinc sulfate	4	g
Copper sulfate	1	g
Molybdic acid	0.5	g

[a] Solution #1 is prepared by adding salts, filling container almost full with water, mixing and finally adding the micronutrient concentrate.

[b] Sequesterene 330 Fe needs to be dissolved in a little water before adding to the dissolved calcium nitrate in solution 2.

[c] The micronutrient concentrate is prepared by adding all salts except boric acid to about 2/3 of the water and dissolving them. The boric acid is added to a separate container containing slightly less than about $1/3$ the water, boiling until it is dissolved, and cooling the solution. The boric acid solution is added to the other micronutrients, diluted to volume, and mixed.

The final solution is prepared by adding 1 part of each stock solution and diluting 200-fold, but the concentrated solutions should not be mixed together without diluting them. A combination of 5 gal each of stock solution 1 and 2, diluted to a final volume of 1000 gal, supplies the following concentrations of the elements in ppm: 119 N, 30 P, 140 K, 100 Ca, 24 Mg, 32 S, 2.5 Fe, 0.25 B, 0.25 Mn, 0.025 Zn, 0.01 Cu, and 0.05 Mo.

TABLE 17.11. Solutions Used for NFT. Rates g per 1000 liters (264 gal)

Salt	Formula	Starting	Topping-up
Calcium nitrate	Ca $(NO_3)_2.4H_2O$	988.0	395.5
Potassium nitrate	KNO_3	658.1	367.6
Magnesium sulfate	$MgSO_4.7H_2O$	496.6	324.3
Monopotassium phosphate	KH_2PO_4	272.0	—
Chelated iron	FeNa EDTA	78.88	32.87
Manganous sulfate	$MnSO_4.H_2O$	6.154	1.539
Boric acid	H_3BO_3	1.714	1.714
Copper sulfate	$CuSO_4.5H_2O$	0.275	0.275
Ammonium molybdate	$(NH_4)\,6Mo7O_2{}^4.4H_2O$	0.092	0.092
Zinc sulfate	$ZnSO_4.7H_2O$	0.308	0.308

Source: Schippers, F. A. 1979. *The Nutrient Flow Technique.* Vegetable Crops Mimeo 212, New York State College of Agriculture, Cornell University, Ithaca, NY, p. 5.

Complete solutions for inert media. To avoid problems of precipitation of nutrients in complete solutions, the ADAS system may use two tanks to separate the calcium nitrate from most other elements and a third solution for nitric acid. If there is enough Ca in the medium, calcium nitrate is omitted and potassium nitrate is increased while potassium sulfate is decreased, reducing the tanks to two (see Table 17.9).

The modified Hoagland solution is prepared from two stock solutions. The first contains most of the macronutrients plus some concentrated micronutrients. The second contains calcium nitrate plus chelated iron. Equal proportions of the two solutions are diluted, mixed, and finally diluted 200-fold (see Table 17.10). Uses of the NFT make up several stock solutions of the different materials and add the dissolved solution to the tank just prior to use (see Table 17.11).

Partial solutions for amended media. The elements provided by the amendments to aggregate cultures can be omitted from the nutrient solution. Adding dolomitic limestone allows the removal of Ca and Mg; adding ordinary superphosphate allows removal of P, Ca, and S; treble superphosphate allows the removal of P; and slow-release micronutrient mixes allow removal of B, Cu, Fe, Mo, and Zn. The removal of these elements greatly simplifies the solution preparation and delivery since needed elements now can be prepared as a concentrated solution in one tank.

Many different ways of producing incomplete nutrient solutions are available. Some of them are listed in Table 17.12. A special solution for use with brackish water is presented in Table 17.13.

TABLE 17.12. Composition of Incomplete Solutions for Use with Fortified Media

Feed ratio N:P_2O_5:K2O	ppm in dilute soln.	Fertilizers	kg 100/L	NH$_4$/NO$_3$ ratio	Use
Media containing superphosphate + limestone + micronutrients					
1:0:3.5	145-0-500	Potassium nitrate Epsom salts**	11.4 2.5	0:100	Low light Slow growth
1:0:2	175-0-350	Potassium nitrate Urea Epsom salts**	10.1 2.16 2.5	30:70	Regular feed
1:0:2	175-0-350	Potassium nitrate Ammonium nitrate Epsom salts**	7.78 2.22 2.5	23:77	Regular feed
1:0:1	250-0-250	Potassium nitrate Urea Epsom salts**	5.56 5.0 2.56	21:79	For more growth
1:0:1	250-0-250	Potassium nitrate Ammonium nitrate Epsom salts**	5.56 5.0 2.5	35:65	For more growth

* To be diluted 1/100. The 100 L make 10,000 L of dilute soln.

** The Epsom salts can be omitted if dolomitic limestone and ordinary superphosphate (0-20-0) are used as amendments. The dolomitic limestone supplies Mg and the ordinary superphosphate supplies SO_4-S. If treble superphosphate (0-46-0) is used, some other source of S would have to be provided.

These solutions suitable for low-light conditions and for plants in the reproductive stage. Higher ammoniacal/nitrate N ratios (50/50 or even 67/33) can be used for starting plants in good light. For poor light, keep ratio of K_2O/N >2/1.

These solutions can be used for media amended with superphosphate and limestone but without micronutrients if a micronutrient solution is added.

Soil Boxes

Containerized growing of plants using the Earth Box and a gradient concept of supplying nutrients has been proposed by C. M. Geraldson (1996) as a means of optimizing nutrient, water, and air inputs.

The gradient concept of adding large quantities of soluble N and K fertilizer in preplaced bands slightly ahead of roots in soil under plastic has been commercially used on a large scale in the field to provide high levels of nutrients relatively free from leaching. The N and K move by diffusion to maintain rapid growth rates; the plastic minimizes water loss by evaporation and fertilizer loss by leaching.

TABLE 17.13. Nutrient Solution Suitable for Crops Grown with Brackish Water*

	Per liter	
Potassium nitrate	900-1100	g
Triple superphosphate	200-300	
Ammonium sulfate	60	
Ferrous sulfate	8-10	(every 2-3 days)
Boric acid	1.5	g
Copper sulfate	0.5	g
Manganese sulfate	1.5	g
Zinc sulfate	0.5	g

* The nutrient solution is satisfactory for producing crops grown with brackish waters having the following composition:

	ppm	
Total salinity	2500-3500	(Approximately 3.9-5.5 mmhos/cm)
Chlorides	700-900	
Sulfates	600-1000	
Calcium	350-450	
Sodium	400-500	
Magnesium	120-180	

Source: Schwartz, M. 1977. *Guide to Commercial Hydroponics,* Fourth Edition. Israel University Press, Jerusalem, Israel.

Combining the gradient system with the Earth Box allows positive control of water and air since the box has an aeration screen that holds soil mix and fertilizer over an air layer below which is a column of water. The Earth Box is roughly 30×70 cm (12×28 inches) at the top, 20×60 cm (8×24 inches) at the base and 30 cm (12 inches) high. The height of the water column is regulated by an overflow, leaving a constant layer of air regardless of the amount of air added via a tube at the surface. The maintenance of a constant water head and air layer ensures large amounts of both for plant growth; the banded fertilizer and that mixed in the soil can provide nutrients for one or two crops.

The concept has produced large crops of tomatoes in regular-sized boxes, and is now being evaluated for much larger containers.

SUMMARY

Sufficient improvements in the air, water, and nutrients of farm soils enabling the satisfactory production of many different crops are made possible by various farm practices outlined in Chapters 9 through 16. But it often is necessary to go beyond these treatments to grow special high-priced plants in field soils because of the crop's special needs, location, or because the soils are subject to severe traffic stress.

A number of materials have been used to improve moisture-holding capacity, air exchange, and supply of nutrients for specialty crops or high-traffic areas in field soils. Such materials as peat moss, composts, sawdust, woodchips, rice and peanut hulls, bagasse, and manures are being used primarily to improve moisture holding capacity and air exchange, although the manures, bagasse, and composts can also supply substantial nutrients. Materials such as sharp sand, vermiculite, perlite, pumice, or calcined clay are added primarily for air exchange. While many of these materials are used for high-traffic areas, most areas rely upon coarse sand, peat moss, and calcined clay to provide sufficient aeration and traffic resistance. Using some crumb rubber for high-traffic areas, at least as a topdressing, is a relatively new approach with considerable promise.

Because of restricted volume and perched water tables, growing plants in containers requires very good aeration, water-holding capacity, and a continuous supply of nutrients. To accomplish these ends, potting soils have been fortified at various times with aerating materials such as arcillite, bagasse, bark, calcined clay, expanded polystyrene flakes, gravel, hadite, leaf mold, composts, manures, peanut hulls, rice hulls, pecan shells, perlite, polyurethane foam, urea formaldehyde resin, vermiculite, and wood chips. Most of these are not used extensively at present, since most commercial operations tend to depend on various combinations of peat moss, coarse sand, perlite, bark, and wood chips.

To provide longer-lasting nutrients, manures or hoof and horn meal have been used extensively in the past, but the trend has been to apply resin-coated materials such as Osmocote, Nutricote, and Procote for the macronutrients and FTE 1, Esmigram, or Perk for the micronutrients. These may be supplemented with nutrient solutions or, in some cases, the long-lasting materials are omitted and the entire nutrition program depends on a rather frequent application of solutions.

Soil in potting mixtures has been almost completely abandoned in favor of combinations of coarse sand, peat moss, or wood compost, vermiculite, or perlite. Recently, the sand has been omitted as lightweight mixes have become the predominant choice for container growing.

It is possible to extend the sides of the fertile triangle by growing plants in (1) air with water containing nutrients sprayed periodically on the roots (aeroponics), (2) water to which nutrients and air are added (liquid hydroponics), or (3) highly aerated substrates to which water and nutrients are added periodically (aggregate hydroponics). In all cases, the air or water sides are increased to maximum lengths while maintaining the other sides near optimum levels.

Aeroponics have had almost no commercial use, but hydroponics has some current use for producing high-valued ornamentals and vegetables. Its use may increase as land values and transportation costs increase.

Nutrient film technique, using shallow films of circulating solutions, has spurred additional use for liquid culture hydroponics, while some new aggregates (Grodan) or new uses for older types of aggregates (peat bolsters, bags of peatlite mixes, bales of straw) have given a boost to aggregate hydroponics.

Many different nutrient solutions have been used over the years for hydroponics. A rather complete nutrient solution containing all essential elements is needed for most hyroponic cultures, but modified ones are better for aggregate cultures or sandy soils that have various amendments or for waters that may have high concentrations of various ions. Complete solutions as well as some of those recommended for amended aggregates or saline waters have been presented.

It has been possible to produce large yields of tomatoes using an Earth Box and the gradient concept of banded fertilizers under a plastic mulch. It remains to be seen whether this concept will work with larger boxes.

Appendix 1

Important Soil, Plant, Water, and Fertilizer Elements; and Ions with Symbols, Atomic, and Equivalent Weights

Element	Symbol	Atomic Wt.[1]	Ion	Symbol and Valence	Atomic Wt.[1]	Equiv- alent Wt.[1]
Aluminum	Al	27	Aluminum	Al^{3+}	27	9
Boron	B	11	Borate	$H_2BO_3^-$	61	61
Calcium	Ca	40	Calcium	Ca^{2+}	40	20
Carbon	C	12	Carbonate	CO_3^{2-}	60	30
			Bicarbonate	HCO_3^-	61	61
Chloride	Cl	35.5	Chloride	Cl^-	35.5	35.5
Copper[2]	Cu	64	Cuprous	Cu^+	64	64
			Cupric	Cu^{2+}	64	64
Fluorine	F	19	Fluoride	F^+	19	19
Hydrogen	H	1	Hydrogen	H^+	1	1
			Hydroxyl	OH^-	17	17
Iron[3]	Fe	56	Ferrous	Fe^{2+}	56	28
			Ferric	Fe^{3+}	56	19
Magnesium	Mg	24	Magnesium	Mg^{2+}	24	12
Manganese[4]	Mn	55	Manganous	Mn^{2+}	55	27.5
			Manganic	Mn^{3+}	55	18
Molybdenum	Mo	96	Molybdate	MoO_4^{2-}	160	80
Nitrogen	N	14	Ammonium	NH_4^+	18	18
			Nitrate	NO_3^-	62	62
			Nitrite	NO_2^-	46	46
			Urea	NH_2^-	16	16
Oxygen	O	16	Oxygen	O^{2-}	16	16
Phosphorus	P	31	Phosphate	$H_2PO_4^-$	97	97
			Phosphate	HPO_4^{2-}	96	48
Potassium	K	39	Potassium	K^+	39	39
Silicon	Si	28	Silicate	$HSiO_3^-$	77	77
Sodium	Na	23	Sodium	Na^+	23	23
Sulfur	S	32	Sulfate	SO_4^{2-}	96	48
Zinc	Zn	65	Zinc	Zn^{2+}	65	32.5

[1] To closest whole or half number.

[2] It is believed that the cupric ion is important for plant nutrition.

[3] Both ferrous and ferric ions are important in soil and plant chemistry but the ferrous form is required by plants.

[4] Manganese can exist in five different valence forms but 2+ and 3+ are the primary manganese ions in soil chemistry processes, the 2+ form being the important one for plants.

Appendix 2

Common and Botanical
Names of Plants

Common name	Botanical name
African violet	*Saintpaula*
Alfalfa or lucerne	*Medicago sativa*
Allamanda	*Allamanda cathartica*
Almond	*Prunus amygdalus*
Alyssum	*Alyssum maratinum*
Apple	
common	*Malus* sp.
kei	*Doyvalis caffra*
sugar	*Syzgium jambos*
Apricot	*Prunus armeniaca*
Apricot, Siberian	*Prunus siberica*
Ash	
black	*Fraxinus nigra*
European	*Sorbus aucuparia*
white	*Fraxinus americana*
Asparagus	*Asparagus officinalis*
Aspen, trembling	*Populus tremuloides aurea*
Aster	
bigleaf	*Aster macrophylius*
seaside	*A. spectabalis*
sky-drop	*A. patens*
stiff	*A. linearifolius*
wave	*A. undulatus*
Avocado	*Persea americana*
Azalea	
Hiryu	*Rhododendron obtusum*
Indian hybrid	*R.* sp. hybrids
pink	*R. periclymenoides*

Common name	Botanical name
Banana	*Musa* sp.
Barley	*Hordaum vulgare*
Basswood or linden	*Tilia* sp.
Bean	
broad	*Vicia Faba*
castor	*Ricinus communis*
field	*V. Faba*
fava or horse	*V. Faba*
lima	*Phaseolus vulgaris*
snap	*P. vulgaris*
wax	*P. vulgaris*
Beech	*Fagus grandifola*
Beet	
mangel wurzel	*Beta vulgaris, crassa* group
sugar	*B. saccharifera*
table	*B. vulgaris, crassa* group
Birch	
black	*Betula nigra*
canoe	*B. papyrifera*
gray	*B. alleghaniensis*
sweet	*B. lenta*
white, American	*B. papyrifera*
wire	*B. populiolia*
yellow	*B. lutea*
Bird of paradise	*Strelitzia reginae*
Bird's-foot trefoil	*Lotus corniculatus*
Blackberry	*Rubus allegheniensis*
Blueberry	
highbush	*Vaccinium corymbosum*
rabbit eye	*V. ashei*
Boxwood	
common	*Buxus sempervirens*
Japanese	*B. macrophylla* var. *japonica*
Boysenberry	*Rubus urainus* var. *longanobassum* cv Boysen
Broccoli	*Brassica oleracea, botrytis* group
Brussel sprouts	*Brassica oleracea, gemnifera* group
Buckwheat	*Fagopyrum esculentum*
Cabbage	*Brassica oleracea, capitata* group
Cabbage, Chinese	*Brassica rapa* L. *chinensis* group
Cajeput	*Umbellisharia california*
Camellia	*Camilla japonica*

Common name	Botanical name
Canna	*Canna* sp.
Cantaloupe or muskmelon	*Cucumis melo* L. *Reticulatus* group
Carambola or starfruit	*Averrhoa carambola*
Carissa	
arduina	*Carissa bispinosa*
natal plum	*C. grandiflora*
Carnation	*Dianthus caryophyllus*
Carrot	*Daucus carota*
Cassava	*Manihot esculenta*
Cassia	*Cinnamomum cassia*
Casuarina	
equisetifolia	*Casuarina equisetifolia*
stricta	*C. ofruta*
Catalpa	*Catalpa* sp.
Cauliflower	*Brassica oleracea, botrytis* group
Cedar, western red	*Thuja plicata* Donn ex D. Don
Celery	*Apium graveolens* var. *dulce*
Celosia	*Agave americana*
Chalkas or orange jasmine	*Murraya paniculata*
Chard, Swiss	*Beta vulgaris, cicla* group
Cherries	
Barbados	*Eugenia uniflora*
black	*Prunus serotina*
choke	*P. virginiana*
pin	*P. pensylvanica*
sour	*P. cerasus*
sweet	*P. avium*
Chrysanthemum	*Chrysanthemum morifolium*
Clover	
alsike	*Trifolium hybridum* L.
bur	*Medicago denticulata*
crimson	*Trifolium incarnatum*
ladino or white	*T. repens*
red	*T. pratense*
subterranean	*T. subterraneum* L.
white sweet	*Mellilotus alba*
yellow sweet	*M. officinalis*
Cocoa	*Theobroma cacao*
Coconut	*Cocos nucifera*
Coco-plum	*Chrysobalanus icaco*
Coffee	*Coffea arabica*

Common name	Botanical name
Collards	*Brassica oleracea, acephala* group
Corn or maize	
common	*Zea mays*
pop	*Z. mays* var. *everta*
sweet	*Z. mays* var. *rugosa*
Cotton	*Gossypium hirsutum*
Cottonwood	*Populus* sp.
Cowpea	*Vigna unguiculata*
Cranberry	*Vaccinium macrocarpon ait*
Crownvetch	*Coronilla varia*
Cucumber	*Cucumis sativus*
Currant	
black	*Ribes nigrum*
red	*R. sativa*
Date	*Phoenix dactylifera*
Desmodium, greenleaf	*Desmodium intortum*
Dogwood	*Cornus alba* and *C. racemosa*
Dracena	
corn plant	*D. fragrans* "Massangeana"
godseffiana	*Dracena surculosa*
Janet Craig	*D. deremensis* "Janet Craig"
reflexa	*D. thaliodes*
Sanders	*D. sanderana*
Warnecki	*D. deremensis* "Warneckii"
Dusty miller	*D. artemisia Stellerana*
Eggplant	*Solanum melongena*
Elder, European red	*Sambucus racemosa*
Elm	*Ulmus americana*
Endive or escarole	*Chichorium endiva*
Eucalyptus	
common	*Eucalyptus deglupta* or *E. grandis*
southern blue-gum	*E. globulus*
Eugenia	*Eugenia* sp.
Euonymus	*Euonymus alatus* (Thunb.) Recel
Euphorbia	*Euphorbia lactea*
Fatsia	*Fatsia japonica*

Common name	Botanical name
Fern	
bird's-nest	*Asplenium nidus*
Boston	*Nephrolepis exalta* "Bostoniensis"
leatherleaf	*Rumohra adiantiformis*
maidenhair	*Adiantum pedatum*
pteris	*Pteris* sp.
Fescue, tall	*Festuca arundinacea*
Fig	*Ficus carica*
Fir	
amabilis	*Abies amabilis*
balsam or Alpine	*Abies balsamea*
Douglas	*Pseudotsuga menziesii*
Flax	*Linum* sp.
Forsythia	*Forsythia intermedia*
Foxtails, meadow	*Alopucurus pratensis*
Gardenia	*Gardenia jasminoides*
Garlic	*Allium sativum*
Geiger	*Cordia sebestena*
Geranium	*Perlargonium* x *hortorum* Bailey
Gerbera or Transvaal daisy	*Gerbera jamesonii*
Gladiolus	*Gladiolus* x *hortulnus*
Gloxinia	*Sinnirgia speciosa*
Grape	*Vitis vinifera, V. labbruska* and hybrids
Grape, muscadine	*Vitis rotundifolia*
Grapefruit	*Citrus* x *paradisi*
Grass	
bahia	*Papsalum notatum*
bent	
colonial	*Agrostis capillaris*
creeping	*A. palustris*
Rhode Island	*A. tenuis*
seaside	*A. palustris*
velvet	*A. canina*
Bermuda	*Cynodon dactylon*
blue	
annual	*Poa annua*
Kentucky	*P. pratensis*
bluejoint	*Calamagrostis canadensis*
brome	*Bromus inermis*

Common name	Botanical name
Grass (*continued*)	
creeping bent	*Agrostis palustris*
crested wheat	*Agropyron desertorum* Schultes
orchard or cocksfoot	*Dactylis glomerata*
pangola	*Digitaria decumbens*
prairie	*Bromus uniloides*
rye	*Lolium perenne*
St. Augustine	*Stenotaphrum secundatum*
sorghum-sudan	*Sorhum sudanese*
sudan	*Sorghum vulgare*
switch	*Planicum virgatum*
tall fescue	*Festuca arundinacea*
zoysia	*Zoysia matrella*
Guava	
cattleya	*Psidium littorale* var. *longipes*
common	*P. guajava*
pineapple	*Feijoa sellowiana*
Hawthorn	*Crategus phaenopyrum*
Hazelnut	*Corylus avellana*
Hemlock, Western	*Tsuga heterophylla*
Hibiscus	*Hibiscus rosa-sinensis*
Holly	
American	*Ilex opaca*
Chinese	*I. cornuta*
English	*I. acquifolium*
Japanese	*I. crenata*
Honey locust, moraine	*Gleditsia triacanthos* var. *inermis*
Honeysuckle	
Cape	*ecomaria capensis*
European fly	*T. canadensis*
Hornbeam, European	*Carpinus betulus*
Horse chestnut	*Aesculus hippocastanum*
Horseradish	*Armoracia rusticana* Mey
Hydrangea	*Hydrangea macrophylla*
Ice plant	*Mesembryanthemum crystallinum*
Iris	
blue flag	*Iris versicolor*
Carolina	*I. caroliniana*
Japanese	*I. kaemferi*
vernal	*I. verna*
Ixora	*Ixora coccinea*

Common name	Botanical name
Ivy	
Boston	*Parthenocissus tricuspidata* var. *veitchii*
English	*Hedera helix* L.
Jaboticaba	*Myrciaria cauliflora*
Jasmine, wax	*Jasminum simplicifolium*
Juniper	
Bar Harbour	*Juniperus horizontalis*
pfitzer compacta	*Pfitzeriana compacta*
Kalanchoe	*Kalanchoe blossfeldiana*
Kale	*Brassica oleracea, acephala* group
Kenaf	*Hibiscus cannabinus*
Kohlrabi	*Brassica oleracea, gongylodes* group
Lantana	*Lantana* sp.
Larch, Japanese	*Larix kaempferi*
Leea or Hawaiian holly	*Leea coccinea*
Leek	*Allium ampeloprasum, porrum* group
Lemon	*Citrus limon*
Lettuce	*Lactuca sativa*
Lilac	
common	*Syrinia vulgaris*
Persian	*S. persica*
Lily	
day	*Hemerocallis* sp.
Easter	*Lilium longiflorum*
Japanese	*L. speciosum*
regal	*L. regale*
tiger	*L. igrinum*
Lime, Persian	*Citrus aurantifolia "tahiti"*
Lipstick plant	*Aeschynanthus pulcher*
Liriope	*Liriope muscari*
Locust	
black	*Robinia psueduacacia*
honey	*Gladirsia triacanthus*
Lupine	
blue	*Lupinus hirsutus*
rose	*L. hartwegi*
white	*L. albus*
yellow	*L. luteus*

Common name	Botanical name
Macadamia	*Macadamia ternifolia*
Magnolia	
saucer	*Magnolia soulangiana*
southern	*M. grandilora*
star	*M. stellata*
Malpighia	*Malpighia* sp.
Mandarin or tangerine	*Citrus reticulata Blanco*
Mandevilla, Dipladenia	*Mandevilla splendens*
Mango	*Magnifera indica*
Maple	
black	*Acer nigrum*
mountain	*A. spicatum*
red	*A. rubrum*
striped	*A. pensylvanicum*
sugar	*A. saccharum*
Marigold	*Targetes* sp.
Millet	*Setaria italica*
Milo, dwarf yellow	*Sorghum vulgare*
Mint	*Menthe arvensis*
Monkey puzzle tree	*Araucaria bidwillii*
Mountain laurel	*Kalmia latifolia*
Mulberry	*Morus* sp.
Mustard	*Brassica juncea*
Myrtle, crape	*Lagerstroemia indica*
Nephytis	*Syngonium podophyllum*
Oak	
black	*Quercus veelutina*
bur	*Q. macrocarpa*
California live	*Q. agrifolia*
English	*Q. robur*
live	*Q. virginia*
pin	*Q. palustris*
red	*Q. rubra*
valonea	*Q. ithaburensis*
white	*Q. alba*
Oat	*Avena sativa*
Olive	
black	*Bucida bucerus*
common	*Olea europaea*
Russian	*Eleagnus augustifolia*

Common name	Botanical name
Onion	*Allium cepa, cepa* group
Orange, navel and Valencia	*Citrus sinensis*
Orchid	
cattleya	*Cattleya* sp.
cymbidium	*Cymbidium* sp.
ladyslipper	*Cypripedium* sp.
phalenopsis or moth	*Phalenopsis* sp.
Palm	
areca	*Chrysalidocarpus lutescens*
bamboo or seifritzii	*Chamaedora erumpens*
cabbage	*Livistona australis*
Canary Island	*Phoenix canariensis*
chamadorea or parlor	*Chamaedorea elegans*
cocus	*Cocus australis*
kentia	*Howea forsterana*
key	*Thrimax microcarpa*
oil	*Elaeis guineesis*
ponytail	*Beaucarnea recurvata*
rhapis or lady	*Rapis excelsa*
robelini	*Phoenix roebelenii*
Washington	*Washington*
Pansy	*Viola tricolor*
Papaya	*Carica papaya*
Parsley	*Petroselinum hortense*
Parsnip	*Pastinaca sativa*
Pea	
English	*Pisum sativum*
field	*P. arvense*
southern or black-eyed	*Vigna unguiculata*
Peach	*Prunus persica*
Peanut	*Arachis hypogaea*
Pear	*Pyrus communis*
Pecan	*Carya illinoensis*
Peony	*Paeonia albiflora*
Peperomia	*Peperomia obtusfolia*
Pepper	*Capsicum annum* var. *annum*
Persimmon, Japanese	*Diospyros kaki*

Common name	Botanical name
Philodendron	
cordatum or heart leaf	*Philodendron scandens oxycardium*
hastatum	*P. hastatum*
panduriforme	*P. panduriforme*
pertussum or monstera	*Monstera deliciosa*
seloum	*Philodendron seloum*
Photinia	*Potinia glaubra*
Pine	
Aleppo	*Pinus halepensis*
Australian	*Araucaria heterophylla, casrino*
Corsican	*Pinus laricio*
jack	*P. banksiana*
Japanese black	*P. thumbergiana*
loblolly	*P. taeda*
lodgepole	*P. contorta "latifolia"*
Norfolk Island	*Araucaria hetrophylla*
radiata	*Pinus radiata*
red	*P. resinosa*
Scotch	*P. sylvestris*
slash	*P. elliottii*
western yellow	*P. ponderosa*
white	*P. strobus*
Pineapple	*Ananus comosus*
Pitomba	*Eugenia luschnathiana*
Pittosporum	*Pittosporum tobira*
Plum or prune	*Prunus domestica*
Podocarpus	*Podocarpus macrophyllus*
Poinsettia	*Euphorbia pulcherrima*
Pomegranate	*Punica granatum*
Poplar	*Populus grandidentata*
Potato	
Irish	*Solanum tuberosum*
sweet	*Ipomea batatus*
Privet	*Ligustrum* sp.
Pumpkin	*Curcubita pepo*
Quince	*Cydonia oblonga*
Radish	*Raphanus sativus*
Rape or canola	*Brassica napus*
Raspberry	*Rubus idaeus*
Rhododendron	*Rhododendron* sp.

Common name	Botanical name
Rhubarb	*Rheum rhaponticum*
Rice	*Oryza sativa* L.
Rose	*Rosa odorata*
Rosemary	*Rosemarinus officinalis*
Rye	*Secale cereale*
Safflower	*Carthamus tinctorius*
Salsify	*Tragopopon porrifolius*
Salvia	*Salvia splendens*
Sansevieria	*Sansevieria laurentii*
Sapote	
black	*Diosyros digyna*
mamey	*Pouteria sapote*
white	*Casimira edulis*
Schefflera or umbrella tree	*Brassaia actinophylla*
Shallot	*Allium cepa, aggregatum* group
Siratro	*Macroptilium atropurpureum*
Snapdragon	*Antirrhinum majus*
sorghum	*Sorghum vulgare*
Sorrel	*Rumex scutatus*
Soybean	*Glycine max*
Spanish bayonet	*Yucca aloifolia*
Spathiphyllum	*Spathiphyllum* sp.
Spider plant	*Chlorophytum comosum*
Spinach	*Spinacia oleracea*
Spirea	*Spirea nipponica*
Spruce	
black	*Picea mariana*
Colorado	*P. pungens*
Engelmann	*P. engelmannii*
Norway	*P. abies*
red	*P. rubens*
Sitka	*P. sitchensis*
white	*P. glauce*
Squash	*Cucurbita pepo* var. *melopepo*
Squawbush	*Viburnum trilobum*
Statice	*Statice limonium* sp.
Strawberry	*Fragaria* sp.

Common name	Botanical name
Stylo	*Stylosanthes humilis*
Sugarcane	*Saccharum officinarum*
Sumac	*Rhus* sp.
Sunflower	*Helianthus angustifolius* and *annus*
Sycamore	*Patanus occidentalis*
Syringa	*Syringa* sp.
Tamarisk	*Tamarix* sp.
Taxus	*Taxus* x *media*
Tea	*Camelia sinensis*
Timothy	*Phleum pratense*
Tobacco	*Nicotiana tabacum*
Tomato	*Lycopersicon esculentum*
Trefoil	
bird's-foot	*Lotus corniculatus*
yellow	*Medicago lupulina*
Turnip	*Brassica raps* var. *rapifera* group
Verbana	*Verbana* x *hybriida*
Vetch, hairy	*Vicia villosa*
Viburnum	*Viburnum acerifolium* or *V. suspensum*
Violet	
African	*Saintpaulia ionantha*
common	*Viola paplionacea*
Walnut	*Juglans regia*
Watercress	*Nasturtium officinale* R. Br.
Watermelon	*Citrullus lanatus*
Wheat	*Triticum aestivum* L.
Willow	*Salix* sp.
Wisteria, Japanese	*Wisteria floribunda*
Yucca	*Yucca elephantipes*
Zamia	*Zamia* sp.
Zinnia	*Zinnia elegans*

References

Chapter 1

Allen, L. H. Jr. 1997. Mechanisms and rates of O_2 transfer to and through rhizomes and roots via aerenchyma. *Soil Crop Sci. Soc. Florida Proc.* 56: 41-54.

Glenn, D. M. and W. V. Welker. 1997. Effects of rhizosphere carbon dioxide on the nutrition and growth of peach trees. *HortScience* 32(7): 1197-1199.

Glinski, J. and J. Lipiec. 1990. *Soil Physical Conditions and Plant Roots.* CRC Press, Inc., Boca Raton, FL.

Glinski, J. and W. Stepniewski. 1985. *Soil Aeration and Its Role for Plants.* CRC Press, Inc., Boca Raton, FL.

Gooch, J. J. 1997. Frosty fields yield early vegetables. *American Vegetable Grower.* 45(11):39-40.

Hewitt, E. J. 1966. *Sand and Water Culture Methods Used in the Study of Plant Nutrition.* Technical Communication No. 22 (Revised Second Edition). Commonwealth Bureau of Horticulture and Plantation Crops, East Malling, Maidstone, England.

Mengel, K. and E. A. Kirby. 1982. *Principles of Plant Nutrition.* International Potash Institute, Worblaufen-Bern, Switzerland.

Snyder, G. H. (Ed.). 1987. *Agricultural Flooding of Organic Soils.* Agricultural Experiment Station, University of Florida Institute of Food and Agricultural Science Bulletin 870 (Technical). Gainesville, FL.

Sposito, G. 1989. *The Chemistry of Soils.* Oxford University Press, New York.

Stolzy, L. 1974. Soil atmosphere. In *The Plant Root and Its Environment,* 335-361. E. W. Carson (Ed). University Press of Virginia, Charlottesville, VA.

Takai, Y., T. Koyama, and T. Kamaru. 1957. Microbial metabolism of paddy soils. *J. Agr. Chem. Soc. Japan,* 31, 211-220.

Turner, F. T. and W. H. Patrick. 1968. Chemical changes in waterlogged soils as a result of oxygen depletion. *Transactions IX Congress International Soil Science Society* 4: 53-65.. Adelaide, Australia.

White, J. W. 1973. Criteria for selection of growing media for greenhouse crops. *Penn. State Agric. Expt. Journal,* Series #4575.

Additional Readings

Rowell, D. L. 1981. Oxidation and reduction. In D. J. Greenland and M. H. B. Hayes (Eds). *The Chemistry of Soil Processes.* John Wiley and Sons, New York.

Chapter 2

Donahue, R. L., R. H. Follett, and R. W. Tulloch. 1995. *Our Soils and Their Management,* Fourth Edition. The Interstate Printers and Publishers, Danville, IL.

FAO. 1971. *Irrigation Practice and Water Management.* Irrigation and drainage paper No. 1. Water Resources and Development Service, Land and Water Development Division, FAO, Rome.

Goldberg, D., B. Gormat, and D. Rimon. 1976. *Drip Irrigation.* Drip Irrigation Scientific Publications, Kfar Schmaryanu, Israel.

Hansen, V. A., O. W. Israelson, and G. E. Stringham. 1980. *Irrigation Practices and Principles.* John Wiley and Sons, New York.

Hiler, E. A. and T. A. Howell. 1983. Irrigating options to avoid critical stress: An overview. In H. M. Taylor, W. R. Jordan, and T. R. Sinclair (Eds.), *Limitations to Efficient Water Use in Crop Production.* American Society of Agronomy, Inc., Crop Science Society of America, Inc., and Soil Science Society of America, Inc., Madison, WI.

Hillel, D. 1988. *The Efficient Use of Water in Irrigation.* World Bank Technical Paper No. 64. The World Bank, Washington, DC.

Lorenz, O. A. 1988. *Knott's Handbook for Vegetable Growers,* Third Edition. John Wiley and Sons, New York.

Miller, R. W. and R. L. Donahue. 1995. *Soils in Our Environment.* Prentice-Hall, Englewood Cliffs, NJ.

Neja, R. A., W. E. Wildman, and L. P. Christensen. 1982. *How to Appraise Soil Physical Factors for Irrigated Vineyards.* Leaflet 2946. Division of Agricultural Science, University of California, Berkeley, CA.

USDA. 1955. Water. In *Yearbook of Agriculture.* USDA, Washington, DC.

Wolf, B. 1996. *Diagnostic Techniques for Improving Crop Production.* The Haworth Press, Inc., Binghamton, NY.

Additional Readings

Jordan, W. R. 1983. Whole plant response to water deficits. In H. M. Taylor and W. R. Jordan (Eds.), *Limitations to Efficient Water Use in Crop Production,* 289-311. American Society of Agronomy, Inc., Crop Science Society of America, Inc., Soil Science Society of America, Inc., Madison, WI.

Mengel, K. and E. A. Kirby. 1982. *Principles of Plant Nutrition.* International Potash Institute, Worblaufen-Bern, Switzerland.

Newman, E. I. 1974. Root and soil water relations. In E. W. Carson (Ed.), *The Plant Root and Its Environment,* 363-440. University Press of Virginia, Charlottesville, VA.

Chapter 3

Bennet, W. F. (Ed.). 1993. *Nutrient Deficiencies and Toxicities in Crop Plants.* APS Press, St. Paul, MN.

Bould, C. and E. J. Hewitt. 1963. Mineral nutrition of plants in soils and in culture media. In F. C. Steward (Ed.), *Plant Physiology.* Academic Press, New York.

Glass, A. D. M. 1989. *Plant Nutrition.* Jones and Bartlett Publishers, Boston/Portla Valley.

International Potash Institute. 1977. *Potassium Dynamics in the Soil.* IPI Extension Guide. International Potash Institute, Worblaufen-Bern, Switzerland.

Mengel, K. and E. A. Kirby. 1982. *Principles of Plant Nutrition.* International Potash Institute, Worblaufen-Bern, Switzerland.

Peck, N. H. 1975. Vegetable crop fertilization. *New York Food and Life Science Bulletin* 52.

Weir, R. G. and G. C. Cresswell. 1993. *Plant Nutrient Disorders 1. Temperate and Subtropical Fruit and Nut Crops.* Inkata Press, North Ryde, Australia.

Weir, R. G. and G. C. Cresswell. 1995. *Plant Nutrient Disorders 2.* Tropical Fruit and Nut Crops. Inkata Press, Melbourne, Australia.

Wolf, B. 1996. *Diagnostic Techniques for Improving Crop Production.* The Haworth Press, Inc., Binghamton, NY.

Wolf, B., J. Fleming, and J. Batchelor. 1985. *Fluid Fertilizer Manual.* National Fertilizer Solutions Association, St. Louis, MO.

Chapter 4

Anonymous. Average Bulk Densities of Different Soil Textural Classes. Information capsule # 137. Midwest Agricultural Laboratories, Inc., Omaha, NE.

Donahue, R. L., R. H. Follett, and R. W. Tulloch. 1995. *Our Soils and Their Management.* The Interstate Printers and Publishers, Danville, IL.

Flegmann, A. W. and R. A. T. George. 1977. *Soils and Other Growth Media.* Avi Publishing Company, Westport, CT.

Miller, R. W. and R. L. Donahue. 1995. *Soils in Our Environment,* Seventh Edition. Prentice Hall, Englewood Cliffs, NJ.

Additional Readings

Chancellor, W. J. 1977. *Compaction of Soil by Agricultural Equipment.* Bull. 1881. Cooperative Extension Division of Agricultural Science, University of California, Berkeley, CA.

Davies, D. B., D. J. Eagle, and J. B. Finney. 1993. *Soil Management.* Farming Press Books and Videos, Ipswich, United Kingdom.

Hartge, K. H. and B. A. Stewart (Eds.). 1995. *Soil Structure.* Lewis Publishers, Boca Raton, FL.

Summer, M. E. and B. A. Stewart. 1992. *Soil Crusting.* Lewis Publishers, Boca Raton, FL.

Taylor, H. M. 1974. Root behavior as affected by soil structure and strength. In E. W. Carson (Ed.), *The Plant Root and Its Environment.* University Press of Virginia, Charlottesville, VA.

Chapter 5

Anonymous. Livestock Manure Folder. Potash and Phosphate Institute and Foundation for Agronomic Research, Norcross, GA.

Collings, G. H. 1947. *Commercial Fertilizers,* Fifth Edition. McGraw-Hill, New York.

Follett, R. H., L. S. Murphy, and R. L. Donahue. 1981. *Fertilizers and Soil Amendments.* Prentice-Hall, Englewood Cliffs, NJ.

Griffith, D. R., J. V. Mannering, and J. E. Box. 1996. Soil and moisture management with reduced tillage. In M. Sprague and G. B. Triplet (Eds.), *No-Tillage and Surface Tillage.* John Wiley and Sons, New York.

Lorenz, O. A. and D. N. Maynard. 1988. *Knott's Handbook for Vegetable Crops,* Third Edition. John Wiley and Sons, New York.

Maples, R. L. 1989-1990. Nitrogen can increase cotton yield and soil organic matter. *Better Crops with Plant Food.* Winter. Potash and Phosphate Institute, Norcross, GA.

Miley, W. N. *Fertilizing Cotton with Nitrogen.* University of Arkansas Coop. Ext. Serv. Leaflet 526.

Obreza, T. and R. Muchovej. 1997. Beneficial use of biosolids in Florida. *Citrus and Vegetable Magazine,* December, pp. 10, 12.

Pieters, A. J. and R. McKee. 1938. The use of cover and green-manure crops. In G. Hambridge (Ed.), *Soils and Men: Yearbook of Agriculture 1938.* United States Government Printing Office, Washington, DC.

Roe, N. E., P. J. Stofella, and D. Graetz. 1997. Composts from various municipal solid waste feedstocks affect vegetable crops. I. Emergence and growth. *J. Am. Soc. Hort. Sci. 122* (3) 427-432.

Russell, E. J. 1966. *Soil Conditions and Plant Growth,* Ninth Edition, Fourth Impression. Longmans, Green, and Co., London.

Tisdale, S. L. and W. L. Nelson. *Soil Fertility and Fertilizers,* Third Edition. Macmillan Publishing, New York.

Webley, D. M. and R. B. Duff. 1965. The incidence in soils and other habitats of microorganisms producing a-ketogluconic acid. *Plant and Soil* 22: 307-313.

Additional Readings

Avnimelich, Y. and A. Cohen. 1989. Use of organic manures for amendment of compacted clay soils. II. Effect of carbon to nitrogen ratio. *Commun. in Soil Sci. Plant Anal.* 20 (15-16): 1635-1644.

McVay, K. A., D. E. Radcliffe, and W. L. Hargrove. 1989. Winter legume effects on soil properties and nitrogen fertilizer requirement. *Soil Sci. Soc. Am. J.* 53: 1856-1862.

Paul, E.A., K.A. Paustian, E. T. Elliott, and C. V. Cole (Eds.). 1996. *Soil Organic Matter in Temperate Agroecosystems.* Lewis Publishers, Boca Raton, FL.

Power, J. F. and J. W. Doran. 1988. Role of Crop Residue Management in Nitrogen Cycling and Use. In W. L. Hargrove (Ed.), *Cropping Strategies for Effi-*

cient Use of Water and Nitrogen. ASA Special Publication No. 51. American Society of Agronomy, Inc., Crop Science Society of America, Inc. and Soil Science Society of America, Madison, WI.

Tester, C. E. 1990. Organic amendment effects on physical and chemical properties of a sandy soil. *Soil Sci. Soc. Am. J.* 54: 827-831.

Unger, P. W., G. W. Langdale, and W. I. Papendick. 1988. Role of crop residues— Improving water conservation and use. In W. L. Hargrove (Ed.), *Cropping Strategies for Efficient Use of Water and Nitrogen.* ASA Special Publication No. 51. American Society of Agronomy, Inc., Crop Science Society of America, Inc., and Soil Science Society of America, Madison, WI.

Wallace, A. 1994. Soil organic matter is essential to solving soil and environmental problems. Soil organic matter must be restored to near original levels. *Commun. Soil Sci. Plant Anal.* 25(1-2): 15-28, 29-35.

Chapter 6

Anonymous. 1978. *Liming Acid Soils.* University of Kentucky Agricultural Extension Service Leaflet AGR-19.

Espinosa, José. 1996. Liming tropical soils—A management challenge. *Better Crops,* 60: 28-311.

Lamotte Chemical Products Co. 1950. *Lamotte Soil Handbook.* Lamotte Chemical Products Co., Towson, MD.

Lorenz, O. A. and D. M. Maynard. 1988. *Knott's Handbook for Vegetable Growers,* Third Edition. John Wiley and Sons, New York.

Lucas, R. E. and J. F. Davis. 1961. Relationships between pH values of organic soils and availability of 12 plant nutrients. *Soil Sci.* 92: 177-182.

National Fertilizer Solutions Association. 1967. Using pH values. *Liquid Fertilizer Manual.* National Fertilizer Solutions Association, Peoria, IL.

Pierre, W. H. 1933. Determination of equivalent acidity and basicity of fertilizers. *Ind. and Eng. Chem., Anal. Ed.* 5(4).

Spurway, C. H. 1941. *Soil Reactions (pH) Preferences of Plants.* Special Bull. 306, Michigan State College, East Lansing, MI.

Tisdale, S. L. and W. L. Nelson. 1966. *Soil Fertility and Fertilizers,* Second Edition. Macmillan, New York.

Additional Readings

Adams, F. 1981. Alleviating chemical toxicities: Liming acid soils. In Arkin, G. F. and H. M. Taylor (Eds.), *Modifying the Root Environment to Reduce Crop Stress.* ASAE Monograph No. 4. American Society of Agricultural Engineers, St. Joseph, MO.

Adams, F. (Ed.), 1988. *Soil Acidity and Liming,* Second Edition. Monograph No. 12. American Society of Agronomy, Inc., Crop Science Society of America, Inc., and Soil Science Society of America, Inc., Madison WI.

Chapter 7

Anonymous. Cation Exchange Capacities of Different Soil Textural Classes. Information Capsule # 137. Midwest Agricultural Laboratories, Inc., Omaha, NE.

Parfitt, R. L. and R. S. C. Smart. 1978. The mechanism of sulfate adsorption on iron oxides. *Soil Sci. Soc. Am. J.,* 42: 48-50.

Additional Readings

Brady, N. C. 1974. *The Nature and Property of Soils,* Eighth Edition. Macmillan, New York.

Harpstead, M. I. and F. D. Hole. 1980. *Soil Science Simplified.* Iowa State University Press, Ames, IA.

Nommik, H. 1965. Ammonium fixation and other reactions involving a nonenzymatic immobilization of mineral nitrogen in soil. In W. V. Batholemew and F. E. Clark (Eds.), *Soil Nitrogen.* American Society of Agronomy, Madison, WI.

Sposito, G. 1989. *The Chemistry of Soils.* Oxford University Press, New York.

Chapter 8

Barick, W. E. 1978. Salt tolerant plants for Florida landscapes. *Proc. Fla. State Hort. Soc.* 91: 82-84.

Bernstein, L. 1964. *Salt Tolerance of Plants.* USDA Bull 283.

Bernstein, L., L. E. Francois, and R. A. Clark. 1972. Salt tolerance of ornamental shrubs and ground covers. *J. Am. Soc. Hort. Sci.* 97: 550-556.

Haehle, R. 1981. *Salt Tolerant Plants.* Broward Co. Agric. Agents Office, Ft. Lauderdale, FL.

Hoffman, G. 1981. Alleviating salinity stress. In G. J. F. Arkin and H. M. Taylor (Eds.), *Modifying the Root Environment to Reduce Crop Stress,* 305-343. ASAE Monograph No. 4. American Society of Agricultural Engineers, St. Joseph, MO.

Joyner, G. 1972. *Salt Tolerant Ornamental Plants for South Florida.* Palm Beach Co. Agric. Agents Office.

Lorenz, O. A. and D. N. Maynard. 1988. *Knott's Handbook for Vegetable Growers,* Third Edition. John Wiley and Sons, New York.

Raeder, L. F. Jr., M. M. White, and C. W. Whittaker. 1943. The salt index—A measure of the effect of fertilizers on the concentration of the soil solution. *Soil Sci.* 55: 210-218.

Rauschkolb, R. and B. Foreman. 1968. USDA Bull. 217.

Russell, E. W. 1966. *Soil Conditions and Plant Growth,* Ninth Edition. Longman, Green and Co., London.

Waters, W. E., J. NeSmith, C. M. Geraldson, and S. S. Woltz. 1972. The interpretation of soluble salt tests and soil analysis by different procedures. *Florida Flower Grower,* 9(4): 5.

Additional Readings

Ayars, R. S. and D. W. Westcott. 1976. *Water Quality for Agriculture.* Irrigation and drainage paper 29. Food and Agriculture Organization of the United Nations, Rome.

Mengel, K. and E. A. Kirby. 1982. *Principles of Plant Nutrition.* International Potash Institute, Worblaufen-Bern, Switzerland.

Chapter 9

Burchett, L. 1981. Conservation tillage here to stay. *Solutions,* March/April.

Eltz, F. L. F. and L. D. Norton. 1997. Surface roughness changes as affected by rainfall erosivity, tillage, and canopy cover. *Soil Sci. Soc. Am. J.* 61: 1746-1755.

Kasper, T. C., D. C. Erbach, and R. M. Cruse. 1990. Corn response to seed-row residue removal. *Soil Sci. Soc. Am. J.* 54: 1112-1117.

Legere, A., N. Samson, R. Rioiux, D. A. Angers, and R. R. Simard. 1997. Response of spring barley to crop rotation, conservation tillage, and weed management intensity. *Agron. J.* 89: 628-638.

Pierce, F. L., M. C. Fortin, and M. J. Staton. 1992. Immediate and residual effects of zone tillage in rotation with no-tillage on soil physical properties and corn performance. *Soil Tillage Res.* 24: 149-165.

Rasmussen, P. E., R. W. Rickman, and B. Klepper. 1997. Residue and fertility effects on yield of no-till wheat. *Agron. J.* 89: 563-567.

Voorhees, W. B., R. R. Allmaras, and C. E. Johnson. 1981. Alleviating temperature stresses. In G. T. Arkin and H. M. Taylor (Eds.), *Modifying the Root Environment to Reduce Crop Stress.* ASAE Monograph No. 4. American Society of Agricultural Engineers, St. Joseph, MI.

Additional Readings

Anonymous. 1997. The benefits of con-tillage. Soil and Water Management. Continuing Education Sells Study Course. *Ag. Consultant,* September: 22-24.

Lal, R., M. McMahon, R. Papendick, and G. Thomas (Eds.). 1996. *Stubble Over the Soil—The Vital Role of Plant Residue in Soil Management.* American Society of Agronomy, Soil Science Society of America, and Crop Science Society of America, Madison, WI.

Sprague, M. A. and G. B. Triplett. 1986. *No-Tillage and Surface Tillage Agriculture.* John Wiley and Sons, New York.

Wagger M. G. and H. P. Denton. 1989. Influence of cover crop and wheel traffic on soil physical properties in continuous no-till corn. *Soil Sci. Soc. Am. J.* 53: 1206-1210.

Chapter 10

Bahe, A. R. and C. H. Peacock. 1995. Bioavailable herbicide residues in turfgrass clippings used for mulch adversely affect plant growth. *HortScience* 30(7): 1393-1395. December.

Farnsworth, R. L., E. Giles, R. W. Frazes, and D. Peterson. 1993. The residue dimension. *Land and Water No. 9,* September. Coop. Extension Service, University of Illinois at Urbana-Champaign.

Griffith, D. H., J. V. Mannering, and J. E. Box. 1986. Soil and moisture management with reduced tillage. In M. R. Sprague and G. L. Triplet (Eds.), *No-Tillage and Surface-Tillage Agriculture.* John Wiley and Sons, New York.

John Deere Tillage Tool: Residue Management Guide. 1991. Datalizer Slide Charts, Inc., Addison, IL.

Liquid Fertilizer Manual. 1980. National Fertilizer Solutions Association, Peoria, IL.

Lucas, R. E. 1982. Organic Soils (Histosols). *Michigan State Research Report #435.*

Salter, R. M. and T. C. Green. 1933. *J. Amer. Soc. Agron.* 25: 622. The American Society of Agronomy, Madison, WI.

Unger, P. W., G. W. Langdale, and R. I. Papendick. 1988. Role of crop residues—Improving water conservation and use. In W. L. Hargrove (Ed.), *Cropping Strategies for Efficient Use of Water and Nitrogen.* ASA Special Publication No. 51, American Society of Agronomy, Crop Science Society of America, Soil Science Society of America, Madison, WI.

Wallace, A. and G. A. Wallace. 1994. A possible flaw in EPA's 1993 new sludge rule due to heavy metal interactions. *Comm. Soil. Sci. Plant Anal.* 25(1&2): 129-135.

Additional Readings

Hartfield, J. L. and B. A. Stewart. 1994. *Crop Residue Management.* Lewis Publishers, Boca Raton, FL.

Lal, R., M. McMahon, R. Papendick, and G. Thomas (Eds.). 1996. *Stubble Over the Soil.* American Society of Agronomy, Soil Science of America, and Crop Science Society of America, Madison, WI.

Magdoff, F. R., M. A. Tabatabi, and E. A. Hamlon Jr. (Eds.). 1996. *Soil Organic Matter: Analysis and Interpretation.* SSSA Special publication No. 46. Soil Science Society of America, Madison, WI.

Chapter 11

Follett, R. H., L. S. Murphy, and R. L. Donahue, 1981. *Fertilizers and Soil Amendments.* Prentice-Hall, Inc., Englewood Cliffs, NJ.

Frank, S. J. and W. R. Fehr. 1983. Band application of sulfuric acid or elemental sulfur for control of Fe-deficient chlorosis of soybeans. *Agron. J.* 75: 451-454.

NCSA Aglime Fact Book. 1986. National Crushed Stone Asociation, Washington, DC.

Wolf, B., J. Fleming, and J. Batchelor. 1985. *Fluid Fertilizer Manual.* National Fertilizer Solutions Association, Peoria, IL.

Additional Readings

Adams, F. (Ed.). 1984. Soil Acidity and Liming, Second Edition. American Society of Agronomy, Madison, WI.

van Breemen, N., J. Mulder, and C. T. Driscoll. 1983. Acidification and alkalinization of soils. *Plant and Soil* 75: 283-308.

Chapter 12

Anonymous. *USDA Handbook 60.* United States Salinity Laboratory, Riverside, CA.

Eaton, F. M. 1966. Chlorine. In H. D. Chapman (Ed.), *Diagnostic Criteria for Plants and Soil*, 98-135. University of California, Division of Agricultural Sciences, Riverside, CA.

Hansen, U. E., O. W. Israelson, and G. E. Stringham. 1979. *Irrigation Practices and Principles.* John Wiley and Sons, Salt Lake City, UT.

Keller, J. 1965. Leaching salt from the soil. Does it matter how it is done? *Farm and Home Sciences* 50, 51 59.

Lunin, J., M. H. Gallatin, C. A. Bauer, and L. V. Wilcox. 1960. *Brackish Water for Irrigation in Humid Regions.* USDA and Virginia Truck Exp't. Sta. Agri. Inf. Bull 213.

Lunt, O. W. 1966. Sodium. In H. D. Chapman (Ed.), *Diagnostic Criteria for Plants and Soils*, 411. University of California, Division of Agricultural Science, Riverside, CA.

Lyerly, P. J. and D. E. Longnecker. 1959. *Salinity Control in Irrigation Agriculture.* Texas Agric. Expt. Sta. Bull. 876.

Richards, L. A. (Ed.). 1934. Diagnosis and improvement of saline and alkaline soils. *USDA Handbook 60,* U.S. Salinity Laboratory, Riverside, CA.

Western Fertilizer Handbook, Seventh Edition. 1985. Interstate Printers & Publishers, Inc., Danville, IL.

Additional Readings

Hillel, D. 1988. *The Efficient Use of Water in Irrigation.* World Bank Technical Paper no. 64, The World Bank, Washington, DC.

Hoffman, G. J. and M. Th. Van Genuchten. 1983. Soil properties and efficient water use: Water management for salinity control. In H. J. Taylor, W. R. Jordan, and T. R. Sinclair (Eds.), *Limitations to Efficient Water Use in Crop Production,* 73-85. American Society of Agronomy, Crop Science Society of America, Soil Science Society of America, Madison, WI.

Chapter 13

Additional Readings

Crane, J. H. and F. S. Davies. 1989. Flooding responses of *Vaccinium* species. *HortScience* 24: 203-210.

Evans, R. and W. Skaggs. 1985. *Operating Controlled Drainage As Subirrigation Systems.* AG-356. North Carolina Agricultural Extension Service, Raleigh NC.

Hansen, V. E., O. W. Israelson, and G. E. Stringham. 1979. Drainage of irrigated lands. *Irrigation Practices and Principles,* 294-310. John Wiley and Sons, New York.

National Handbook of Conservation Practices. 1977. USDA, Soil Conservation Service, Washington, DC.

Chapter 14

Hewitt, E. J. 1966. *Sand and Water Culture Methods Used in the Study of Plant Nutrition.* Technical Communication No. 22 (Revised Second Edition). Commonwealth Bureau of Horticulture and Plantation Crops, Commonwealth Agricultural Bureaux, Farnham Royal, England.

Langan, T. D., J. W. Pendleton, and E. S. Oplinger. 1986. Peroxide coated seed emergence in water-saturated soil. *Agron. J.* 78: 769-772.

Schwarz, M. 1977. *Guide to Commercial Hydroponics.* Fourth Printing. Israel University Press, Jerusalem.

Steiner, A. 1968. Soilless culture. *Proceedings of the 6th Colloquium of the International Potash Institute,* Florence, Italy, 324-341. International Potash Institute, Worblaufen-Bern, Switzerland.

Winsor, G. W., R. G. Hurd, and D. Price. 1979. Nutrient film technique. Growers Bulletin No. 5. Glasshouse Crops Research Institute, Littlehampton, England.

Additional Readings

Jones, J. B. Jr. 1983. *A Guide for the Hydroponic & Soilless Culture Grower.* Timber Press, Braverton, OR.

Kratky, B. A. 1989. Non-circulating hydroponic systems: An affordable alternate. *Am. Vegetable Grower,* April, 82, 84.

Chapter 15

Boswell, M. J. 1990. *Micro-Irrigation Design Manual.* J. H. Hardie Irrigation, El Cajon, CA.

Bouyoucos, G. J. 1956. Improved soil moisture meter. *Agricultural Engineering* 37(4): 261-262.

Bucks, D. A., E. S. Nakayama, and R. G. Gilbert, 1979. Trickle irrigation water quality and preventative maintenance. *Agricultural Water Management* 2: 149-162.

Goldhammer, D. A. and R. L. Snyder (Eds.). 1989. *Irrigation Scheduling.* Division of Agriculture and Natural Resources Publication 21454, University of California, Oakland, CA.

Irrigation Scheduling. 1977. Coop. Ext. Serv., University of Nebraska, Lincoln, NE.

Miyomota, S. and J. L. Stroelein. 1975. Sulfuric acid. *Progressive Agriculture in Arizona* 27: 13-16.

Rhoads, F. M. 1981. Plow layer soil water management and program fertilization on Florida utisols. *Proc. Soil and Crop Scientific Soc. of Florida* 40: 12-16.

Schwab, D. *Irrigation Water Measurement.* OSU Extension Facts 1502. Cooperative Extension Service, Division of Agriculture, Oklahoma State University, Stillwater, OK.

Scott, V. H. and C. E. Houston. 1981. *Measuring Irrigation Water.* Leaflet 2956. Division of Agricultural Science, University of California, Davis, CA.

Stroehlein, J. L. and A. D. Halderman. 1975. *Sulfuric Acid for Soil and Water Treatment.* Arizona Agric-File Q-357.

Wolf, B. 1996. *Diagnostic Techniques for Improving Crop Production.* The Haworth Press, Inc., Binghamton, NY.

Additional Readings

Ayars, R. S. and D. W. Westcott. 1976. *Water Quality for Agriculture.* Food and Agriculture Organization of the United Nations, Rome.

Farnham, D. S., R. F. Hasek, and J. L. Paul. 1985. *Water Quality: Its Effects on Ornamental Plants.* University of Calif. Division of Agriculture and Natural Resources Leaflet No. 2995.University of California, Berkeley, CA.

Hillel, D. 1988. *The Efficient Use of Water in Irrigation.* World Bank Technical Paper No. 64. The World Bank, Washington, DC.

Irrigation Practice and Water Management. 1971. Food and Agriculture Organization of the United Nations. Rome.

Kidder, G. and E. A. Hanlon. 1985. *Neutralizing Excess Carbonates.* University of Florida. Notes in Soil Science No. 18. University of Florida, Gainesville, FL.

Kovach, S. P. 1984. *Determination of Water Requirements for Vegetable Crops.* Fla. Coop. Ext. Serv. Circular 607, IFAS, University of Florida, Gainesville, FL.

Parchomchuk, P., C. S. Tan, and R. G. Berard. 1997. Practical use of time domain reflectometry for monitoring soil water content in microirrigated orchards. *HortTech* 7(1), January-March, 17-22.

Smajstrala, A. G., D. A. Harrison and L. P. Parsons. 1983. Irrigation scheduling with evaporative pans. *The Citrus Industry* 14, 17-18.

Chapter 16

Collings, G. H. 1955. *Commercial Fertilizers,* Fifth Edition. McGraw-Hill, New York.

Dinauer, R.C. (Ed). 1971. *Fertilizer Technology and Use.* Soil Science Society of America, Inc., Madison WI.

Follett, R. A., L. S. Murphy, and R. L. Donahue. 1981. *Fertilizers and Soil Amendments.* Prentice-Hall, Upper Saddle River, NJ.

Gascho, G. C. 1985. Use of fluid fertilizer in irrigated multi-cropping systems. *Fluid Fertilizer Foundation Symposium Proceedings.*

Hignett, T. P., E. Frederick, and B. Halder (Eds.). 1979. *Fertilizer Manual.* International Fertilizer Development Center, Muscle Shoals, AL.

Lorenz, O. A. and D. N. Maynard. 1988. *Knott's Handbook for Vegetable Growers,* Third Edition. John Wiley and Sons, New York.

Mortvedt, J. J. 1990. Micronutrients: Unlocking agronomic potential. *Solutions* 64(6): 30-31.

Roberts, T. L. and J. T. Harapiak. 1997. Fertilizer management in direct seeding systems. *Better Crops* 81(2): 18-20.

Wolf, B. 1982. An improved universal extracting solution and its use for diagnosing soil fertility. *Commun. Soil Sci. Plant Anal.* 13(12): 1005-1033.

Wolf, B. 1996. *Diagnostic Techniques for Improving Crop Production.* The Haworth Press, Inc., Binghamton, NY.

Wolf, B., J. Fleming, and J. Batchelor. 1985. *Fluid Fertilizer Manual.* National Fertilizer Solutions Association, Peoria, IL.

Additional Readings

Anonymous. Fertilizer and chemical applications through irrigation systems. A & L Mid West Agricultural Laboratories, Inc., Omaha, NE.

Chapter 17

Bunt, A. C. 1976. *Modern Potting Composts.* George Allen & Unwin Ltd., London.

Cooper, A. 1978. *Commercial Applications of NFT.* Growers Books, London.

Geraldson, C. M. 1996. The containerized gradient concept—Potential for undiminishing nutritional stability. *Soil Crop Sci. of Florida Proc.* 55: 20-22.

Guide Notes on the Nutrient Film Technique for Tomatoes. 1981. Booklet G1, Ministry of Agriculture, Fisheries, and Food, London.

Jensen, M. H. *New Developments in Hydroponic Systems: Descriptions, Operating Characteristics, Evaluations.* Environmental Research Laboratory, University of Arizona, Tucson, AZ.

Jensen, M. H. 1997. Hydroponics. *HortScience* 32: 1018-1021.

Johnson Jr., H. *Soilless Culture of Greenhouse Vegetables.* Cooperative Extension, Leaflet 21402. University of California, Riverside, CA.

Johnson, P. and A. Johnson. 1968. *Horticultural and Agricultural Uses of Sawdust and Soil Amendments.* Technical Bulletin. Compton, CA.

Kaufman, P. B., T. L. Mellichamp, J. Glimn-Lacy, and J. D. LaCroix. 1985. *Practical Botany.* Reston Publishing, Reston, VA.

Larsen, J. E. 1980. *Peat-Vermiculite Mix for Growing Transplants and Vegetables in Trough Culture.* Texas Agric. Ext. Serv., Texas A&M University, College Station, TX.

Matkin, O. A. and P. A. Chandler. 1957. The U.C.-type mixes. In Baker, K. F. (Ed.). *The U. C. System for Producing Healthy Container-Grown Plants,* 68-85. Manual 23. University of California Division of Agricultural Sciences, Agricultural Experiment Station, Extension Service, University of California, Berkeley, CA.

McCoy, E. L. 1992. Quantitative physical assessment of organic materials used in sports turf rootzone mixes. *Agron. J.* 84: 375-381.

Poole, R. T., C. A. Conover, and J. Joiner. 1981. *Foliage Plant Production.* Prentice-Hall, Englewood Cliffs, NJ.

Poole, R. T. and W. E. Waters. 1972. *Florida Nurseryman* 17(3): 12-13.

Schippers, P. A. 1979. *The Nutrient Flow Technique.* Vegetable Crops Mimeo 212. New York State College of Agriculture, Cornell University, Ithaca, NY.

Schwartz, M. 1977. *Guide to Commercial Hydroponics,* Fourth Edition. Israel University Press, Jerusalem.

Spomer, L. A. 1973. Soils in plant containers: Soil amendment, air and water relationships. Illinois Research, Illinois Agricultural Experiment Station 15: 3, 16-17. University of Illinois, Urbana, IL.

Steiner, A. 1968. Soilless culture. *Proc. 6th Colloquium of the International Potash Inst., Florence, Italy.* International Potash Institute, Worblaufen-Bern, Switzerland.

Tomatoes in Peat. 1980. Growers Guide No. 8. Growers Books, London.

Ward, G. W. 1975. *Fertilizer Schedule for Greenhouse Tomatoes and Cucumbers in Southwestern Ontario.* Publication 1562. Information Division, Canada Department of Agriculture, Ottawa.

Western Fertilizer Handbook. 1990. Horticultural Edition. The Interstate Printers and Publishers, Danville, IL.

Winsor, G. W., R. G. Hurd, and D. Price. 1979. *Nutrient Film Technique.* Growers' Bulletin No. 5., Glasshouse Crops Research Institute, Littlehampton, England.

Wittwer, G. H. and S. Honma. 1979. *Greenhouse Tomatoes, Lettuce and Cucumbers.* Michigan State University Press, East Lansing, MI.

Additional Readings

Anonymous. 1983. *Genesis Rooting System.* Publication BR-01-83. Genesis Technology, Boulder, CO.

Arkin, G. F. and H. M. Taylor (Eds.). 1981. *Modifying the Root Environment to Reduce Crop Stress.* ASAE Monograph No. 4. American Society of Engineers, St. Joseph, MO.

Davis, W. B., J. L. Paul, J. H. Madison, and L. Y. George. 1970. *A Guide to Evaluating Sands and Amendments Used for High Trafficked Turfgrass.* University of California Agricultural Extension. AXT-n113. Davis, CA.

Sheldrake, R. and J. W. Boodley. *Commercial Production of Vegetables and Flower Plants.* Information Bulletin 82. New York State College of Agriculture and Life Sciences, Cornell University, Ithaca, NY.

Vanini, J. T. and J. N. Rogers. 1995. Topdressing with crumb rubber from used tires in turfgrass areas. *65th Ann. Michigan Turfgrass Conf. Proc.* 24: 235-240.

White, J. W. 1973. *Criteria for Selection of Growing Media for Greenhouse Crops.* Journal Series No. 4574. The Pennsylvania State University Agricultural Experiment Station, University Park, PA.

Willumsen, J. (Ed.). 1988. *Symposium on Horticultural Substrates and Their Analysis.* ISHS Technical Communication No. 221, ISHS, Wageningen, Netherlands.

Index

Order Your Own Copy of
This Important Book for Your Personal Library!

THE FERTILE TRIANGLE
The Interrelationship of Air, Water, and Nutrients in Maximizing Soil Productivity

_____ in hardbound at $69.95 (ISBN: 1-56022-878-4)

COST OF BOOKS_____

OUTSIDE USA/CANADA/
MEXICO: ADD 20%_____

POSTAGE & HANDLING_____
(US: $3.00 for first book & $1.25
for each additional book)
Outside US: $4.75 for first book
& $1.75 for each additional book)

SUBTOTAL_____

IN CANADA: ADD 7% GST_____

STATE TAX_____
(NY, OH & MN residents, please
add appropriate local sales tax)

FINAL TOTAL_____
(If paying in Canadian funds,
convert using the current
exchange rate. UNESCO
coupons welcome.)

☐ **BILL ME LATER:** ($5 service charge will be added)
(Bill-me option is good on US/Canada/Mexico orders only;
not good to jobbers, wholesalers, or subscription agencies.)

☐ Check here if billing address is different from
shipping address and attach purchase order and
billing address information.

Signature_____

☐ **PAYMENT ENCLOSED: $**_____

☐ **PLEASE CHARGE TO MY CREDIT CARD.**

☐ Visa   ☐ MasterCard   ☐ AmEx   ☐ Discover
☐ Diner's Club

Account #_____

Exp. Date_____

Signature_____

Prices in US dollars and subject to change without notice.

NAME _____

INSTITUTION _____

ADDRESS _____

CITY _____

STATE/ZIP _____

COUNTRY _____ COUNTY (NY residents only) _____

TEL _____ FAX _____

E-MAIL_____

May we use your e-mail address for confirmations and other types of information? ☐ Yes   ☐ No

Order From Your Local Bookstore or Directly From
The Haworth Press, Inc.
10 Alice Street, Binghamton, New York 13904-1580 • USA
TELEPHONE: 1-800-HAWORTH (1-800-429-6784) / Outside US/Canada: (607) 722-5857
FAX: 1-800-895-0582 / Outside US/Canada: (607) 772-6362
E-mail: getinfo@haworthpressinc.com
PLEASE PHOTOCOPY THIS FORM FOR YOUR PERSONAL USE.

BOF96

TO ORDER: CALL: 1-800-HAWORTH / FAX: 1-800-895-0582 (outside US/Canada: + 607-771-0012) / E-MAIL: getinfo@haworthpressinc.com

☐ YES, please send me **Horticulture as Therapy**
____ in hard at $79.95 ISBN: 1-56022-859-8
(Outside US/Canada/Mexico: $96.00)

- Individual orders outside US, Canada, and Mexico must be prepaid by check or credit card.
- Discounts are not available on 5+ text prices and not available in conjunction with any other discount.
- Discount not applicable on books priced under $15.00.
- 5+ text prices are not available for jobbers and wholesalers.
- Postage & handling: in US: $4.00 for first book, $1.50 for each additional book.
  Outside US: $5.00 for first book; $2.00 for each additional book.
- Canadian residents: please add appropriate sales tax after postage & handling.
  NY, MN, and OH residents: please add appropriate sales tax after postage & handling.
- Payment in UNESCO coupons welcome.
- If paying in Canadian dollars, use current exchange rate to convert to US dollars.
- Please allow 3-4 weeks for delivery after publication.
- Prices and discounts subject to change without notice.

Signature _____

☐ **BILL ME LATER** ($5 service charge will be added):
(Minimum order: $15.00. Bill-me option is not available to individuals outside US/Canada/Mexico. Service charge is waived for jobbers/wholesalers/booksellers.)
☐ Check here if billing address is different from shipping address and attach purchase order and billing address information.

☐ **PAYMENT ENCLOSED $** _____
(Payment must be in US or Canadian dollars by check or money order drawn on a US or Canadian bank.)

☐ **PLEASE BILL MY CREDIT CARD:**

☐ Visa ☐ MasterCard ☐ American Express ☐ Discover ☐ Diners Club

Account Number _____

Expiration Date _____

Signature _____

FAX

THE HAWORTH PRESS, INC., 10 Alice Street, Binghamton, NY 13904-1580 USA

Please complete the information below or tape your business card in this area.

NAME _____

INSTITUTION _____

ADDRESS _____

CITY _____

STATE _____ ZIP _____

COUNTRY _____

COUNTY (NY residents only) _____

E-MAIL _____

May we use your e-mail address for confirmations and other types of information?
( ) **Yes** ( ) **No.** We appreciate receiving your e-mail address and fax number. Haworth would like to e-mail or fax special discount offers to you, as a preferred customer. We will never share, rent, or exchange your e-mail address or fax number. We regard such actions as an invasion of your privacy.

☐ **YES, please send me Horticulture as Therapy (ISBN: 1-56022-859-8)** to consider on a 60-day examination basis. I understand that I will receive an invoice payable within 60 days, or that **if I decide to adopt the book, my invoice will be cancelled.** I understand that I will be billed at the lowest price. (Offer available only to teaching faculty in US, Canada, and Mexico.)

Signature _____

Course Title(s) _____

Current Text(s) _____

Enrollment _____

Semester _____

Office Tel _____ Hours _____

Decision Date _____

(32) 04/99 [BIC99]

DATE DUE

DEC 1 5 2000			

Demco, Inc. 38-293